PERGAMON INTERNATIONAL LIBRARY
of Science, Technology, Engineering and Social Studies

*The 1000-volume original paperback library in aid of education,
industrial training and the enjoyment of leisure*

Publisher: Robert Maxwell, M.C.

LANGUAGE
OF THE
EARTH

THE PERGAMON TEXTBOOK
INSPECTION COPY SERVICE

An inspection copy of any book published in the Pergamon International Library
will gladly be sent to academic staff without obligation for their consideration for
course adoption or recommendation. Copies may be retained for a period of 60 days
from receipt and returned if not suitable. When a particular title is adopted or
recommended for adoption for class use and the recommendation results in a sale
of 12 or more copies the inspection copy may be retained with our compliments.
The Publishers will be pleased to receive suggestions for revised editions and new
titles to be published in this important international Library.

Pergamon Titles of Related Interest

LANGUAGE OF THE EARTH

EDITED BY
FRANK H. T. RHODES
RICHARD O. STONE

PERGAMON PRESS
New York Oxford Toronto Sydney Paris Frankfurt

Pergamon Press Offices:

U.S.A.   Pergamon Press Inc., Maxwell House, Fairview Park,
                     Elmsford, New York 10523, U.S.A.

U.K.   Pergamon Press Ltd., Headington Hill Hall,
                     Oxford OX3 0BW, England

CANADA   Pergamon Press Canada Ltd., Suite 104, 150 Consumers Road,
                     Willowdale, Ontario M2J 1P9, Canada

AUSTRALIA   Pergamon Press (Aust.) Pty. Ltd., P.O. Box 544,
                     Potts Point, NSW 2011, Australia

FRANCE   Pergamon Press SARL, 24 rue des Ecoles,
                     75240 Paris, Cedex 05, France

FEDERAL REPUBLIC Pergamon Press GmbH, Hammerweg 6, Postfach 1305,
OF GERMANY   6242 Kronberg/Taunus, Federal Republic of Germany

Library of Congress Cataloging in Publication Data

Main entry under title:

    Language of the earth.
  1. Geology--Addresses, essays, lectures. I. Rhodes,
Frank Harold Trevor. II. Stone, Richard O., 1920-1978.
QE35.L28   550   80-23088
ISBN 0-08-025981-2
ISBN 0-08-025980-4 (pbk.)

Printed in the United States of America

Contents

Preface

One of the problems with our conventional styles of teaching and conventional patterns of learning at the introductory undergraduate level is that the "subject"—whatever it may be—all too easily emerges as given, frozen, complete, canned. Add to this quizzes, multiple-choice exams, and a single textbook, and the pursuit of knowledge sometimes becomes a kind of catechism—the recital of prepared answers to a limited set of questions ("You lost one point because you missed the hardness of hornblende," or some other equally inconsequential fragment of information). It is against such a limited view of knowledge that students so frequently react, and so they should.

But it need not be so. In every department there are still successful teachers who can win the interest and enthusiasm of once uninterested and unenthusiastic students. They do so partly by the example of their own commitment to learning, and partly by revealing knowledge as a continuing personal quest. For no knowledge, and least of all scientific knowledge, exists as a finished corpus of categorized facts. Knowledge exists because there are people; it is the accumulated personal experience of our race. It becomes meaningful, useful and intelligible as we grasp, not only its content, but also its basis, its implications, its relationship and its limitations. Its coherence and significance lie in its relatedness to the whole of the rest of our human experience.

The aim of this book is to provide such a context for the particular category of knowledge which we identify as earth science. Our intention is not didactic. We make no attempt to cover the ground in the sense of describing the current content of each area of earth science. Our aim is to illustrate the scope and range of the science and to convey its flavor and style, rather than catalog its contents; to display all our knowledge as provisional rather than infallible, as refinable rather than complete and finished; to show the inspiration and sweeping implications of earth science, rather than representing it as an isolated area of study.

We hope that by using the form of an anthology, based as it is on the writings of authors representing many countries, periods and viewpoints, we can show something of the individuality which lies at the heart of science. The categorization of science as less humane than other areas of human knowledge and as dehumanizing in its effects is a verdict reached too readily by many of our contemporaries. For scientific knowledge, in spite of its public verifiability, grows by private insight and personal intuition. "If you want to know the essence of scientific method," Einstein is said to have remarked, "don't listen to what a scientist may tell you. Watch what he does." So this book is a window on the world of people 'doing'—not just geologists—but sculptors and soldiers, artists and aviators, politicians and poets, prophets and prospectors, novelists and naturalists.

The thread that gives continuity is their common concern with the Earth. Nor is the span of time and place any less comprehensive than that of occupation. We have deliberately selected writings from the time of the 5th or 6th century B.C., when the book of Job was written, to the heyday of the Victorian era, when William Buckland fascinated and dominated the Oxford scene; from the medieval ages of wonder in the created universe, to the novels of John Fowles written in the 1970's. In the most literal sense, geology involves the comprehension of the whole Earth. Our authors' accounts range from dripping caves, deep below the Pyrenees, to the deserts of North Africa, from the Alps to the Andes, from mountain tops to the depths of the ocean, from the Antarctic ice sheet to the surface of the moon. Their writings concern both social and scientific topics, as well as personal and political goals. All reveal some aspect of geology in the life of mankind.

We have arranged our selections in various categories, in order to give some structure and coherence to the book, and also to allow the reader to use it more selectively. These categories are generally self-evident, but a few articles could have fitted into some other category just as easily as the one in which they stand. Most of the articles that we include are mere fragments of much longer articles, essays, or books, and even the extracts themselves, we have generally abridged. We regret the need to make these abbreviations, and we have tried very carefully to preserve the sense and style of the original authors. Our reluctance to make the abridgement is exceeded only by our conviction that diversity and variety are of more importance than comprehensive quotation of a smaller number of authors. In the book as it now stands, our task of selection has been painfully difficult; we have collected many outstanding articles for which we have been unable to find space. With each extract, we have retained the essential footnotes and references, though we have shortened or deleted footnotes and references where these were not essential to the main argument. We hope that our readers will wish to explore the new worlds of information represented in some of the references that are quoted, just as we also hope that they may develop a sufficiently strong taste for some of the present authors to encourage them to explore their other writings.

We hope that the book will provide useful supplementary reading for those who are enrolled in introductory earth science courses. We see it—not merely as a collection of required readings—but as an anthology for browsing. We hope that the book might even stand by itself, and have some interest for those who, though having no formal concern with courses or teaching in earth science, possess a curiosity about the planet which is our home, and about our varying responses to it. This response is of more than casual importance. It is, of course, of vital concern as a basis for the vocational skills and technical knowledge required of those pursuing courses in earth science. But, in the far wider sense, it is an important component in adding to the richness and exuberance of life. We have attempted to use a concern with the Earth to introduce the reader to a growing range of experiences and involvements ranging from sculpture to literature, and from architecture to history. Through such interests, it is possible to develop a new perspective which enriches the quality of human experience.

Our response to the Earth is a factor of profound significance. We share our over-crowded, polluted, plundered planet with 3.5 billion other members of our race. The next 30 years will call for critical decisions in the problems of conservation, energy supplies, mineral and water resources, land use, atmospheric protection and utilization of the oceans. The terms of our survival will depend in large part on our careful comprehension of the language of the earth and our stewardship of this rare planet. Earth's fitness as a home for humans is the end product of its long and elaborate history. Soil, air, and water, food and fuels, minerals and microbes are the products of earth processes acting over incalculable periods, and of the life, activity, and death of countless organisms. Each of us, in his own turn, absorbs, utilizes and recreates this sustaining environment.

"Speak to the Earth, and it shall teach thee," Job was told 25 centuries ago. To those with the patience to master its language, the Earth still responds. To those with the insight to comprehend its moods and reflect on its mysteries, the Earth still teaches.

Dick Stone and I began work on this book in 1972. We worked steadily at it, fitting it in as best we could between other pressing commitments. Most of the writing and selection of materials was finished by 1976.

Because of increasing publishing costs, we were then faced with the need to reduce the length of the book. This took much longer than either of us had anticipated. Dick Stone became ill with a disease against which he fought with courage and hope, but from which he died on July 23, 1978. I have completed the book, but its essential form and much of its content are as we both planned it.

Frank H. T. Rhodes

Cornell University
Ithaca, New York

Acknowledgments

I am most grateful to several individuals whose willing cooperation and enthusiastic interest have contributed significantly to the completion of this book. My greatest debt is to my late colleague, Dick Stone, who derived great pleasure from the selection and preparation of material for the book. The breadth of the topics we cover in the book reflects the scope and quality of his own interests.

To those authors and publishers who have given permission to reproduce their work, to extract from it and otherwise abridge it, I have a particular debt of gratitude. In some cases, this permission was given without the requirement of payment of copyright fees. All the original sources of information are fully acknowledged.

Mrs. Jean Schleede and Mrs. Margaret Gillingham typed earlier versions of the manuscript, while Mrs. Marcia Parks, Mrs. Clara Pierson and Mrs. Joy Wagner assisted in its final preparation. Their careful typing, checking and verification of sources have been especially important, and I am most grateful for their help.

Frank H. T. Rhodes

Part I

GEOLOGIC KNOWLEDGE

Geologic knowledge, as it appears in the pages of a textbook, represents an abstraction and a synthesis, a distillation and a summary of the personal experiences of individual geologists from their various encounters with the Earth. There is a widespread supposition that scientific knowledge, especially in its methods and conclusions, differs in some way from the methods and conclusions of other areas of human experience. In two important respects these do differ in their emphasis: deliberate abstraction and pervasive quantification are very characteristic features of science, though they are certainly not unknown in other human activities. For science is a statement of human experience, differing from others chiefly in its exclusive concern with the material world and in its degree of generality.

Science does indeed concern itself with what most of us assume to be a real world and an actual universe, but that does not provide it with some official imprimatur, certifying its authenticity by the authority of the cosmos at large. Science exists because there are people. It presents no final statement of truth, binding us and the rest of the universe to observe its dictates and its ordained boundaries. It is tentative, provisional, refinable, subject not only to minor correction and clarification, but also to the kind of wholesale restatement, partial rejection and sweeping reinterpretation which succeed such revolutions as those produced by the discovery of the heliocentric nature of the solar system, organic evolution and sea floor spreading. Yet even these revolutions, which change the whole context and direction of scientific inquiry, are the results of the deeply personal insights and individual experiences of a Copernicus, a Darwin, a Hess or a Vine. Knowledge grows neither by public polls nor by majority votes at scientific conferences, but by individual perception and by personal insight. It is public in its interest, verifiability,

1

and utility; but it is private in its origin and source. It is not committees that lead us into new truth; it is individuals.

The selections that follow deal with various aspects of geologic knowledge, including especially the characters, motives and attitudes of some individuals whose work has contributed to the sum total of our present understanding. Other sections describe typical field geologists encountering the phenomena of the earth, the role of controversy in geology, and some of the distinctive philosophical problems involved in geology itself. Concluding sections deal with the biography of life, and some current problems and controversies in geology.

Time and Thought

In the whole history of thought no transformation in men's attitude to Nature—in their 'common sense'—has been more profound than the change in perspective brought about by the discovery of the past. Rather than take this discovery for granted, it is almost preferable to exaggerate its significance.

—**Stephen Toulmin and June Goodfield**

CHAPTER I

Geologists Are Also Human

Introduction

Because knowledge is rooted in the knower, science is rooted in the scientist just as surely as art is rooted in the artist. Scientific knowledge, therefore, represents not only an enclosing edifice, in which each of the component parts blends into a more or less harmonious whole, but an edifice which is the product of individual insight. It is these individual insights, hard-won from patient reflection, and prolonged encounter with the natural world, which are the real elements of scientific knowledge. Yet the insights are achieved, not by white-coated technocrats, the high priests of some particular terrestrial vision, but by geologists, who are also very human. It is in that sense that scientific knowledge is both private and public, for when a geologist is not practicing his craft, he engages in the same activities and shares the same hopes and prejudices as his neighbors, digging the garden, listening to Mozart, voting in elections and walking the dog. For this reason, the character of the individual observer is inevitably involved in subtle ways in the character of the observations which he makes and the conclusions which he draws. The individual's frame of reference, the commitment to devote weeks or years to the study of some isolated and perhaps obscure phenomenon, the determination to journey to remote and often inhospitable regions to gain firsthand access to certain phenomena, all these reflect the relationship between individual character and the scientific work which the individual produces.

Nor is the history of geology confined to an account of the data accumulated by professional geologists during the last two hundred years. It stretches back as far as early man, who, in collecting flints for weapons, and using fossils to decorate the graves of his dead, asserted his belief in the dignity and persistence of life. It stretches back to Pliny the Elder, Roman naturalist, philosopher and sailor, who lost his life during the great eruption of Vesuvius. His nephew, Gaius Plinius (Pliny the Younger), records in a letter to his

3

friend Tacitus that Pliny was at Misenum, in command of the Roman fleet, when at one o'clock on the 24th of August A.D. 79, his attention was drawn to a great cloud which "seemed of importance and worthy of nearer investigation to the philosopher." He ordered a light boat to be got ready "but as he was about to embark, Pliny received a plea from Rectina, wife of Caesius Bassus, living on the other side of the bay, begging him to rescue her from what she thought was certain death." This converted his plan of observation into a more serious purpose. "He steered directly for the dangerous spot whence others were flying, watching it so fearlessly as to be able to dictate a description and be able to take notes of all the movements and appearances of this catastrophe as he observed them." After landing, but being unable, because of violent seas, to embark, he encouraged and reassured the community. When day came, they found his body perfect and uninjured, seeming by his attitude to be asleep, rather than dead.

In various ways, the world view of particular individuals had profound effects upon their commitment to science. In the case of Nicholas Steno, for example, the 17th Century Danish priest and philosopher, and the first to formulate the essential principles of historical geology, his conversion to Catholicism led to the end of his scientific career, for his efforts thereafter were devoted wholly to the service of the Catholic Church. In other cases, such as that of James Hutton, religious commitment was a catalytic influence in scientific thought, as shown by Hutton's commitment to the idea of providence in Nature, which was a major influence in his determination to develop his theory of the Earth, even though the work was bitterly attacked by some of his theological critics. In still other cases, chance encounters with the Earth while following some other career have been all important, as in the case of William Smith, the self-taught surveyor and engineer, whose occupation in constructing canals in southern England involved trenching through stratified rocks and, thus, allowed him the opportunity to formulate fundamental principles of stratigraphy.

There is also an interaction in the other direction, however, because our sense of the Earth on which we live may, in turn, influence our world view. At least one Hindu tribe, for example, living in an area subject to earthquakes, regarded the earth as supported by elephants standing on a giant turtle (an incarnation of the God Vishnah) which, in turn, rested on a cobra, the symbol of water. This floating concept of the earth fitted well with their experience of earthquakes, which were believed to be caused by movements of the elephants, adjusting their stance in response to the load of the earth.

The selections that follow range from the 19th century to the present, reflecting the different ways in which working geologists and others respond to the challenge of unraveling the history of the Earth. Several accounts describe the ways in which individuals were first attracted to geology and the course of events which led them to devote their lives to it.

The short extract written by George F. Sternberg, was originally published in the Journal of the Kansas State Teachers' College at Hays. *It describes a fossil hunting expedition in the early years of the 20th century, which was directed by the writer's father, C. H. Sternberg, in the richly fossiliferous Cretaceous strata of Wyoming.*

THRILLS IN FOSSIL HUNTING

George F. Sternberg

Will you go with me for a few moments into Converse County, Wyoming, where I was fortunate enough to discover a remarkable skeleton of a duck-billed dinosaur wrapped in its skin impression, which at the time of discovery was said to be the finest example of a dinosaur known. It was in July, 1908, that I drove our outfit into the rough sheep and cattle country north of Lusk, Wyoming, along the breaks of the Cheyenne River to search for dinosaur remains. My father, C. H. Sternberg, was in charge of the party. We were assisted by my two younger brothers, Charles and Levi. Father was the only one of the party who had ever collected a dinosaur bone. But we boys were looking forward to doing great things. We trudged over the rough exposures day after day and week after week only to return to our humble camp, pitched some sixty miles from Lusk where we must go for our supplies. As time went on we began to realize that finding dinosaur specimens was not as easy as we had expected. We wished that we were back in Kansas where the sunflowers grew. Camp supplies were getting low, and a trip to Lusk for more food must be made. We had gathered together a few unimportant specimens which could be hauled in, and it was decided that my younger brother, Levi, and I should remain in camp while Charles, who was the official camp cook, should accompany father on the trip.

Charles informed me that if I would ride to a sheep wagon some seven miles away and get some baking powder, there would be enough food in camp to last the two of us until they could return three days later. This I did; and upon arriving on the spot where the sheep wagon had been located for some time, we found it just ready to move some eight or ten miles farther away. We were informed that the dry weather had caused the water to give out. The herder also told us that he was running short of grub. However, I secured the necessary baking powder and returned to camp thinking all was well.

The evening before, I had accidently found some bones of a duckbilled dinosaur protruding from the base of a high sandstone ledge. Although it looked as if something worthwhile might be buried there, still the amount of work necessary to uncover the prospect did not lend any encouragement to us boys. Besides, the quarry was a mile and a half from our camp, and we had only one saddle horse between us to ride back and forth. Nevertheless, we went to work in earnest, and after two days of hard digging we laid bare a floor large enough to trace out most of the skeleton. Then with tools we began to follow the bones into the bank starting in the region of the hips. Slowly and carefully we worked. The removal of each bit of dirt revealed more and more of the skeleton. We were beginning to get hopeful. "But what if the head were gone?", we began to think. For it so often happens that after finding an articulated skeleton one finds the head has been severed from the body before burial and is nowhere to be found. This practically destroys the skeleton for museum purposes.

About this time our food gave out except for a sack of potatoes. I began to realize the baking powder for which I had ridden fourteen miles was of little use to us as there was only enough flour for two days. Had I been told that there was practically no salt in camp I might have borrowed some from the sheep herder. But the wagon should be back almost any time now, so rather than leave our interesting specimen and ride to the nearest ranch for food we decided to "go it on spuds." And "spuds" it was for three whole days. It would not have been so bad if we had had some salt to season them.

Finally by the evening of the third day, I had traced the skeleton to the breast bone, for it lay on its back with the ends of the ribs sticking up. There was nothing unusual about that. But when I removed a rather large piece of sandstone rock from over the breast I found, much to my surprise, a perfect cast of the skin impression beautifully preserved. Imagine the feeling that crept over me when I realized that here for the first time a skeleton of a dinosaur had been discovered wrapped in its skin. That was a sleepless night for me. Had I missed my regular cup of coffee or eaten too many potatoes for supper?

We were loathe to leave our treasure and go for food though we knew there was not a human being for miles to disturb it. Few cattle men ever rode this way as this was a high and dry region where cattle were seldom found ranging. The sheep men came here with their flocks only when there was melting snow to furnish water.

It was about dusk on the evening of the fifth day when we saw the wagon loaded with provisions roll into camp. "What luck?" was my father's first question. And before he could leave his seat I had given him a vivid sketch of my find; for we had found every bone in place except the tail and hind feet which had protruded from the rock and had washed away. The head lay bent back under the body. Traces of the skin were to be seen everywhere. "Let's go and see it," he shouted as he jumped from his seat on the wagon. I grabbed some food from the boxes on the wagon and away the two of us went, leaving the others to prepare a meal for us. Darkness was nearly upon us when we reached the quarry and there laid out before us was the specimen. One glance was enough for my father to realize what I had found and what it meant to science. Will I ever forget his first remark as we stood there in the fast approaching twilight? It thrills me now as I repeat it. "George, this is a finer fossil than I have ever found. And the only thing which would have given me greater pleasure would have been to have discovered it myself."

I do not remember what we had for supper that night. I do not remember what news was brought from home and loved ones. I do remember, however, that is was another restless night for not only myself but for my father as well. I could hear him roll in his bed and cough or make some noise which told me he too was spending a sleepless night. The day was beginning to break before I finally dozed off, only to awake as the sun came up over the hill. There was much to be done to collect the specimen. But by the time it was boxed and ready for shipment, Professor Osborn of the American Museum of Natural History, New York City, had sent a man to see it and had arranged to secure it for his great institution.

It was not until 1911 that I again had the privilege of seeing my prize specimen. There it lay in that great institution just as it was when I found it resting on its back with the head bent down under its body and its fore limbs stretched out on either side. It had been so skillfully prepared that the impression of the skin had been preserved over nearly all parts of the body. In Professor Osborn's description of the specimen he has called it the "Mummy Dinosaur." Could there be a greater thrill in the life of a man than to know he has been able to discover one of these buried treasures of bygone days and to have it placed before the world as an everlasting memorial? Is a fossil hunter's life worthwhile? Are there any thrills in it?

Beauty

"It was when I was happiest that I longed most. . . . The sweetest thing in all my life has been the longing. . . . to find the place where all the beauty came from."

—C.S. Lewis

Hans Cloos (1886–1951) was born in Magdaburg, Germany, and received his geological training at the University of Freiburg. His early studies concerned the tectonics of the Jura Mountains, and, after receiving his doctorate, he worked as a mining geologist in Southwestern Africa, and later in petroleum exploration in Indonesia. Cloos later returned to Germany, and joined the faculty of the University of Marburg. He subsequently occupied the Chair of Geology at the University of Breslau and later the University of Bonn. An outstanding field worker, and a major contributor to our understanding of intrusion and tectonic processes, Cloos was especially well known for his work in experimental modeling, by which he sought to simulate geologic structures and their formation. He was awarded the Penrose Medal of the Geological Society of America in 1948. The present extract concerns his journey out to Africa to take up his first geological appointment. Taken from his book Conversation With the Earth, *it describes the impression made upon him by Stromboli and Aetna.*

SHIP'S WAKE

Hans Cloos

The present is a distraction; the future a dream; only memory can unlock the meaning of life.

—DESMOND MACCARTHY

The next day was gloomy and rainy. The gray triangle of Stromboli, most punctual volcano on earth and Italy's weather prophet, slid past. Three times I saw the clouds of smoke which Stromboli has been exhaling at regular intervals since the days of antiquity. These intervals decrease slightly with increasing barometric pressure. That afternoon Sicily came into view, with Mt. Aetna dominating, its snowy head ringed with little clouds of steam, and Messina at its base. There had just been a severe earthquake, and through field glasses I saw the ruins of hotels on the quay. Again the earth's two faces showed themselves—the black of the tenebrous depths and the light one of the non-volcanic crust. For although Messina had not been destroyed by the volcano which towers over the city, the crust had been rent asunder by great tremors, and the volcano's action showed that the fissures must reach down into the still fluid depths of the earth.

A little later the ship turned left, and henceforth sailed straight through the open Mediterranean for the mouth of the Nile.

The next morning, like countless seafarers before and after me, I made a discovery: on ahead, where the steady bow of the ship cut through the blue sea as if it were molten glass, one looks into the future; astern, at the taffrail, above the noisy propellers, images of the past seem to emerge from the

greenish-white wake. One peers into the restless Charybdian whirlpool boiling behind the ship; unfathomable masses of water and endless horizons
recede before one's gaze, and dreams turn to the past.

While I stood there, looking back at the gulls as they swooped this way
and that, becoming smaller and smaller, then racing after the ship from which
they got their food, a traveling companion joined me. He too for a time
stared silently at the water, and finally observed that geology must be a
curious profession. How had I happened to take it up? How could anybody?
Every geologist has been asked that question a few times. I told him about
my own intellectual development, and about the evolution of a geologist. I
told him that my earliest childhood memory—which goes back almost to the
age of two—involved an attempt to catch a sparrow-hawk that had flown
into the room. I told him how my relationship with the earth had developed
out of many experiences, and I told him about this dramatic one:

The land had been drenched by a cloudburst. The gardener of the estate
carried me—then aged ten—across the flooded village street which had become a stream. When the water had subsided, I followed its course. There
was a fresh, deep washout in the ravine in the little wood, and a bridge had
been covered with gravel. Beyond and below the bridge, at the mouth of the
ravine, limestone debris and mud had spread out over the meadows, forming
a new layer which raised the level of the ground. Yesterday's meadow was
now covered and obliterated. Did the earth grow like a tree, adding a ring
at a time? Could the land change its shape from one day to the next? Even
at that early age I had been half-aware of these questions.

A little later I developed an interest in fossils and minerals. A deep gorge,
or "Klamm," was cut into the "Muschelkalk" (shell-limestone) near my home.
The stony creek bed and adjacent fields were full of fossils like ammonites,
fluted bivalves, and smooth terebratula as large as nuts—all petrified into
yellowish-white limestone and washed free of the surrounding rock. Today
this region has been much picked over and it may take a long time before
it is again a good spot for collectors. I inspected the limestone fields and the
stone piles at their boundaries, and collected a rich booty. Small finds I kept
in a cigar box; large slabs found a place on the garden wall. Since that time
some selected samples have accompanied me wherever I have gone. A splendid ceratite that my father found, resembling a curled snake and weighing
several pounds, now decorates my desk. At the beginning my unbridled
imagination led me to fancy that I had found fossilized snakes, skulls, heads,
and worms. But gradually I learned the limits of fossil preservation and at
the same time began to understand the strange historical character of these
extinct forms of life, which evolved under entirely different conditions from
those of the present.

Never again did nature seem so permeated with wonder, and never again
was my relationship with it so strong and direct as during these years of

transition from my childhood when I was stimulated by fairy-tales and play, to the period of sober and objective scientific observation of the adult. The former taught me to see the richness of forms, the latter added color and life to diversity. Bluebells bloomed in the spring woods, the perfume of daphne was sweet on the air. Sand-yellow vipers sunned themselves in the summer's heat, ringed snakes glided across the path. Fat emerald lizards clung to the warm walls.

During those years I understood the voices of animals and stones as did Siegfried after he had tasted dragon's blood. Later, when I was able to feel less strange in distant lands, deserts, and jungles, and when I learned something of the strange languages of the wide, wild world, I was thankful for those summers of communion with nature during my boyhood. The language of nature must be learned like the mother tongue. Anyone who begins only at the age of twenty has little prospect of mastering that language.

It was somewhat later when I became enchanted by the brightly colored minerals and sparkling crystals. Even now I can feel my astonishment when I first found gypsum in the gray wall of an abandoned quarry. It was clear as glass, yet soft as wax or wood. To think there should be anything like that in the rocks of the ground! When a well-house was hewn out of the limestone, little cavities were exposed; tiny caves evolved like doll house rooms. In the dumps of an ore mine I found multi-colored crystals, growing side by side and on top of one another like flowers. A summer vacation in the Hunsrück led me to the old agate cutting works of the Idar Valley. For days I would dig in the tailings, looking for purple amethysts, red ribbon agates, fibrous and yellow cat's-eyes. Even today a certain odor, apparently that of the oil used by the cutters, is ineradicably linked to my early impressions of this dead and yet so colorful and multiform world.

Then suddenly schooldays were over, and the choice was between architecture and geology. Was that not vacillating, indicative of a lack of character? At the time I reproached myself for this indecision, and I realized only many years later that architecture can be applied to the study of the earth's crust and that, indeed, during my entire life I have done little else.

And now my own choice, geology! For me, geology was far superior to all other fields, and that is as it should be. In geology, too, there were books, numbers and other ballast for my memories. Like the rest, I wore out books and filled blank papers with notes. But by far the most important books for geology students were the quarries and clay pits, the cliffs and creek beds, the road and railroad cuts in woods and fields. Our words and letters were the imprints of plants and animals in stone, the minerals and crystals, and our vast, inexhaustible, incorruptible, and infallible library was nature itself.

And to this end we went out into the field, time and time again. Often at brief intervals we returned to the same outcrop where our Pythia, nature, opened her mouth from time to time to utter her equivocal oracles. Time and again we studied the same stratification, or the same interpenetration of rocks, and yet each time advanced one step further, because of what we had learned on the last visit, because the previous impression had had time to settle, or because this time our eyes were a little keener and now observed what hitherto had escaped them.

The day came when I was far enough along with my work to dare to ask the experienced teacher for a project of my own. Now real research began. Every good geological doctorate dissertation should be based on a first modest research expedition, and countless theses really were.

The first independent piece of research had an educational effect: I stood alone, face to face with nature. In that chosen small portion of the earth's surface there were no reference books to depend on, and no professors to turn to for help. The student had to depend on his own eyes and head, and on his legs and hobnailed boots as well. As a rule nature was very reserved about her revelations. She gave very little freely. Ask, and ask intelligently and persistently, ask again and again, and then listen carefully: perhaps nature will answer.

Then the knapsack was packed a bit more carefully than usual and I got underway. For the first few weeks, I groped around uncertainly. Evenings I returned to the village inn, either in despair, or filled with hope or pride, but invariably in possession of many errors! Yet slowly I began to understand, and during ensuing days and weeks the future geologist commenced to evolve.

Toward the end of the first summer in the field, the professor came out for two or three days. He looked over the exposures, the mapping, and the drawings. He listened to what we had to say, and drank some of the good wine that had grown on his student's layered rocks. He praised the wine and the stratified rocks, and returned to his comfortable Institute, where a thousand finished theses looked down from the shadowy bookshelves. Then I was on my own again, and had to figure out how to muddle through.

After the final examination I was completely at a loss. The tasks I had worked on for two solid years was completed; what I had learned in four years had been divulged in two hours. Then I had some luck, without which the best man in the world cannot succeed. Even so, I had to open the door, and reach out for it. Four months later I was sitting in the Freiburg-Basel-Olten-Lucerne express, passing swiftly through my first modest field of geological enterprise, and on toward the second and vastly larger one. By evening of that day I was already in Milan on the south side of the Alps, the following day I was in Florence, at the western base of the Apennines, and a few days later in Rome and Naples. And now the mighty limestone walls of Crete were sliding by to the left over the smooth blue of the sunny Mediterranean. They are the extreme southerly margin of the Eurasian belt of mountain ranges, on whose northern border at Freiburg I had just passed my doctoral examination. Now it was a case of: "Hic Rhodus, hic salta!";

that is, "Let's see what you have learned, my boy!" The actual Rhodes, of course, after Crete the most easterly pier in the marginal wall of Europe, remained hidden somewhere to the left beyond the horizon. But already the new continent, the black continent with the big blank places on its map, was blowing its glowing desert air towards me. Africa . . . !

Fossil Collecting

As a warning against overenthusiastic collecting, I will close with a little experience of my lamented friend Gilbert Van Ingen. While still working with the New York Survey, 'Van' walked one rainy day through the main street of Granville, New York. On the wet sidewalk he noticed many fine specimens of the Cambrian jellyfish *Dactyloises bulbosus*, and although he was to return that day to Albany, he set out enthusiastically to procure some slabs. He sought the manager of the quarry and asked him to take up some of the slabs and replace them with new ones. The manager kindly agreed and asked him to mark the desired slabs. So 'Van' marked the slabs with a red cross and went home. Within a week or two a letter announced that a freight car loaded with the whole sidewalk of the main street was ready for shipment and that the bill was $600.00 Doctor Clarke had neither funds nor inclination to meet the expense and the wires were kept hot that day between the perspiring Gilbert Van Ingen and the kind manager. It happened that 'Van' had put the paint and brush behind a near-by barn before leaving, and a little boy who had watched him had continued the job of marking the sidewalk.

—**Rudolf Ruedemann**

R. A. Bartlett, Professor of History at the University of Oklahoma, is the author of the eminently readable Great Surveys of the American West, *from which the next extract is taken. Clarence Rivers King (1842–1901) graduated from Yale in 1862 and spent the following winter studying under Alexander Agassiz. In 1863 he followed the North Platte and the Humbolt rivers in the west, and, after losing all his belongings in a fire, was forced to cross the Sierra Nevada on foot. For the three following years he served as assistant geologist to the Geological Survey of California. Returning to Washington in 1866, it was largely his persistence that led to the appropriation by Congress of funds to finance a survey of the 40th parallel. He was only 25 years old at the time he made the expedition, but he devoted the next seven years of his life to developing the work. The survey, which began from California, collected over 5,000 rock samples, and published its results as a series of articles on systematic geology.*

It was King who later convinced Congress to establish the U.S. Geological Survey, and he became its first director. He occupied the position for only two years, however, before returning to work as a consulting mining geologist and engineer. His death in 1901 was hastened by pneumonia, which he contracted while working in mining properties in Butte, Montana. He remains one of the most colorful, if controversial, figures of North American geology.

KING'S FORMATIVE YEARS

R. A. Bartlett

Clarence King was born at Newport, Rhode Island, on January 6, 1842, the newest addition to a long line of transplanted Englishmen. Tradition had it that his great grandfather, Samuel King, of Newport, had been an artist of some ability and a friend of Benjamin Franklin's, even helping the good doctor with some of his electrical experiments. And throughout the American part of their history, the Kings had enjoyed a generous sprinkling of gentlemen, scholars, and men of wealth.

King died in China, at the port of Amoy, June 2, 1848. He left his fortune, what there was of it, invested in the trading firm, and when it went bankrupt in 1857, the widow King and her young son were left nearly penniless.

But the widowed Mrs. King was a remarkably intelligent young woman who was determined to help her little son in every way possible. In the pre-Darwinian era, fossils and pretty rocks were strange things indeed for a small boy to collect, but it is related that she shared his enthusiasms. Once, when Clare was just seven, she let him take her by the hand and lead her over a mile of snow to show her what was perhaps his first geological discovery, a fossil fern in a stone wall. This so aroused the boy's interests that from then on the room he and his mother inhabited became "a veritable museum." The close mother-son relationship was to last throughout King's life. In manhood, he found in his mother his closest intellectual companion, and even on his death bed his closest thoughts were of her.

In the summer of 1862, King was graduated with the first class to receive bachelor of science degrees from Sheffield Scientific School. For a summer vacation he rowed up Lake Champlain and down the St. Lawrence to Quebec, accompanied by three friends including Gardner. That winter of 1862–63 found him studying glacial geology with Alexander Agassiz and even dabbling in the study of the Pre-Raphaelite school of art. Not until the spring of 1863 did the emergency arise that determined the course of his life.

It concerned the health of a rather frail little fellow, of considerably smaller stature than King, who was nevertheless his best friend. James Terry Gardner had known King since they were both lads of fourteen, and, as Gardner once testified before a congressional committee, he and King had lived together "on terms of intimacy closer than those of most brothers." During their summers together they had learned the joys of hunting and fishing and botanizing, and when the time came for them to choose their professions, both had chosen science. Gardner had attended the Polytechnic School (Rensselaer) at Troy and then had gone on to the Sheffield School, where he and King had renewed their friendship. But the cold climate and the hard work had taxed Gardner's constitution, and the two close friends, King, handsome and robust, and Gardner, haggard and exhausted, decided to head for California.

One wonders at the sad partings from home. What did King's mother have to say about her favorite child leaving for the Far West? What did Jamie's parents say when their frail son, harboring an insistent cough, proudly announced that California was the place for *him?* But it was 1863. That summer witnessed Gettysburg and the fall of Vicksburg. The argument that going to California was safer then enlisting in the Union Army had a telling effect. The young men won out, and, with a third companion named Jim Hyde, they hit the road for Kansas, the great jumping-off place for the long trek to the land of the setting sun.

At St. Joseph, Missouri, they met a man named Speers who was leading a wagon train to California, and they joined up with him. They bought horses, gathered their frontiersman's gear together, and set out. Hardly had they hit the trail before their adventures began.

One day, still in June, the wagon train came to a place among the hills where the earth had been grooved by thousands of wagon wheels, and as they coasted through it, they realized that they had just crossed South Pass. It was quite a letdown, for they were not in mountains, and there was not a pine tree in sight. Southward lay high bluffs, however, and far to the northwest, rising abruptly in the summer haze, lay the ominous hulks of the

Wind River Mountains. Straight west lay the cool, shallow waters of the Green River. The divide was crossed, and the true West lay beyond: Mormons, the desert, the Comstock Lode at Virginia City, the high Sierras, and California. With renewed energy they clattered on.

Westward they traveled, to Fort Bridger, crowded with nine hundred Indians held prisoners; to Echo Canyon and the Weber Valley; and finally to Salt Lake City, where they encamped on a square in the center of the city and feasted on lettuce and green onions.

But California still called. The wagon train worked its way around the northern shores of the Great Salt Lake, then westward across the parched sagebrush and alkali flats to the Humboldt River in Nevada. In this valley the three young men turned away from the main party and toward Virginia City. Until then, they had followed the approximate route of the future railroad and had included in their journey much of the area which would later be studied by the Fortieth Parallel Survey. The trip so far had taken them three months, and each of the three thought of himself a completely different man than he had been just three months before back in New York.

As they tramped up into the Sierras (they had sold their horses), they paused occasionally for breath and turned their eyes eastward, surveying the land over which they had passed. It was hot and dry out there, but a few clouds shared the hills, and the delicate blues of the distant mountains would have defied the painting abilities of a great artist. Both King and Gardner had fallen in love with the Great Basin, and when Jamie wrote home to his mother, he told her so. "But seriously," he said, "before we left the plains we had become so fascinated with the life and so interested in the vast loneliness in those deserts . . . that I would gladly have turned around and traveled right back over the same road."

They trekked across the divide and down through the Mother Lode country, analyzing the lay of the land as they went. Soon they were at Sacramento, where they took passage on a river steamer for San Francisco. The steamer was filled with men from the mines, rough, sunburned men in flannel shirts and high boots, wearing belts and revolvers. But one of them, though dressed much the same way, had a different face. Gardner walked past him, then sat down opposite him and pretended to be reading a newspaper. "An old felt hat, a quick eye, a sunburned face with different lines from the other mountaineers, a long weather-beaten neck protruding from a coarse gray flannel shirt and a rough coat, a heavy revolver belt and long legs made up the man," Gardner recalled. "And yet," he felt, "he is an intellectual man: I know it." The "intellectual" turned out to be William H. Brewer, assistant to Josiah Whitney on the California Geological Survey. Brewer had worn down his exploring party to a single packer, and he eagerly hired King to go northward with him to Lassen Peak and Mount Shasta. While Gardner went on to San Francisco and took work with the U.S. Topographical Engineers, King got in his first large scale geological work and had an opportunity for a first-hand study of volcanism—a branch of geology in which he later became the foremost authority in the United States.

For the next four years, King's and Gardner's stars were hitched to the California Geological Survey. Once during this period, King took time off to help in a geological survey of the huge Mariposa estate, owned by John C. Frémont. On another occasion, he and Gardner accompanied General McDowell on a reconnaissance of the desert regions of southern California and part of Arizona. Gardner had joined the survey in April, 1864. The bracing San Francisco climate had not improved his health, and he had resigned his job rather than work on the Sabbath. He had about decided to return home when Professor Whitney invited him to go into the southern Sierras with King, Brewer, and Charles F. Hoffman, the survey topographer. Gardner was to be an "assistant," furnishing his own horse and blankets, for a trip of about four months. As the party progressed, Gardner helped Hoffman and thus learned the secrets of the good topographical methodology which he put to good use during his service with the Fortieth Parallel Survey and with Hayden.

The importance of the association of King and Gardner with the California Geological Survey cannot be overestimated. Josiah Whitney was one of the most respected geologists then at work in the United States, and his appointment as head of the survey had been urged by such eminent scientists as the Sillimans, Dana, Marsh, Leidy, and Meek. His associate, Professor Brewer, like King and Gardner, was a product of the Sheffield Scientific School. Charles F. Hoffman, the topographer, was a well-trained young German who became one of the most valuable men on the survey.

The California Survey had been created by an act of the state legislature in 1860, and Whitney was appointed state geologist. He was supposed to make a complete geological survey of the state and furnish a complete scientific report. From the first, however, he failed to placate the state legislature, and that old bogey appropriations—or lack of them—brought the Survey to an end in 1872, although Whitney continued as state geologist until 1874.

Whitney was a scholar and a scientist, and he never understood the necessity of coming down from the clouds of scientific speculation and doing some earthly lobbying. American legislators tend to be pragmatic, and California legislators in the 1860's were no exception. They had to have an idea of a practical, industrial application for the results of a geological survey, or else they would reject the survey entirely. Whitney's overall plan embraced a preliminary topographical and geological survey as preparation for more specialized studies which ultimately would have practical significance. But the legislature could neither understand this nor wait for the transformation from lofty speculation to concrete and applicable data. When a lawmaker brought one of the survey publications, a fossil-like tome on paleontology, onto the floor of the legislature and proceeded to read from it, the California Geological Survey was as doomed as the dinosaur.

King learned both good and bad things from his experiences with the California Survey. To his benefit were nearly four years with some of the nation's leading geologists, four years of working with scientific field expeditions, and four years of roughing it in the unexplored mountains of the Sierras. Then there was Whitney's idealism. His men, wrote Brewster, "were about the only persons in California who were concerned with the earth, and were not trying to make money out of it." Nor could any member of the survey make a penny for himself.

Such insistence upon personal integrity was of course good. But the personal aspect embraced the policy of the survey as a whole: *it* reflected no interest in the mining industry. It was engaged in pure science, something very close to a scientist's heart. But its appropriations came from a less intellectual body, and so the survey failed for lack of funds. King's concept of a survey of the fortieth parallel reflected in some degree this impracticality, this interest in pure science, which was a product of his years with Whitney.

These were the formative years for Clarence King, and happy ones, and he left more reminiscenses of his work with the California Survey than he did for all his years on the fortieth parallel project. Most of his stories are gathered in a single volume, *Mountaineering in the Sierra Nevada*, which is for the most part a compilation of sketches which had previously appeared in the *Atlantic Monthly*. The book is filled with striking word pictures of the western country. Take, for example, King's description of the desert as seen from far up on its western rim:

> Spread out below us lay the desert, stark and glaring, its rigid hillchains lying in disordered grouping, in attitudes of the dead. The bare hills are cut out with sharp gorges, and over their stone skeletons scanty earth clings in folds, like shrunken flesh; they are emaciated corpses of once noble ranges now lifeless, outstretched as in a long sleep. Ghastly colors define them from the ashen plain in which their feet are buried. Far in the south were a procession of whirlwind columns slowly moving across the desert in spectral dimness. A white light beat down, dispelling the last tract of shadow, and above hung the burnished shield of hard, pitiless sky.

The leading American critic of the time, William Dean Howells, expressed the greatest admiration for King's literary accomplishment. "He has brilliantly fixed forever a place of the Great West already vanished from actuality," Howells said. "In one glowing picture he has portrayed a sublime mood of nature, with all those varying moods of human nature which best give it relief."

Yet King rightfully looked upon these pleasant years in the Sierras as a period of preparation for more serious duties, as perhaps a last golden experience in the happy days of youth and a prelude to the more profound professional work of adulthood. "There are turning-points in all men's lives which must give them both pause and retrospect," he wrote.

> In long Sierra journeys the mountaineer looks forward eagerly, gladly, till pass or ridge-crest is gained, and then, turning with a fonder interest, surveys the scene of his march, letting the eye wander over each crag and valley, every blue hollow or pine-land or sunlit gem of alpine meadow.... With a lingering look he starts forward, and the closing pass-gate with its granite walls shuts away the retrospect, yet the delightful picture forever after hangs on the gallery wall of his memory. It is thus with me about mountaineering; the pass which divided youth from manhood is traversed, and the serious service of science must hereafter claim me.

For Clarence King, the year 1866 marked his crossing of the pass from youth to manhood.

Moses

> At the watershed the track divided suddenly down by short zigzags into the head of Wadi Musa, till it ended abruptly in a block of big boulders.
> This is Ain Musa, the rock which Moses is said to have struck: from under it there flows a stream that runs on down the valley and in the old days fed Petra. In all the country between here and Suez one is impressed with the personality of Moses, as perpetuated in many Arab names and traditions. He is the one local man whose memory has stuck, even from 3000 years or more ago.

—R. A. Bagnold

*Haroun Tazieff gives a vivid account of his first close contact with vol-
canoes. While working as a mining geologist in the Belgian Congo, he
was suddenly summoned to drive north to investigate an active volcano.
The detailed account that he gives, taken from his book* Craters of Fire,
conveys something of the flavor of this encounter.

THE MAKING OF A VOLCANOLOGIST

Haroun Tazieff

I returned to camp one evening and found a native runner who had arrived
shortly before. Squatting serenely on his heels with a pot of *bukari* in front
of him, at regular intervals he deftly flicked down his throat little pellets of
this porridge made of cassava flour, which he had previously rolled between
his fingers. He got up and placidly wiped his hands on the seat of his khaki
shorts, leaving a smear of grease on them.

'Yambo, Bwana.' ('Hail, Master.')

He produced from his pocket the official envelope that he had brought
me. The telegram was dated several days earlier. It had taken the message
several hours to travel by wireless across the expanses of the immense Congo,
as far as Léopoldville, and then, having been left lying about in various
government offices in the capital of the province, it had ended up by coming
to me at the steady jog-trot of this bureaucratised native. The news it contained
rejoiced my heart. My distant chief instructed me to hasten with all possible
speed to the north of Lake Kivu, in the territory of the Albert National Park,
to observe a volcanic eruption in the Virunga Range.

This unexpected mission promised me weeks and perhaps months of a life
of freedom and variety, taking me to new scenery and, I imagined, wild
mountain air, on an undertaking where everything would have to be impro-
vised. The difficulties, which I could guess at, the prospect of rough living
spiced, no doubt, with still more or less undefined dangers—all these were
additional attractions.

Already, while I was doing honour to the grilled banana served by my
'boy' Paya, I saw looming up before me the mass of the eight volcanoes that
form the Virunga Range: eight giants reared up on a vast plateau of lava at
which the jungle nibbled. I knew that six of them, with altitudes of between
10,800 and 14,800 feet, were considered definitely extinct, but that two
others, at the western end of the range, were still active. Which of the two
had suddenly burst into renewed virulence? Was it Nyamlagira, 9,840 feet
in height, which had been dormant since 1940, after an eruption lasting two
years? Or was it Niragongo, that mighty truncated cone towering to a height
of almost 11,500 feet over the blue meanderings of Lake Kivu? A plume of
steam formed a permanent crown to it. In the course of one of my trips I had

seen this plume aglow in the night, lit up by the flames of lava in the immense crater. That night again, althought I was so far away, it went on glowing for a long time in my half-sleep . . .

It is about 140 miles from Bukavu, nestling right down at the southernmost point of the lake, at the far end of a pretty bay fretted with the foliage of eucalyptus trees, to Goma, a tiny European centre on the northern shore. They are connected by a mountain road, which is both winding and bad. On the average, the journey takes seven hours.

As I drove, I began to consider my new job.

A geologist who goes in for volcanoes, I said to myself, must be called a *volcanologist* . . . The idea of the mission pleased me more and more, but I began to feel some uneasiness when I thought of how little knowledge I had of the subject. Sitting on a bench in a lecture-room, I had paid precious little attention to these pustules on the body of the earth. Our teachers, extremely learned when it was a matter of phenomena dating from several millions of years ago or of what must be going on several thousands of leagues down under the earth, were rather disdainful about these all-too-contemporary manifestations on the surface. And so, realising my ignorance, I had spent the previous afternoon rummaging through the official library at Coste, in an effort to find a handbook on volcanology. But in vain. Despite its proximity to volcanoes, the library seemed equally determined to ignore the subject. In despair of finding what I wanted, I had had to fall back on the 'Volcanoes' chapter in a general work on geology; and I must say it was excellent as far as it went. Altogether I was feeling slightly at a loss. Still, I thought to myself, while the truck was contending with the most wanton corkscrew bends in the road, had it not always been thus in any colony? There everyone learns for himself, by sheer trial and error, how to do what, in theory, he knows nothing about, and nobody does the thing he has been trained for. The mechanic builds houses, the prospector nurses the sick, the retired sea-faring man plants coffee . . . and in the long run everyone manages somehow. A geologist who has done some surface-mining should be able to get along quite well in volcanology!

As far as actual volcanoes went, all I had ever really seen was the peak at Teneriffe, that superb bulk of dark indigo outlined against the liquid gold of the sunset sky. Superb, of course, but not much to go on in the way of scientific information . . . The ship had kept steadily on her course, without calling there. For the rest, there had been certain exceptionally clear nights when from Bukavu I had been able to look across sixty miles of lakewater to where Niragongo sent up the lurid glow that had just been haunting my vigil. Some time previously I had even crossed that Virunga Range towards which we were at this moment driving, but the weather had been too thick for us to see much of our surroundings.

Naturally enough, like everyone else I knew that there were *active* volcanoes, which may be either dormant or in eruption, and those that are *extinct*. I also knew that they are all situated on important cracks in the crust of the earth, cracks that are produced in places where the crust is weak, generally on the edge of deep oceanic faults. And furthermore I knew that certain volcanoes are violently explosive, whereas others behave in a manner that is essentially effusive; some send out viscous lava, others fluid lava.

And finally, I knew that although the general form of the volcano is the conic mountain with a crater opening at the summit, there are also others: the 'shield' volcano, the volcano with a 'plug dome', the 'strato-volcano', which has for a crater a vast 'sink-hole', a sort of caved-in cylindrical tank pierced by one or more shafts, also cylindrical, and again the kind called *caldera* (the Spanish for 'cauldron')—immense craters with a more or less horizontal floor covered with smaller cones, as well as pits and fractures.

But what did I know of the *why* behind volcanic phenomena, the force driving to the surface the magma from the depths of the earth? Of all the aspects of my ignorance, the most serious, in my eyes, was ignorance of the volcanologist's *job*. There was nothing definite in my mind but my haste to get there, my good will, and a curiosity ready for anything. Doubtless this was a great deal, but it was, perhaps, scarcely enough. I tried, without much success, to imagine what was waiting for me and to work out the plan of attack. These reflections filled hour after hour of my journey, jolting along through the darkness.

The Longest Chapter

When the work of the geologist is finished and his final comprehensive report is written, the longest and most important chapter will be upon the latest and shortest of the geological chapters.

—Grove Karl Gilbert

Sir William Edgeworth David (1858–1934) was born in Wales, and was educated at Oxford, intending to enter holy orders. His health broke down, however, and he emigrated to Australia, first joining the Geological Survey of New South Wales, and subsequently, in 1891, becoming Professor of Geology at the University of Sydney. He served as leader of the Royal Society's Funafuti expedition in 1897, and was chief of the scientific staff to Shackleton's Antarctic Expedition of 1907–09. Though 50 years old when he reached the Antarctic, he led the climb of Mt. Erebus in 1908 and was one of the small party that reached the South Magnetic Pole in 1909. The events described in the present extract took place on that memorable journey.

WITH SHACKLETON IN THE ANTARCTIC

M. E. David

On January the 28th, the Professor's fiftieth birthday, they sighted the top of the cone of Mount Erebus. He said:

The excitement grew on board as we drew nearer and nearer to the marvellous mass of land which formed such a dramatic termination to Ross' ever memorable voyage in 1841.

Little by little the giant peak of Erebus and its smoke-cloud were shut out from our view by the sister peak of Mount Terror. . . . At midnight of January 28th the sun was still shining brightly at the back of Mount Terror, and the scene was one of exquisite beauty. . . . In the middle distance, across a strip of purplish-grey sea, were the ice-cliffs of the heavily crevassed glaciers creeping seawards down the slopes of Mount Terror, with here and there immense masses of black lava breaking the even line of white. To the left, the slopes of Terror showed the most delicate tints of bluish-purple and pale blue violet, while to the west, where the sunlight at the back of Mount Terror was reflected down by a dense cumulus cloud, the snow-slopes glowed with a soft golden radiance. . . . In the distance was the skyey peak of Erebus with its great steamcloud flushed with pink below and all dark grey above. One understood then in some measure what must have been the feelings of Ross when he discovered his wonderful land.

While working steadily at the daily tasks that lay about him, his eyes were often turned towards that majestic mountain, Erebus, which, rising rapidly from sea-level, rears itself aloft from near the western side of Ross Island to a height of over 13,000 feet. To the Professor this volcano,

. . . its flanks and foothills clothed with spotless snow, patched with the pale blue of glacier ice, its active crater crowned with a spreading smoke-cloud and overlooking the vast white plain of the Barrier to the east and south, is one of the fairest and most majestic sights that Earth can show. . . . For us, living under its shadow, the longing to climb it and penetrate the mysteries beyond the veil, soon became irresistibly strong.

Though several expeditions had been in its neighbourhood, it had never been climbed, and when the suggestion to do so had been made various difficulties had to be considered by the leader before he gave his consent to the project.

> In the first place, the only party who had ascended the foothills of Erebus had found their path barred by heavily crevassed ice. . . . Then, too, the winter was fast approaching, bringing with it blizzards, and temperatures likely to be specially low at high altitudes. . . . After careful consideration Lieutenant Shackleton decided a reconnaissance in the direction of Mount Erebus might be made, and that, if the risk did not appear to be too great, an attempt might be made to reach the summit of the mountain.

The date for starting was fixed for the following morning, March the 5th, and immediately all was excitement and bustle at the winter-quarters, which 'literally rang with the clang of preparation.' It was past midnight before this ceased.

At nine the following morning a party of six, consisting of the summit party and a supporting party, set out with an eleven-foot sledge. The summit party consisted of David, Mawson, and Dr. Mackay, and was under the Professor's leadership. The supporting party was composed of Lieutenant Adams, Dr. Marshall, and Sir Philip Brocklehurst, and was led by Adams. The summit party was provisioned for eleven days, and the supporting party for six.

All that day and the following night they were marooned in their sleeping-bags by the blizzard. As the tent-poles had been left in the sledge they merely doubled the tents over the top of the sleeping-bags. Adams, Marshall, and Brocklehurst, who shared a three-man sleeping-bag, had a narrow escape during the day, Brocklehurst being blown away down the gully when he emerged from the bag, and Adams, who had accompanied him, was also blown down. They managed, by crawling on their hands and knees, and with the utmost difficulty, to make their way back to the sleeping-bag, where they arrived so numb and exhausted that they could hardly struggle into it. The party managed to get some sleep that night, and when they awoke at 4 a.m. the following morning were devoutly thankful to find that the blizzard was over.

This experience upholds the old saying that 'God tempers the wind to the shorn lamb,' for to live in sleeping bags without the shelter of a tent for two nights and a day in a blizzard, at an altitude of 8,750 feet above sea-level, without any ill effects, is, to say the least, unusual in polar travel. It also speaks well for the endurance of the party, particularly for the fifty-year-old member of it, who provided his own warmth in a one-man sleeping-bag. As Shackleton afterwards said, they 'were winning their spurs not only on their first Antarctic campaign, but in their first attempts at serious mountaineering.'

They started the ascent again at 5:30 a.m. The gradient was now a rise of 1 in 1½.

> Burdened as we were with our forty-pound loads, and more or less stiff after thirty continuous hours in our sleeping-bags, and beginning besides to find respiration difficult as the altitude increased, we felt exhausted while we were still 800 feet below the rim of the main crater. Accordingly we halted at noon, thawed some snow with the primus, and were soon revelling in cups of delicious tea, hot and strong, which at once invigorated us. Once more we tackled the ascent.

Mackay, underestimating the effects of altitude on a man carrying a heavy load, took a short cut up a dangerously steep snow slope, cutting steps with his ice-axe instead of following the more gradual rocky route taken by the others. He was soon heard calling for help, so the rest of the party hurried to the top of the ridge and Dr. Marshall and the Professor dropped down to his assistance. Mackay fainted on reaching safety, but soon recovered.

On arriving at the rim of the main crater they found themselves on the brink of a massive precipice, 80 to 100 feet high, of black rock forming the inner edge of the vast depression. Beyond this was an extensive snow-field, the floor of the old crater, which sloped up to the lip of the active cone and crater at its south end, the latter emitting great volumes of steam.

Having chosen for their camping-place a little rocky gully on the slope of the main cone, they prepared a meal, while Dr. Marshall examined Brocklehurst's feet, which had felt numb for some time. They were found to be so badly frost-bitten that the doctor decided that Brocklehurst must stay in camp in his sleeping-bag. The party lunched at 3 p.m. and all but the invalid set off to explore the snow slope leading up to the active crater.

They arrived at the camp soon after 6 p.m., and as they sat on the rocks enjoying their tea, they had a glorious view to the west.

> While the foothills of Erebus flushed rosy red in the sunset, a vast rolling sea of cumulus cloud covered all the land from Cape Bird to Cape Royds. . . . Far away the western mountains glowed with the purest tints of greenish-purple and amethyst. That night we had nothing but hard rock rubble under our sleeping-bags, and quite expected another blizzard; nevertheless, 'weariness can snore upon the flint,' and thus we slept soundly couched on Kenyte lava.

> The following morning had two surprises for us; first, when we arose at 4 a.m. there was no sign of a blizzard, and next, while we were preparing breakfast, someone exclaimed, 'Look at the great shadow of Erebus!' and a truly wonderful sight it was. All the land below the base of the main cone, and for forty miles to the west of it, across MacMurdo Sound, was a rolling sea of dense cumulus cloud. Projected obliquely on this, as on a vast magic lantern screen, was the huge bulk of the giant volcano. The sun had just risen, and flung the shadow of

Erebus right across the Sound and against the foothills of the Western Mountains. Every detail of the profile of Erebus, as outlined on the clouds, could be readily recognized. There to the right was the great black fang, the relic of the first crater; far above and beyond that was to be seen the rim of the main crater, near our camp; then further to the left and still higher rose the active crater with its canopy of steam faithfully portrayed on the cloud screen. Still further to the left the dark shadow dipped rapidly down into the shining fields of cloud below. All within the shadow of Erebus was a soft bluish grey; all without was warm, bright, and golden. Words fail to describe a scene of such transcendent majesty and beauty.

At 6 a.m. they left camp and made all speed to reach the crater summit. Roped together, they climbed several slopes formed by alternating beds of hard snow and vast quantities of large and perfect felspar crystals mixed with pumice.

A little further on we reached the foot of the recent cone of the active crater; here we unroped, as there was no possibility of any crevasses ahead of us.

Our progress was now painfully slow, as the altitude and cold combined to make respiration difficult. . . . A shout of joy and surprise broke from the leading files when, a little after 10 a.m., the edge of the active crater was at last reached. We had travelled only about two and a half miles from our camp, and had ascended just 2,000 feet, and yet this had taken us, with a few short halts, just four hours.

The scene that now suddenly burst upon us was magnificent and awe-inspiring. We stood on the edge of a vast abyss, and at first could neither see to the bottom nor across it, on account of the huge mass of steam filling the crater and soaring aloft in a column 500 to 1,000 feet high. After a continuous loud hissing sound, lasting for some minutes, there would come from below a big dull boom, and immediately afterwards a great globular mass of steam would rush upwards to swell the volume of the snow-white cloud which ever sways over the crater. These phenomena recurred at intervals of a few minutes during the whole of our stay at the crater. Meanwhile the air around us was redolent of burning sulphur.

Presently a gentle northerly breeze fanned away the steam cloud, and at once the whole crater stood revealed to us in all its vast extent and depth. Mawson's measurements made the depth 900 feet, and the greatest width about half a mile. . . . While at the crater's edge we made a boiling point determination with the hypsometer. . . . As the result of averaging aneroid levels, together with the hypsometer determination at our camp at the top of the old crater, calculations made by us showed that the summit of Erebus is probably about 13,370 feet above sea-level.

As soon as our measurements had been made, and some photographs had been taken by Mawson, we hurried back towards our camp, as it was imperatively necessary to get Brocklehurst down to the base of the main cone that day, and this meant a descent in all of nearly 8,000 feet.

On arrival in camp they had a hurried meal, shouldered their heavy packs, and started down the steep mountain slope.

As the Professor in his 'Narrative' does not relate all that followed, Mawson's account of this episode is now quoted:

> I was busy changing photographic plates in the only place where it could be done—inside the sleeping-bag. . . . Soon after I had done up the bag, having got safely inside, I heard a voice from outside—a gentle voice—calling:
> 'Mawson, Mawson.'
> 'Hullo!' said I.
> 'Oh, you're in the bag changing plates, are you?'
> 'Yes, Professor.'
> There was silence for some time. Then I heard the Professor calling in a louder tone:
> 'Mawson!'
> I answered again. Well, the Professor heard by the sound I was still in the bag, so he said:
> 'Oh, still changing plates, are you?'
> 'Yes.'
> More silence for some time. After a minute, in a rather loud and anxious tone:
> 'Mawson!'
> I thought there was something up, but could not tell what he was after. I was getting rather tired of it and called out:
> 'Hullo. What is it? What can I do?'
> 'Well, Mawson, I am in a rather dangerous position. I am really hanging on by my fingers to the edge of a crevasse, and I don't think I can hold on much longer. I shall have to trouble you to come out and assist me.'
> I came out rather quicker than I can say. There was the Professor, just his head showing, and hanging on to the edge of a dangerous crevasse.

Mawson had brought an ice-axe with him. He now inserted the chisel edge of this into a hole which he chipped in firm ice at the crevasse's edge, and holding on to the pick-point, swung the handle towards the Professor, who seized it and managed to clamber on to the solid ice, after which he went off to continue his sketching.

LIFE, TIME, AND DARWIN

Frank H. T. Rhodes

The extracts quoted from the writing of Charles Darwin concern his boyhood years and the voyage of the Beagle, but it is important to put that voyage into perspective. At a meeting of the Linnean Society of London on 1 July 1858, Charles Darwin and Alfred Russell Wallace presented a joint paper entitled "On the Tendency of Species to form Varieties; and on the Perpetuation of Varieties and Species by Natural Means of Selection." The association of the two authors was remarkable. In February 1858, Wallace lay stricken with fever at Ternate, in the jungles of the Moluccas. As his mind wandered over the problem of the development of species, a subject that had exercised his attention for a number of years, he suddenly recalled an *Essay on Population* by the Rev. Robert Malthus, which he had read twelve years before. Malthus argued that the human race would increase in a geometric progression were it not for the fact that many of its members failed to survive and to reproduce. In a sudden flash of insight, Wallace realized the applicability of this to the organic world as a whole and conceived the idea of natural selection in the development of species. Within a week he sent to Darwin a summary of his conclusions under the title "On the Tendency of Varieties to depart indefinitely from the Original Type." Wallace wrote that the idea expressed seemed to him to be new, and asked Darwin, if he also thought it new, to show it to Sir Charles Lyell.

Darwin received the essay with astonishment and dismay, for Wallace's hypothesis was identical with that which he himself had formulated. Darwin, who curiously enough had also been much influenced by Malthus's essay, had devoted the previous twenty years to the patient accumulation of evidence which he proposed to publish as a book. Year after year Darwin accumulated more and more data, and slowly the enormous treatise took form, although he had, in fact, prepared an outline of his theory as far back as 1842, and a more lengthy account two years later. Of these, he wrote to Lyell, "I never saw a more striking coincidence; if Wallace had my MS. sketch written out in 1842, he could not have made a better short abstract! Even his terms now stand as heads of my chapters." It was under such circumstances that Sir Charles Lyell and Sir Joseph Hooker suggested a joint presentation of papers announcing the theory. Darwin and Wallace readily agreed, and the joint publication included Wallace's essay and an extract from Darwin's manuscript of 1844, together with an extract from one of his letters to Asa Grey written in October 1857 "in which," (as Lyell and Hooker noted in their accompanying letter of presentation) "he repeats his views, and which shows that these remained unaltered from 1839 to 1857." The paper was calmly received and few of those who heard it could have predicted the way in which this new theory of evolution was soon to shatter the tranquillity of Victorian thought.

The following year Darwin completed a brief abstract (as he called it) of his researches and on 24 November 1859, there was published the most important book of the century—*The Origin of Species*. On the day of its publication, the first edition of 1,250 copies was sold out.

The effect of *The Origin* upon public opinion was immediate and profound, but its effect upon the natural sciences was revolutionary. It provided the key that not only integrated and interpreted the maze of biological data but also gave new impetus and urgency to every avenue of research.

To Darwin, as to Wallace, the ultimate solution of the problem of the origin of species came suddenly: "In October 1838," he wrote ". . . I happened to read for amusement Malthus on *Population,* and being well prepared to appreciate the struggle for existence which everywhere goes on, from long-continued observation of the habits of animals and plants, it at once struck me that under these circumstances favourable variations would tend to be preserved, and unfavourable ones to be destroyed. . . . I can remember the very spot in the road whilst in my carriage, when to my joy the solution came to me." Yet the path by which Darwin reached this conclusion was a long and laborious one, and certainly his early years gave little hint of the development of genius. He was born at Shrewsbury in 1809, the son and grandson of physicians. His grandfather, Erasmus Darwin, achieved considerable recognition for his poetical exposition of evolutionary views on the origin of species, similar to, but achieved independently of those of Lamarck. Darwin's school days made little impression upon him, although he became an avid collector of minerals and insects.

He himself has described these years in his autobiographical notes.

"In the summer of 1818 I went to Dr. Butler's great school in Shrewsbury, and remained there for seven years . . . Nothing could have been worse for the development of my mind than [this] school, as it was strictly classical, nothing else being taught, except a little ancient geography and history. The school as a means of education to me was simply a blank . . . Towards the close of my school life, my brother worked hard at chemistry, and made a fair laboratory with proper apparatus in the tool-house in the garden, and I was allowed to aid him as a servant in most of his experiments . . . The subject interested me greatly, and we often used to go on working till rather late at night. This was the best part of my education at school . . . but I was once publicly rebuked by the headmaster . . . for thus wasting my time on such useless subjects. . . .

As I was doing no good at school, my father wisely took me away at a rather earlier age than usual, and sent me [October 1825] to Edinburgh University with my brother, where I stayed two years. . . . My brother was completing his medical studies . . . and I was sent there to commence mine. But soon after this period I became convinced . . . that my father would leave me property enough to subsist on with some comfort . . . [and] my belief was sufficient to check any strenuous effort to learn medicine . . . During my second year at Edinburgh I attended Jameson's lectures on Geology and Zoology but they were incredibly dull. The sole effect they produced on me was the determination never as long as I lived to read a book on Geology, or in any way to study the science . . .

After having spent two sessions in Edinburgh, my father perceived . . . that I did not like the thought of being a physician, so he proposed that I should become a clergyman . . . and [I] went to Cambridge after the Christmas vacation, early in 1828. . . . During the three years which I spent at Cambridge my time was wasted, as far as the academical studies were concerned, as completely as at Edinburgh and at school . . . [but] by answering well the examination questions in Paley, by doing Euclid well, and by not failing miserably in Classics, I gained a good place among the οι πολλοι, or crowd of men who do not go in for honours."

Yet in spite of his own somewhat melancholy assessment, Darwin's Cambridge years were to mark the turning point of his career, for there grew up a lasting friendship with J. S. Henslow, the Professor of Botany. It was Henslow who urged him to pursue the despised science of geology, which he did under Adam Sedgwick for an extra term at Cambridge after his graduation. It was also Henslow who arranged for Darwin to accompany H. M. S. Beagle (a 240-ton, ten-gun brig) on a survey voyage to South America and thence round the world. Henslow's parting gesture to the young Darwin was to suggest that he study carefully Lyell's newly published first volume of the *Principles of Geology*, "but on no account accept the views therein advocated". The five-year voyage was, for Darwin, a time of diligent observation and collecting, which gradually opened up a new world of study before him, a world in which most of the treasures were geological and a world which (in spite of Henslow's warning) he saw through the eyes and studied by the methods of Lyell. "I clambered over the mountains of Ascension with a bounding step", he wrote, "and made the volcanic rocks resound under my geological hammer."

The following account concerns Darwin's voyage with the Beagle *and describes his visit in October, 1835, to the Galapagos Islands, a group of ten volcanic islands lying on the equator some 500 miles west of Ecuador. It was here that Darwin was brought into contact with "that mystery of mysteries—the first appearance of new beings on earth." The extract indicates his meticulous observations, and the astonishing gift he had of comparison and synthesis, by which he was able from isolated phenomena to draw wider conclusions. This capacity was, itself, shaped by the voyage. Darwin later wrote "The Voyage of the 'Beagle' has been by far the most important event in my life and has determined my whole career. . . . I have always felt that I owe to the voyage the first real training or education of my mind."*

THE VOYAGE OF THE BEAGLE

Charles Darwin

October 8th.—We arrived at James Island: this island, as well as Charles Island, were long since thus named after our kings of the Stuart line. Mr. Bynoe, myself, and our servants were left here for a week, with provisions and a tent, whilst the Beagle went for water. We found here a party of Spaniards, who had been sent from Charles Island to dry fish, and to salt tortoise-meat. About six miles inland, and at the height of nearly 2000 feet, a hovel had been built in which two men lived, who were employed in catching tortoises, whilst the others were fishing on the coast. I paid this party two visits, and slept there one night. As in the other islands, the lower region was covered by nearly leafless bushes, but the trees were here of a larger growth than elsewhere, several being two feet and some even two feet nine inches in diameter. The upper region being kept damp by the clouds, supports a green and flourishing vegetation. So damp was the ground, that there were large beds of a coarse cyperus, in which great numbers of a very small water-rail lived and bred. While staying in this upper region, we lived entirely upon tortoise-meat: the breast-plate roasted (as the Gauchos do *carne con cuero*), with the flesh on it, is very good; and the young tortoises make excellent soup; but otherwise the meat to my taste is indifferent.

One day we accompanied a party of the Spaniards in their whaleboat to a salina, or lake from which salt is procured. After landing, we had a very rough walk over a rugged field of recent lava, which has almost surrounded a tuff-crater, at the bottom of which the salt-lake lies. The water is only three or four inches deep, and rests on a layer of beautifully crystallized, white salt. The lake is quite circular, and is fringed with a border of bright green succulent plants; the almost precipitous walls of the crater were clothed with wood, so that the scene was altogether both picturesque and curious. A few years since, the sailors belonging to a sealing-vessel murdered their captain in this quiet spot; and we saw his skull lying among the bushes.

During the greater part of our stay of a week, the sky was cloudless, and if the trade-wind failed for an hour, the heat became very oppressive. On two days, the thermometer within the tent stood for some hours at 93°; but in the open air, in the wind and sun, at only 85°. The sand was extremely hot; the thermometer placed in some of a brown colour immediately rose to 137°, and how much above that it would have risen, I do not know, for it was not graduated any higher. The black sand felt much hotter, so that even in thick boots it was quite disagreeable to walk over it.

The natural history of these islands is eminently curious, and well deserves attention. Most of the organic productions are aboriginal creations, found nowhere else; there is even a difference between the inhabitants of the different islands; yet all show a marked relationship with those of America, though separated from that continent by an open space of ocean, between 500 and 600 miles in width. The archipelago is a little world within itself, or rather a satellite attached to America, whence it has derived a few stray colonists, and has received the general character of its indigenous productions. Considering the small size of the islands, we feel the more astonished at the number of their aboriginal beings, and at their confined range. Seeing every height crowned with its crater, and the boundaries of most of the lava-streams still distinct, we are led to believe that within a period geologically recent the unbroken ocean was here spread out. Hence, both in space and time, we seem to be brought somewhat near to that great fact—that mystery of mysteries—the first appearance of new beings on this earth.

Of land-birds I obtained twenty-six kinds, all peculiar to the group and found nowhere else, with the exception of one lark-like finch from North America (Dolichonyx oryzivorus), which ranges on that continent as far north as 54°, and generally frequents marshes. The other twenty-five birds consist, firstly, of a hawk, curiously intermediate in structure between a buzzard and the American group of carrion-feeding Polybori; and with these latter birds it agrees most closely in every habit and even tone of voice. Secondly, there are two owls, representing the short-eared and white barn-owls of Europe. Thirdly, a wren, three tyrant-flycatchers (two of them species of Pyrocephalus, one or both of which would be ranked by some ornithologists as only varieties), and a dove—all analogous to, but distinct from, American species. Fourthly, a swallow, which though differing from the Progne purpurea of both Americas, only in being rather duller colored, smaller, and slenderer, is considered by Mr. Gould as specifically distinct. Fifthly, there are three species of mocking thrush—a form highly characteristic of America. The remaining land-birds form a most singular group of finches, related to each other in the structure of their beaks, short tails, form of body and plumage: there are thirteen species, which Mr. Gould has divided into four sub-groups.

All these species are peculiar to this archipelago; and so is the whole group, with the exception of one species of the sub-group Cactornis, lately brought from Bow Island, in the Low Archipelago. Of Cactornis, the two species may be often seen climbing about the flowers of the great cactus-trees; but all the other species of this group of finches, mingled together in flocks, feed on the dry and sterile ground of the lower districts. The males of all, or certainly of the greater number, are jet black; and the females (with perhaps one or two exceptions) are brown. The most curious fact is the perfect gradation in the size of the beaks in the different species of Geospiza, from one as large as that of a hawfinch to that of a chaffinch, and (if Mr. Gould is right in including his sub-group, Certhidea, in the main group) even to that of a warbler. Seeing this gradation and diversity of structure in one small, intimately related group of birds, one might really fancy that from an original paucity of birds in this archipelago, one species had been taken and modified for different ends. In a like manner it might be fancied that a bird originally a buzzard, had been induced here to undertake the office of the carrion-feeding Polybori of the American continent.

The Ocean of Truth

I do not know what I may appear to the world, but to myself I seem to have been only like a boy playing on the sea-shore, and diverting myself in now and then finding a smoother pebble or a prettier shell than ordinary, whilst the great ocean of truth lay all undiscovered before me.

What Descartes did was a good step. You have added much several ways, & especially in taking ye colours of thin plates into philosophical consideration. If I have seen further it is by standing on ye shoulders of Giants.

—Sir Isaac Newton

Sir Archibald Geikie (1835–1924) was one of the great scientist-statesmen of the late 19th and early 20th centuries. He had a distinguished career as a geologist with what was then the Geological Survey of the United Kingdom, ultimately becoming its director. He also served as the first Murchison Professor of Geology at the University of Edinburgh. Some of his early work was carried out jointly with his fellow-Scot, Sir Roberic Murchison, with whom in 1862 he published one of the earliest geologic maps of Scotland. An outstanding field geologist, he made major contributions to the study of geomorphology, vulcanism, and the geology of Scotland. His textbooks of geology were widely read, but he was also a perceptive and literate author of more general works, as the following account illustrates.

In the present extract taken from his autobiography A Long Life's Work, *published in 1924, Geikie described the discovery of fossils made early in his career and reflected upon his own personal commitment to geology. The passage also includes a fascinating account of a meeting with Hugh Miller. The events described took place in the autumn of 1856.*

A LONG LIFE'S WORK

Sir Archibald Geikie

A new interest in the valleys and ravines of the Pentland Hills arose when I had the good fortune to discover among them, underneath conglomerates and massive sheets of lava of Old Red Sandstone age, a group of shales crowded with fossils. Some twenty years earlier Charles Maclaren had found a shell (*Orthoceras*) there, but nothing further had been done in searching the strata for more organic remains which would determine their geological age. I disinterred a good many species of brachiopods, lamellibranchs and other fossils, well preserved and identical with familiar organisms in the Upper Silurian part of the Geological Record. When subsequently the Survey Collector was put on the ground he obtained a number of additional species.

Such a "find" as this gave special emphasis to a thought which from the beginning had often been in my mind. The work on which I was now engaged, and to which I had dedicated my life, was not merely an industrial employment; the means of gaining a livelihood; a pleasant occupation for mind and body. It often wore to me an aspect infinitely higher and nobler. It was in reality a methodical study of the works of the Creator of the Universe, a deciphering of His legibly-written record of some of the stages through which this part of our planet passed in His hands before it was shaped into its present form. A few of the broader outlines of this terrestrial history had been noted by previous observers in the Lothians, but many other features still remained to be recognised and interpreted, while the mass of complicated detail, so needful for an adequate comprehension of the chronicle as a whole, was practically still unknown. I felt like an explorer entering an untrodden land. Every day the rocks were yielding to me the story of their birth, and thus making fresh, and often deeply interesting, additions to my stock of knowledge. Lucretius, in a well-known passage of his great poem on Nature, has given expression to the enthusiasm wherewith a mind may be filled by the contemplation of a region of philosophical thought which no human foot has yet entered, and he warmly pictures the joy of the first comer who approaches and drinks from springs which no human lips have yet tasted. It is true that the poet confessed that he felt goaded onward by the sharp spur of fame, and by the hope that from the new flowers of the unexplored realm he would gather a splendid coronal, such as had never yet graced the head of any mortal. I had no such ambitious dreams, though I looked forward to rising in my profession to something higher than the "surveyor's drudge or Scottish quarryman" of Macculloch's sarcasm. Meanwhile the discovery of new facts in the ancient history of the land gave me in itself an ample store of pleasure, augmented by the delight of fitting them into each other and extracting from them the connected narrative which they had to reveal. I remember the indignation with which, about this time, in reading Cowper's poems, as I loved to do, I came upon his ignorant and contemptuous denunciation of astronomers, geologists and men of science in general whose work in life was described as

the toil
Of dropping buckets into empty wells
And growing old in drawing nothing up.

More tolerable was Wordsworth's condescending description of the geologist, going about with a pocket-hammer, smiting the "luckless rock or prominent stone," classing the splinter by "some barbarous name" and hurrying on, thinking himself enriched, "wealthier, and doubtless wiser than before."

Samuel Johnson is recorded by his biographer to have expressed his belief that "there is no profession to which a man gives a very great proportion of his time. It is wonderful, when a calculation is made, how little the mind is actually employed in the discharge of any profession." Whether the active clergyman, lawyer or medical man would subscribe to this statement may be questioned. But I found that the profession which I had chosen demanded all the time and thought which I could give to it, both with body and mind.

The discovery of a large group of organic remains, previously unknown to exist where I found them, might have been made by anybody, and conferred no particular credit on the finder. But it made a profound impression on my mind, as its full meaning in the unravelling of the geological history of the district gradually revealed itself. I lost no time in including it, with an account of my recent doings in Skye, in a letter to my kind friend Hugh Miller. His note in reply, from its interest in reference to his own field-observations, which he never lived to publish, may be inserted here.

Wednesday evening 9th October, 1856

MY DEAR SIR

Could you drop in upon me on the evening of Saturday first, and share in a quiet cup of tea. I am delighted to hear that you have succeeded in reading off the Liassic deposits of Skye, and shall have much pleasure in being made acquainted with the result of your labours. Like you, I want greatly a good work on the English Lias, but know not that there is any such. My explorations this season have been chiefly in the Pleistocene and the Old Red. I have now got Boreal shells in the very middle of Scotland, about equally removed from the eastern and western seas. But the details of our respective explorations we shall discuss at our meeting.

Yours ever

HUGH MILLER.

Having been fortunately able to accept this invitation, I spent a couple of hours with him in his home at Portobello. He was interested in the discovery of so large a series of well-preserved fossils in the basement rocks of the Pentland Hills, and not less pleased with the details of my work in Skye. He had spread out on the table his trophy of northern shells, which enabled him to affirm that, at a late date in geological time, Scotland was cut in two by a sea-strait that connected the firths of Forth and Clyde. He had found marine shells at Buchlyvie in Stirlingshire, on the low ground between the two estuaries. Finding me not quite clear as to the precise location of Buchlyvie, he burst out triumphantly with the lines which Scott puts at the head of the immortal chapter of *Rob Roy* wherein are told the adventures at the Clachan of Aberfoyle:

> Baron o' Buchlyvie,
> May the foul fiend drive ye,
> And a' to pieces rive ye,
> For building sic a toun,
> Where there's neither horse-meat, nor man's meat, nor a chair
> to sit doun.

*Hugh Miller (1802–1856) was a Scottish author, philosopher and geol-
ogist. After a rather wild youth, he was apprenticed to a stonemason at
17 and became a journeyman mason, following his trade all over Scotland.
He later became an accountant in a local bank, and embarked on a writing
career. This included some early poetry, and antiquarian and natural
history studies of northern Scotland, and his skills subsequently brought
him the editorship of a biweekly political newspaper,* Witness. *A man of
deep religious conviction, Miller pursued geology as a hobby. His best-
known work,* The Old Red Sandstone, *from which our extract is taken,
appeared first in serial form in* Witness. *His self-taught technical skill in
describing the fossil fishes of the Old Red Sandstone was highly praised
by Huxley. Amongst his dozen or so other books,* Footprints of the Creator
was a vigorous attack on Chamber's evolutionary Vestiges of Creation.
*His later years were marred by lung complications, induced by his early
quarrying work.*

THE OLD RED SANDSTONE

Hugh Miller

It was twenty years last February since I set out, a little before sunrise, to
make my first acquaintance with a life of labour and restraint; and I have
rarely had a heavier heart than on that morning. I was but a slim, loose-
jointed boy at the time, fond of the pretty intangibilities of romance, and of
dreaming when broad awake; and, woful change! I was now going to work
at what Burns has instanced, in his "Twa Dogs," as one of the most disagreeable of all employments,—to work in a quarry. Bating the passing
uneasiness occasioned by a few gloomy anticipations, the portion of my life
which had already gone by had been happy beyond the common lot. I had
been a wanderer among rocks and woods, a reader of curious books when
I could get them, a gleaner of old traditionary stories; and now I was going
to exchange all my day-dreams, and all my amusements, for the kind of life
in which men toil every day that they may be enabled to eat, and eat every
day that they may be enabled to toil!

The quarry in which I wrought lay on the southern shore of a noble inland
bay, or frith rather, with a little clear stream on one side, and a thick fir
wood on the other. It had been opened in the Old Red Sandstone of the
district, and was overtopped by a huge bank of diluvial clay, which rose over
it in some places to the height of nearly thirty feet, and which at this time
was rent and shivered, wherever it presented an open front to the weather,
by a recent frost. A heap of loose fragments, which had fallen from above,
blocked up the face of the quarry, and my first employment was to clear
them away. The friction of the shovel soon blistered my hands, but the pain

was by no means very severe, and I wrought hard and willingly, that I might see how the huge strata below, which presented so firm and unbroken a frontage, were to be torn up and removed. Picks, and wedges, and levers, were applied by my brother-workmen; and, simple and rude as I had been accustomed to regard these implements, I found I had much to learn in the way of using them. They all proved inefficient, however, and the workmen had to bore into one of the inferior strata, and employ gunpowder. The process was new to me, and I deemed it a highly amusing one: it had the merit, too, of being attended with some such degree of danger as a boating or rock excursion, and had thus an interest independent of its novelty. We had a few capital shots: the fragments flew in every direction; and an immense mass of the diluvium came toppling down, bearing with it two dead birds, that in a recent storm had crept into one of the deeper fissures, to die in the shelter. I felt a new interest in examining them. The one was a pretty cock goldfinch, with its hood of vermilion, and its wings inlaid with the gold to which it owes its name, as unsoiled and smooth as if it had been preserved for a museum. The other, a somewhat rarer bird, of the woodpecker tribe, was variegated with light blue and a grayish yellow. I was engaged in admiring the poor little things, more disposed to be sentimental, perhaps, than if I had been ten years older, and thinking of the contrast between the warmth and jollity of their green summer haunts, and the cold and darkness of their last retreat, when I heard our employer bidding the workmen lay by their tools. I looked up and saw the sun sinking behind the thick fir wood beside us, and the long dark shadows of the trees stretching downwards towards the shore.

This was no very formidable beginning of the course of life I had so much dreaded. To be sure, my hands were a little sore, and I felt nearly as much fatigued as if I had been climbing among the rocks; but I had wrought and been useful, and had yet enjoyed the day fully as much as usual. It was no small matter, too, that the evening, converted, by a rare transmutation, into the delicious "blink of rest" which Burns so truthfully describes, was all my own. I was as light of heart next morning as any of my brother-workmen. There had been a smart frost during the night, and the rime lay white on the grass as we passed onwards through the fields; but the sun rose in a clear atmosphere, and the day mellowed, as it advanced, into one of those delightful days of early spring which give so pleasing an earnest of whatever is mild and genial in the better half of the year. All the workmen rested at mid-day, and I went to enjoy my half-hour alone on a mossy knoll in the neighbouring wood, which commands through the trees a wide prospect of the bay and the opposite shore. There was not a wrinkle on the water, nor a cloud in the sky, and the branches were as moveless in the calm as if they had been traced on canvas. From a wooded promontory that stretched half-way across the frith there ascended a thin column of smoke. It rose straight as the line of a plummet for more than a thousand yards, and then, on reaching a thinner stratum of air, spread out equally on every side, like the foliage of a stately tree. Ben Wyvis rose to the west, white with the yet unwasted snows of winter, and as sharply defined in the clear atmosphere as if all its sunny slopes and blue retiring hollows had been chiselled in marble. A line of snow ran along the opposite hills: all above was white, and all below was purple. I returned to the quarry, convinced that a very exquisite pleasure may be a very cheap one, and that the busiest employments may afford leisure enough to enjoy it.

The gunpowder had loosened a large mass in one of the inferior strata, and our first employment, on resuming our labours, was to raise it from its bed. I assisted the other workmen in placing it on edge, and was much struck by the appearance of the platform on which it had rested. The entire surface was ridged and furrowed like a bank of sand that had been left by the tide an hour before. I could trace every bend and curvature, every cross hollow and counter ridge, of the corresponding phenomena; for the resemblance was no half resemblance,—it was the thing itself; and I had observed it a hundred and a hundred times, when sailing my little schooner in the shallows left by the ebb. But what had become of the waves that had thus fretted the solid rock, or of what element had they been composed? I felt as completely at fault as Robinson Crusoe did on his discovering the print of the man's foot in the sand. The evening furnished me with still further cause of wonder. We raised another block in a different part of the quarry, and found that the area of a circular depression in the stratum below was broken and flawed in every direction, as if it had been the bottom of a pool recently dried up, which had shrunk and split in the hardening. Several large stones came rolling down from the diluvium in the course of the afternoon. They were of different qualities from the sandstone below, and from one another; and, what was more wonderful still, they were all rounded and water-worn, as if they had been tossed about in the sea or the bed of the river for hundreds of years. There could not, surely, be a more conclusive proof that the bank which had enclosed them so long could not have been created on the rock on which it rested. No workman ever manufactures a half-worn article, and the stones were all half-worn! And if not the bank, why then the sandstone underneath? I was lost in conjecture, and found I had food enough for thought that evening, without once thinking of the unhappiness of a life of labour.

The immense masses of diluvium which we had to clear away rendered the working of the quarry laborious and expensive, and all the party quitted it in a few days, to make trial of another that seemed to promise better. The one we left is situated, as I have said, on the southern shore of an inland bay,—the Bay of Cromarty; the one to which we removed has been opened in a lofty wall of cliffs that overhangs the northern shore of the Moray Frith. I soon found I was to be no loser by the change. Not the united labours of a thousand men for more than a thousand years could have furnished a better section of the geology of the district than this range of cliffs. It may be regarded as a sort of chance dissection on the earth's crust. We see in one place the primary rock, with its veins of granite and quartz, its dizzy precipices of gneiss, and its huge masses of hornblende; we find the secondary rock in another, with its beds of sandstone and shale, its spars, its clays, and its nodular limestones. We discover the still little-known but highly interesting fossils of the Old Red Sandstone in one deposition, we find the beautifully preserved shells and lignites of the Lias in another. There are the remains of two several creations at once before us. The shore, too, is heaped with rolled fragments of almost every variety of rock,—basalts, ironstones, hyperstenes, porphyries, bituminous shales, and micaceous schists. In short, the young geologist, had he all Europe before him, could hardly choose for himself a better field. I had, however, no one to tell me so at the time, for Geology had not yet travelled so far north; and so, without guide or vocabulary, I had to grope my way as I best might, and find out all its wonders for myself. But so slow was the process, and so much was I a seeker in the dark, that the facts contained in these few sentences were the patient gatherings of years.

In the course of the first day's employment I picked up a nodular mass of blue limestone, and laid it open by a stroke of the hammer. Wonderful to relate, it contained inside a beautifully finished piece of sculpture,—one of the volutes, apparently, of an Ionic capital; and not the far-famed walnut of the fairy tale, had I broken the shell and found the little dog lying within, could have surprised me more. Was there another such curiosity in the whole world? I broke open a few other nodules of similar appearance,—for they lay pretty thickly on the shore,—and found that there might. In one of these there were what seemed to be the scales of fishes, and the impressions of a few minute bivalves, prettily striated; in the centre of another there was actually a piece of decayed wood. Of all Nature's riddles, these seemed to me to be at once the most interesting and the most difficult to expound. I treasured them carefully up, and was told by one of the workmen to whom I showed them, that there was a part of the shore about two miles farther to the west where curiously-shaped stones, somewhat like the heads of boarding-pikes, were occasionally picked up; and that in his father's days the country people called them thunderbolts, and deemed them of sovereign efficacy in curing bewitched cattle. Our employer, on quitting the quarry for the building on which we were to be engaged, gave all the workmen a half-holiday. I employed it in visiting the place where the thunderbolts had fallen so thickly, and found it a richer scene of wonder than I could have fancied in even my dreams.

What first attracted my notice was a detached group of low-lying skerries, wholly different in form and colour from the sandstone cliffs above or the primary rocks a little farther to the west. I found them composed of thin strata of limestone, alternating with thicker beds of a black slaty substance, which, as I ascertained in the course of the evening, burns with a powerful flame, and emits a strong bituminous odour. The layers into which the beds readily separate are hardly an eighth part of an inch in thickness, and yet on every layer there are the impressions of thousands and tens of thousands of the various fossils peculiar to the Lias. We may turn over these wonderful leaves one after one, like the leaves of a herbarium, and find the pictorial records of a former creation in every page: scallops, and gryphites, and ammonites, of almost every variety peculiar to the formation, and at least some eight or ten varieties of belemnite; twigs of wood, leaves of plants, cones of an extinct species of pine, bits of charcoal, and the scales of fishes; and, as if to render their pictorial appearance more striking, though the leaves of this interesting volume are of a deep black, most of the impressions are of a chalky whiteness. I was lost in admiration and astonishment, and found my very imagination paralysed by an assemblage of wonders that seemed to outrival in the fantastic and the extravagant even its wildest conceptions. I passed on from ledge to ledge, like the traveller of the tale through the city of statues, and at length found one of the supposed aerolites I had come in quest of firmly imbedded in a mass of shale. But I had skill enough to determine that it was other than what it had been deemed. A very near relative, who had been a sailor in his time on almost every ocean, and had visited almost every quarter of the globe, had brought home one of these meteoric stones with him from the coast of Java. It was of a cylindrical shape and vitreous texture, and it seemed to have parted in the middle when in a half-molten state, and to have united again, somewhat awry, ere it had cooled enough to have lost the adhesive quality. But there was nothing organic in its structure; whereas the stone I had now found was organised very curiously indeed. It was of a conicle form and filamentary texture, the filaments radiating in straight lines from the centre to the circumference. Finely-marked veins like white threads ran transversely through these in its upper half to the point; while the space below was occupied by an internal cone, formed of plates that lay parallel to the base, and which, like watchglasses, were concave on the under side and convex on the upper. I learned in time to call this stone a belemnite, and became acquainted with enough of its history to know that it once formed part of a variety of cuttle-fish, long since extinct.

My first year of labour came to a close, and I found that the amount of my happiness had not been less than in the last of my boyhood. My knowledge, too, had increased in more than the ratio of former seasons; and as I had acquired the skill of at least the common mechanic, I had fitted myself for independence. The traditional experience of twenty years has not shown me that there is any necessary connection between a life of toil and a life of wretchedness; and when I have found good men anticipating a better and a happier time than either the present or the past, the conviction that in every period of the world's history the great bulk of mankind must pass their days in labour, has not in the least inclined me to scepticism.

One important truth I would fain press on the attention of my lowlier readers: there are few professions, however humble, that do not present their peculiar advantages of observation; there are none, I repeat, in which the exercise of the faculties does not lead to enjoyment. I advise the stonemason, for instance, to acquaint himself with Geology. Much of his time must be spent amid the rocks and quarries of widely-separated localities. The bridge or harbour is no sooner completed in one district than he has to remove to where the gentleman's seat or farm-steading is to be erected in another; and so, in the course of a few years, he may pass over the whole geological scale.

Should the working man be encouraged by my modicum of success to improve his opportunities by observation, I shall have accomplished the whole of it. It cannot be too extensively known, that nature is vast and knowledge limited, and that no individual, however humble in place or acquirement, need despair of adding to the general fund.

". . . that looks like goald"

Monday 24th this day some kind of mettle was found in the tail race that looks like goald first discovered by James Martial, the Boss of the Mill. Sunday 30th Clear and has been all the last week our metal has been tride and proves to be goald and it is thought to be rich we have pict up more than a hundred dollars worth last week.

—A Diary

Raphael Pumpelly (1837–1923) was a geologist and mining engineer who worked in Japan, China, Mongolia, and Turkestan, as well as the United States. The present extract gives a glimpse of student life at Freiberg in 1857, when he was a student at the University there. Of particular interest is his comparison between the professor-student relationship in Germany and that in the United States at that time.

LIFE AT FREIBERG

Raphael Pumpelly

Weissbach was an excellent teacher, quick, impetuous, often sarcastic. Once, in explaining the instruments, he set up a theodolite, pointed the telescope on an object, looked hastily into it, and called the students one by one to see the cross-hair and focus it. Each one would look and look again, and answer the impatient question: "Well, haven't you focussed it yet?" with a frightened hesitating "ye-es Herr Professor." But when one student resolutely answered "No, Herr Professor," there came the quick exclamation: "Stupid!" Then looking himself, and finding that the cap had all the time been covering the objective, he laughed and said: "It was I, after all, who was the stupid one!"

During the years I was at Freiberg we had abundant practice in surveying on the surface and underground.

Breithaupt was already old; he was, however, one of the fathers of Mineralogy, and an inspiring lecturer. He taught crystallography without mathematics, using wooden models, of which he had one for every possible combination, and for twins. The models were two to three inches in diameter, to be visible to the whole class. He had a long knife blade which he held against the face he was describing. One could never think of the man, the model or the knife separately; they seemed a trinity. He created in his students an interest in crystal forms and systems, that I did not find later in the mathematical treatment under Alvin Weissbach.

Cotta took us on frequent excursions in connection with subjects in his lectures. During the years I was in Freiberg, these trips covered not only the neighboring region, but extended to Bohemia and Thuringia, sometimes lasting a week or even longer. In some of these, in the holidays, several of the other professors joined, and the whole became a jovial picnic. These men, old and middle-aged, were quite as good comrades as any of us. Outside the lecture room they were as young as we were; we liked to mingle with them at the restaurant tables, drinking beer, smoking, and telling stories, or in serious talk; and they joined us in the same spirit. I hold this to have been an important element, in education. It is a phase that seems to be lacking in our universities, and is only partially represented in our conferences on special subjects. Professor Shaler of Harvard was the only American instructor I can recall who was a representative of the type of these men in this respect. Bless them and him: for they are remembered with affection and gratitude.

I remember one excursion in my second year, which Cotta and a fellow student—von Andrian—and I made for two weeks in Thuringia. The scenery and historical associations were most interesting, the inns and the beer and wines were good. Andrian and I were helping Cotta in work on the geology of the region.

One evening during a talk on paleontology Andrian had said: "There are so many contradictory determinations of species that it seems as though paleontologists often made them *nach dem Gefühl*."

Now, *Gefühl* may mean either feeling, touch, state of mind, or sentiment. Both of my companions spoke English. The discussion at once lost its seriousness. Cotta said: "It is true that *das Gefühl* in some of its senses may influence the paleontologist; it enters into most decisions, even in a wager; for instance, I wager you can't swallow two eggs on an empty stomach; that is an instance where the decision rests only on *Gefühl*."

Von Andrian said at once: "I can't take that bet because I *feel* it is impossible."

"What is the wager?" I asked.

"Suppose we make it a bottle of champagne," they both answered.

"Good," I responded, "I *feel* that I can do it, so I accept the bet."

"Order the bottle," said Cotta, "for you have lost through your *state of mind*."

"But why?"

"Because when you have swallowed the first egg your stomach is no longer empty."

"But suppose I take both down at one swallow."

They said that would be impossible, but they would let it be decided by experiment.

"That's your *sentiment*," I answered, "order the eggs."

I had my doubts as to success. I broke the eggs into a glass, and mixed them thoroughly; then, under close, and somewhat embarrassing observation, with a successful gulp, I got those two potential chicks simultaneously into my stomach. I had won.

"Now," I said, "I used *Gefühl* in all its senses. I liked eggs; that was *feeling*. I was hungry: that was a *state of mind*. I knew how they would feel slipping down my throat; that was *touch*. But the decision was based on experiment, and therefore on induction. Order the bottle!"

They confessed that it had all been planned beforehand to get me to take the bet.

The champagne went down more pleasantly, though, than the eggs.

On this excursion we visited a wonderfully interesting cave. The rooms were small, as I remember them, but the walls and roof were covered with a mass of great crystals of gypsum instead of stalactites. No torches had blackened them with smoke. As large as one's arm, and several feet long, transparent, white with pearly luster and silken texture, these glistening crystals projected out from the intertwined ground-mass far into the open, and formed a scene of dreamlike beauty.

One day, I think it was near Ilmenau, I had the good luck to find several specimens of crystals showing an unrecorded law of twining. They were pseudomorphs after a feldspar. I gave them later to Professor Breithaupt, who published them, giving me credit. I imagine that this was the first time my name had appeared in print, except perhaps when I was lost in Corsica and being looked for through Europe.

At the end of the excursion, as we were walking to the station, in Eisenach I think, and in tramp-worn clothes, we saw two ladies ahead of us. One of them had dropped a handkerchief. Andrian picked it up and, as we passed them, raising his hat, politely handed it to one of the ladies. We heard one of them say: "Ach! how very polite for a common laborer!"

After we were out of hearing, Andrian laughed and said: "Those ladies are the Duchesse Hélène, daughter of Louis Phillipe, and her daughter. I used to see her when as a boy I was a page at the French Court." Andrian was an Austrian baron.

I looked for Adam

No one stirred up so much controversy in 19th century England as Charles Darwin with his theory of evolution. At his death, however, Britain acknowledged his genius and buried him with much pomp in Westminster Abbey. After the cermony, pallbearer Thomas Huxley was asked by a worried-looking peer, "Do you believe that Darwin was right?" "Of course he was right," explained Huxley. His lordship surveyed the Abbey with a pained expression, then observed in a low tone, "Couldn't he have kept it to himself?"

—Herbert Wendt

CHAPTER 2

Eyewitness Accounts of Geological Phenomena

Science, it is generally asserted, is concerned with facts. But ultimately there is nothing in Nature labeled "fact." Facts represent human abstractions, and our recognition and understanding of facts are based upon individual perception and experience. To this extent geology includes both contrived encounters, which we call experiments, which are based on studies such as those in the laboratory where conditions and materials are carefully controlled, and also other encounters and experiences that can be gained only by firsthand contact with the Earth as it is, which generally means the rocks of the Earth's crust. In this sense, the work of the field geologist provides the basic link between our knowledge and the materials and processes on which it is based.

The role of experiment on Earth Science is limited by the problems of scaling down the Earth to appropriate models, the problem of physical states existing in the Earth but unattainable in the laboratory, and the time element, which is rarely attainable under laboratory conditions. For these and other reasons, laboratory experimentation plays a lesser role in geology than in most other sciences, although there is a rather different kind of "experimentation" which is also possible in the field, especially with such natural geological phenomena as the various agents of erosion and deposition, hot springs, geysers, and volcanoes, to take some obvious examples.

Geology has a twofold concern. It is concerned first with the present configuration and composition of the earth, and second with the interpretation of the past history of the Earth. Field observations are of critical importance, because the recognition of earlier events in the history of the Earth depends heavily on an analysis of present composition and configuration of Earth materials, and the identification of the action of processes and laws observable in our existing environment. Here, again, time and scale are important elements. Catastrophes do indeed take place, but most processes that influence the crust of the Earth act almost imperceptibly. Unfortunately, nearly all the processes that we can observe (as opposed to those we can infer), are rapid ones, which play only a local or limited role in the overall development of the Earth's crust. In spite of this limitation, such events and processes are still of great significance because they allow us to observe and compare multiple sequences of events in which various components and conditions may be analyzed, and in which a measure of prediction and experimentation is possible. Even the speed and extent of these processes has great significance.

Because most centers of population are deliberately established in areas remote from scenes of frequent terrestrial violence and instability, field observation frequently takes the geologist into distant and isolated areas. The areas included in the present accounts range from the Sahara Desert to the Antarctic, and the accounts themselves present fascinating differences in character and style.

Song of the Dredge

Down in the deep, where the mermen sleep
    Our gallant dredge is sinking;
Each finny shape in a precious scrape
    Will find itself in a twinkling!
They may twirl and twist and writhe as they wist,
    And break themselves into sections,
But up they all, at the dredges call
    Must come to fill collections.

—Edward Forbes

Landslides are rapid down-slope movements of rock debris. They represent a particular example of the universal down-slope movement of weathered material. This "mass wasting" takes place under the influence of gravity, and involves down-slope movement ranging from sudden rockfalls to slow hill creep, which proceeds almost imperceptibly over many years. Landslides are facilitated by certain structural rock conditions, such as faults, or bedding planes inclined parallel to the slope of the ground, interbedding of shales with more brittle strata, excessively heavy rainfall which may lubricate joints and bedding planes, and by artificial disturbance of the geologic environment. Two examples of these slides are given.

The first took place in 1903 near the small mining village of Frank, Alberta, and killed 70 people. It is estimated that about 40 million cubic yards of rock avalanched down the face of the 3,000 feet high Turtle Mountain at a speed reaching 60 miles per hour, and struck the valley bottom with such force that it spread across the two-mile width of the floor and was carried 400 feet up the other side. Turtle Mountain consisted of jointed limestone, in which, although the stratification was perpendicular to the face of the mountain, the joints were parallel to it. The opening of these joints by weathering facilitated the movement. The account that follows is taken from the official report, published by the Canadian Department of the Interior soon after the event took place.

THE TURTLE MOUNTAIN SLIDE

R. G. McConnell and R. W. Brock

At dawn, on April 29, 1903, a huge rock mass, nearly half a mile square and probably 400 to 500 feet thick in places, suddenly broke loose from the east face of Turtle mountain and precipitated itself with terrific violence into the valley beneath, overwhelming everything in its course. The great mass, urged forward by the momentum acquired in its descent, and broken into innumerable fragments, ploughed through the bed of Old Man river, and carrying both water and underlying sediments along with it, crossed the valley and hurled itself against and up the opposite terraced slopes to a height of 400 feet. Blocks of limestones and shale, mingled with mud, now cover the valley to a depth of from 3 to probably 150 feet, over an area of 1.03 square miles.

The number of people killed by the slide is not known exactly, but it is given at about 70. The property destroyed includes the tipple and plant at the mouth of the Canadian American Coal and Coke Company's mine, the company's barn and seven cottages at the east end of the town of Frank, half a dozen outlying houses, with some shacks and camps, besides a considerable number of horses and cattle and a couple of ranches. The track of the Crow's

Nest Railway was hopelessly buried for a distance of nearly 7,000 feet and the lower mile of the Frank and Grassy Mountain Railway met a similar fate. The people occupying the houses in the track of the slide were all swept away with it and destroyed, with the exception of a few near the edge of the slide, who escaped in some almost miraculous way, which they themselves cannot explain.

The slide occurred about 4:10 a.m., at a time when most of the inhabitants of the valley were asleep and before full daylight. The statements of the few eye-witnesses throw little light on the character of the slide, but the following notes obtained from them are not without interest:—

Karl Cornelianson was awakened by the noise. He rushed to the door of his house, which looked out over the first terrace flat and the base of the second. His first thought was that there had been an explosion at the mine and his first look was in that direction. Seeing nothing there, he glanced round to the terrace flat in time to see the rock debris hurl itself against the slope of the second terrace, and its momentum spent, fall back to the lower level. His impression was that an explosion had taken place directly in front of him. The edge of the slide was about a quarter of a mile from his door.

Mr. McLean, who kept a boarding house in Frank, was already up. Hearing the noise, he rushed to the door in time to see the slide rush by only a few feet in front of him. The passage of the slide was so instantaneous that he thought an eruption had taken place directly in front of him.

A freight train was shunting on the mine siding at the time of the disaster. They had taken some cars of coal from the tipple at the mouth of the tunnel and were slowly backing up for another load, when the engineer heard the rocks breaking away from the mountain above. He immediately changed to full speed ahead and ran out of danger. The conductor saw the men at the tipple become alarmed and start to run, but they were overtaken by the slide and perished. Immediately after everything was shrouded in a cloud of dust.

Mr. Warrington, who was sleeping in one of the cottages destroyed, was awakened by a noise which he thought was caused by hail. He jumped out of bed and then realized that it was something more serious, but before he had time to become alarmed the house began to rock, and the next thing he was conscious of was finding himself in the lee of some rocks, forty feet from where the house had stood. His bed was some twenty feet farther on. His thigh was broken and he was otherwise injured by small fragments of debris being forced into his body. He pulled himself out with his arms, and was trying to work his way to some children, whose cries he heard, when the first rescue party arrived.

The report by Gordon Gaskill of the 1963 dam failure at the Vajont Dam in the Italian Alps, describes one of the most terrible disasters of modern times. Some three years before the disaster, a great dam of 858 feet high had been built across the narrow valley to impound a deep reservoir. Repeated, but very small earth movements were recorded in the ensuing period, but on the night of October 9, after torrential rains, a rockslide of some 600 million tons occurred high in the valley above the reservoir and threw a huge volume of rock material into the reservoir, creating a great surge of water that rose 300 feet above the top of the dam. As it swept through the the valley below, some 2,000 people lost their lives.

Even in the absence of the reservoir, the rock fall was of such proportions that it would probably have caused considerable damage, but the rock fall itself was facilitated by the construction of the dam. The bedrock consisted of interbedded limestones and clays, which had been weakened by deformation, and further weakened by water petrolating through it from the reservoir.

THE NIGHT THE MOUNTAIN FELL

Gordon Gaskill

They say the animals knew. In that last peaceful twilight—Wednesday, October 9, 1963—hares grew suddenly bold and, oblivious of passing men and automobiles, raced silently, intently, down the paved road—away from the lake. As darkness gathered, cows milled uneasily in their stalls, dogs whimpered, chickens stirred in their pens, unwilling to sleep. A couple watching television was irritated by the unnatural, noisy fluttering of their caged canary. Then the fluttering abruptly stopped: in its strange panic the bird had caught its head in the cage bars and strangled to death. Husband stared at wife: "Something's going to happen! The dam . . . ?"

Life or death this night in the small town of Longarone, in northeast Italy, would turn on a simple, single fact: how high up the hillsides of the river valley below the Vajont Dam you happened to be. All but those in the highest parts would soon die.

An engaged couple, due to marry in six days, had a slight difference of opinion. Giovanna wanted to go to the movies at Belluno, the provincial capital about 12 miles away, but her fiancé, Antonio, felt too tired and begged off. They separated for the night, he to his higher home, she to a lower one. Next morning he would be digging in the muddy waste where Giovanna, her family and home had disappeared, repeating endlessly, "If only I had taken her to the movies. . . . If only I had taken her to the movies."

A teen-age boy astride his motorbike fidgeted with embarrassment as, from a window, his mother tried to talk him out of riding off to another village to see a girl. But from inside the house his father, remembering his own salad days, called out indulgently, "Oh, let him go!" The mother sighed, gave in and the boy rode off to safety—never to see home or parents again.

Visiting Longarone were three Americans of Italian descent, all staying in a low-lying little hotel. One of them, John De Bona, of Riverside, Calif. retired to his room—and never would be seen again. Two others, Mr. and Mrs. Robert De Lazzero, of Scarsdale, N.Y. had panted uphill nearly 150 steps to have dinner with his great-aunt Elisabetta, and two cousins. Shortly before ten, dinner over, they were about to walk back down to the hotel when Aunt Elisabetta said, "Don't go yet. See, I've saved a special bottle of wine for you." Somewhat reluctantly, they stayed on a little longer. They were lucky.

For the clocks of Longarone would never strike 11 on this night. Just before that hour Longarone and the hamlets clustering near it would be erased from the earth, and more than 2000 people would die in perhaps the world's most tragic dam disaster.

For four years the great new Vajont Dam had been both the pride and the fear of people living around Longarone, a sub-Alpine town not far south of the Austrian border. Flung across the nearby Vajont gorge—so deep and narrow that sunlight touched its bottom only fleetingly at noon—the dam was the highest arch dam in all the world, a showpiece for visiting foreign engineers, a magnet for tourists. Its graceful curving wall of tapered concrete rose 858 feet above its base—nearly five times as high as Niagara Falls, 132 feet higher than America's highest dam, Hoover. Its impounded lake, not yet full, would provide enormous amounts of electricity to bring industry, jobs and prosperity to mountain folk for miles around.

Still, many feared it.

Ever since 1959 a swelling chorus of protest had demanded that the dam be stopped or that absolute assurance be given that it was safe. But the site had been approved and all work had been carefully supervised by some of Italy's most respected geologists and engineers. Chief among them was the very father of the Vajont project, Dr. Carlo Semenza, an internationally known engineer who had built dams in several countries. He and others concluded that there might be a little minor land slippage at first—as with nearly all new artificial lakes—but nothing to worry about.

People living nearby weren't so sure. Above all, they distrusted the stability of Monte Toc, which anchored the dam's left shoulder and hung nearly 4000 feet above the new lake. They gave it an ominous nickname, *la montagna che cammina*—"the mountain that walks." The village of Erto, just above the lake, felt itself especially menaced and made most of the early protests.

With construction having begun in 1956, all was ready for a partial test-filling of the lake by March 1960. The results were worrisome. Even modest amounts of water produced, on November 4, 1960, an alarming crack in the earth high up on Monte Toc—a gaping split more than a foot wide and about 8000 feet long. At the same time, a half-million tons of earth and rock slipped into the still-small lake, churning up waves six feet high.

Disappointed, Società Adriatica di Elettricità (SADE), the power company building the dam, lowered the lake and, its timetable wrecked, turned to two years of expensive testing. Also, it performed extensive remedial work: strengthening the dam, digging a large bypass tunnel, sealing suspected rock fractures with pressurized concrete.

But, less than six months after the warning slide, Dr. Semenza himself began to lose hope. In an April 1961 letter, not divulged until after the disaster, he wrote to an engineering friend: "The problems are probably too big for us, and there are no practical remedies to take." He died six months afterward. Yet neither he nor others sharing his worries ever dreamed of any serious danger to *human life*. They feared only that landslides might so clog the basin as to make it useless for water storage.

One test (No. 19) on SADE's elaborate scale model (1/200 actual size) of the entire dam and basin was to have especially disastrous consequences. The formal report on this test said that *if the lake's level was 75 feet below maximum*, this would be "absolutely safe, even in the face of the most catastrophic landslide that can be foreseen." Such a fantastic slide, said the test, would churn up dangerous waves about 80 feet high on the lake.

The safety measures to counteract this seemed obvious: if a slide seemed imminent, (a) get the lake down at least 75 feet below maximum, and (b) evacuate everybody from the shoreline belt so that the expected waves could spend their rage harmlessly. As for people in Longarone, 1½ miles *below* the dam, there was no reason to worry. With the lake down to this "safety level," only about five feet of water, a harmless trickle, could possibly go *over* the dam.

In April 1963 the situation seemed ripe for a new test-raising of the water. By now a new factor had entered the picture. SADE and many other private power companies had been forcibly nationalized under a new state power board, ENEL (Ente Nazionale per l'Energia Elettrica). The exit valves were closed, and the water started up again.

As the waters crept higher, the old disheartening signals appeared. Another long, frightening crack split the earth high on Monte Toc. From July to September, small earth tremors shook the area. Strange rumbling came from deep in the earth; the lake water "boiled up" ominously.

SADE earlier had implanted dozens of sentinel bench marks on the mountain's flank. These, watched regularly by optical instruments so sensitive that they noted even a hairbreadth of movement, would signal any tendency of the earth to slip downward. And signal they did. Quiet for months, they suddenly began reporting an ever-rising tendency of Monte Toc's earthen flank to slip . . . 6 . . . 8 . . . 12 . . . 22 millimeters per 24 hours—edging up toward the highest "danger reading" of 40, registered three years earlier at the time of the first slide.

Perturbed, ENEL-SADE halted the test-raising with the waters still 41 feet below maximum, hoping that the earth would settle down to a new stability. Unfortunately, it did not. And, to make matters worse, thrice-normal rains, the heaviest in 20 years, had made the earth unusually sodden.

Nino A. Biadene, ENEL-SADEL deputy director-general for technical matters, from the head office in Venice, now declared that there would be no question of letting the water go any higher, even though it was authorized. On the contrary, he would order the water lowered if the alarm signals continued.

They did continue—and got worse. Thus, on September 26—with disaster 13 days away—Biadene gave the emergency order: "Take the water down!" Instantly, the great exit valves were opened, and water began rushing out. But not *too* fast, for this would too quickly remove the water-cushion supporting Monte Toc's soaked earth and make it even more unstable.

Next day was golden and merry in Longarone. Most of the harvesting had been done. The year's business had been good: new factories had been coming in; ever more tourists had come to visit the dam; everybody had jobs and fatter paychecks. Best of all, Longarone's famed ice-cream makers, who fanned out over Europe every March to make and sell their delicacy, were now streaming home to spend the winter with their families. It was a gay time of homecoming, of matchmaking and marrying, of meeting old friends in the many bar-cafés which served as clubs, of swapping stories of the season's business over a glass of *vino*.

True, the great dam above cast a certain shadow on the merriment. Word had leaked out that the slippage rate was high today. A truck driver named Antonio Savi said that he had driven over the paved road on Monte Toc and that it was buckling so much he refused to go there again. But this seemed familiar stuff somehow.

Hundreds of men stayed at the bar-cafés later than usual that evening. They were ardent soccer fans, and at 9:55 began a television relay from Madrid of an important game between the Spanish Real Madrid team and the Scottish Glasgow Rangers. Most would never live to know who won the game (the Spaniards, 6–0), and later it was bitterly cursed for luring so many to death. In fact it probably made little difference in the final death toll; it doomed some, but saved others.

Meanwhile, up on Monte Toc the sentinel bench marks had gone wild. Far in the past was the old danger-reading of 40. Now the signals were reading 190 . . . 200!

Just before 9 p.m.—with disaster barely 100 minutes away—Biadene in Venice decided that it would be wise to institute *below* the dam and around

Longarone some of the same precautions already in force up around the lake itself. By phone he ordered a subordinate engineer in Belluno to have police bar all traffic on the below-dam roads in and around Longarone. And from the dam went out a series of phone calls to people and establishments down by the Piave—to the sawmill, the spinning mill, the quarry, a tavern—passing on the message: "Maybe a little water over the dam tonight . . . nothing to get alarmed about."

At 10:39 the mountain fell. Not all of it, but a greater single mass than has fallen in Europe since prehistoric times—with a shock so great that it simulated genuine earthquake effects on seismographs in five countries. About 600 million tons came down—roughly equivalent to a football field piled with earth and rock to a height of *40 miles*. It didn't fall slowly, by inches, as predicted, disintegrating as it went. Instead, the mountain split away cleanly, as if cut by a knife, and fell straight into the lake.

Don Carlo Onorini, parish priest of Casso village perched high on a mountain just across the lake, happened to be watching. In the bright glare of the floodlights he saw the mountainside suddenly slip loose "with a sound as if the end of the earth had come." A muddy flood leaped up toward him and he saw, just below him, the mountainside clawed away, a church and some lower houses vanish, before the flood fell back into the valley. An enormous blue-white flash filled the sky as the great 20,000-volt high-tension lines shorted, fused and broke, plunging the valley into darkness.

All around the lakeshore the tormented water raced—not 80 feet high, but in places, clawing up to *800 feet* above lake level. It thundered at the dam— and the dam held. But the water went over the dam not five feet high but up to 300 feet high, and smashed to the bottom of the gorge 800 feet below. There it was constricted as in a deadly funnel, and its speed fearfully increased. It shot out of the short gorge as from a gun barrel and spurted across the wide Piave riverbed, scooping up millions of deadly stones. Ahead of it raced a strange icy wind and a storm of fragmented water, like rain, but flying *upward*. By now it was more than a wave, more than a flood. It was a tornado of water and mud and rocks, tumbling hundreds of feet high in the pale moonlight, leaping straight at Longarone.

In the next minutes—about six—it thundered far up the hillside where Longarone stood, then recoiled into the Piave Valley with a fearful sucking noise, as if a mile-wide sink were emptying. In those six minutes Longarone vanished from the earth.

Almost none of the survivors—even those watching from so high up that they were never in danger—could give any coherent account of what they saw. One man remembers "a great milky cloud settling over our town." Another, "a huge grayish-silverish mass that seemed so big it hardly seemed to move at all. Then I saw things swirling in the mass—bodies, lumber, automobiles." Most remember the strangely cold wind and the horrible noise "like a thousand express trains rushing on us . . . a noise so great the ears refused to hear it."

At one bar, somebody yelled, "The dam's broken! Run for your lives!" Those spry enough made it; those too old or too dazed died. In another bar, those who jumped out uphill windows made it; those who went out the front door did not.

A girl of 22, Maria Teresa Galli, was just closing her balcony shutters when she felt a great cold wind, and somehow the house seemed to dissolve around her. Some great force, part wind, part water, picked her up and whirled her along as she thought dazedly: "I'm flying . . . walking . . . swimming!" Two hundred yards away, an old couple, Arduino Burrigana and his wife Gianna, watched from the top floor as the flood invaded the ground floor—and dropped some dark bundle which emitted a groan. It was Maria Teresa Galli, fainting with shock, bruised, but little harmed.

A paralyzed man, helpless in his chair, called out in panic to his wife, "What is it? What is it?" She stepped out on the balcony to look. He felt the house tremble to some great shock, and called out, "Where are you?" She never answered. A passing edge of the wave had flicked her away.

The visiting American couple from Scarsdale had nearly finished the special bottle of wine when the roar came. One of the cousins pulled open the door, stared out, slammed it shut again and cried, "We're all dead!" Water poured over them, and the man remembers dazed thoughts passing through his mind like, "What's the use . . . with a thousand feet of water over us?" Yet in a moment, miraculously, the water retreated. He was safe, and so were his wife and cousins—except for fractures—but it had been too much for Aunt Elisabetta. She lay dead.

It took some time before the outer world learned how enormous the tragedy had been. The flood had isolated Longarone in a sea of mud. The first reporter waded in about 2:30 a.m., and well before dawn in came 1000 of Italy's famed mountain troops, the *Alpini*, the vanguard of nearly 10,000 rescue workers—soldiers, police, firemen, Red Cross, Boy Scouts, volunteers of all kinds.

The great flood of death was matched by an equal flood of compassion and help from all Italy. While no single Italian admitted to any blame for the disaster, somehow Italy felt responsible.

All around the world the news produced horror and help. From Australia to Canada, many Italian communities—often led by local Italian-language newspapers—gave money. James Bez, 58, an unemployed worker in Stamford, Conn., who had been born in Longarone and lost nearly all his family there, personally collected $350 in cash plus 40 boxes of clothing, which a Greek ship carried to Italy free.

One day recently, along with an engineer who had no connection with the dam, I climbed a narrow, muddy road, still being rebuilt, up to the village of Casso—where, on that terrible night, the parish priest had looked across to see the mountain fall. Here, 800 feet above the lake, the whole scene lay clear before us.

Below us, to the right stood the great dam, still intact save for a few minor bruises at the top. Directly below us lay the huge mass of the slide, looking as if it had been there forever, its trees and bushes still growing, already being called Monte Nuovo—"the new mountain." It nudges up against the dam and is, in effect, a new natural dam about 1½ miles thick with earth and rock, towering nearly 300 feet higher than the man-made dam, now forever useless.

The lake has shrunk to about half its former size. But Italy will soon need all the electric power it can get, and there is an unspoken hope that later, when fear and feeling have cooled, some absolutely safe way may be found to go on using what is left of the lake for precious water storage.

As we stood at Casso looking down, my engineer companion pointed with his pipestem to the great dam, still a proud work of man. "In building things these days," he said, "man can calculate stresses and strains just about 100 percent—as that dam proves. But so far, not even the best experts, using the best equipment, can be absolutely sure what goes on deep down in the earth. These days engineering is pretty much an exact science. Geology isn't—not yet."

Secrets of the Earth

Happy the man whose lot it is to know
The secrets of the earth. He hastens not
To work his fellows' hurt by unjust deeds
But with rapt admiration contemplates
Immortal Nature's ageless harmony
And how and when her order came to be,
Such spirits have no place for thoughts of shame.

—Euripides

Earthquakes are movements of the earth's crust generally caused by movement of rock masses relative to one another. They range in intensity from minor tremors, so slight that they can be detected only by sensitive seismographs, to earthquakes so violent that major areas are destroyed. The great loss of life caused by some earthquakes is often the result of secondary causes, such as fires from broken gas mains (which may run out of control because of broken water mains), great tidal waves (tsunamis) from submarine earthquakes and flooding caused by landslides. Some 900,000 are said to have perished in the Shantzee earthquake of 1556, for example.

The two accounts of the Lisbon earthquake are quite different. While neither represents an eyewitness description—and both are reflective, as well as descriptive—the account by Voltaire is taken from Candide, *a novel which parodied contemporary virtue and philosophy. Dr. Pangloss, Candide's tutor, and a "brutal sailor" suffer storm and shipwreck and are finally cast ashore in Portugal, only to find themselves then involved in the great Lisbon Earthquake. Voltaire wrestles with the problem of death, including death resulting from "acts of God" (the drowning of the Anabaptist) and the injustice of death at the hands of other men. Pangloss, still arguing not only for a "train of sulphur running underground from Lima to Lisbon," but also for the ultimate harmony of things, ("It is impossible that things should not be what they are; for all is well.") is hanged by the Portuguese because of his comments on the earthquake. "Must I see you hang without knowing why" pleads Candide of his dead master. "Was it necessary that you should be drowned in port?" he groans over his dead Anabaptist friend. In these questions, Voltaire echoes a baffled and angry agony that swept Europe after the Lisbon Earthquake of 1755.*

CANDIDE

Voltaire

Half the enfeebled passengers, suffering from that inconceivable anguish which the rolling of the ship causes in the nerves and in all the humors of bodies shaken in contrary directions, did not retain strength enough even to trouble about the danger. The other half screamed and prayed; the sails were torn, the masts broken, the vessel leaking. Those worked who could, no one cooperated, no one commanded. The Anabaptist tried to help the crew a little: he was on the main deck; a furious sailor struck him violently and stretched him on the deck; but the blow he delivered gave him so violent a shock that he fell head-first out of the ship. He remained hanging and clinging to part of the broken mast. The good Jacques ran to his aid, helped him to climb back, and from the effort he made was flung into the sea in full view of the sailor, who allowed him to drown without condescending even to look at him. Candide came up, saw his benefactor reappear for a moment and then be engulfed for ever. He tried to thrown himself after him into the sea; he was prevented by the philosopher Pangloss, who proved to him that the Lisbon roads had been expressly created for the Anabaptist to be drowned in them. While he was proving this *a priori,* the vessel sank, and everyone perished except Pangloss, Candide and the brutal sailor who had drowned the virtuous Anabaptist; the blackguard swam successfully to the shore and Pangloss and Candide were carried there on a plank. When they had recovered a little, they walked toward Lisbon; they had a little money by the help of which they hoped to be saved from hunger after having escaped the storm. Weeping the death of their benefactor, they had scarcely set foot in the town when they felt the earth tremble under their feet; the sea rose in foaming masses in the port and smashed the ships which rode at anchor. Whirlwinds of flame and ashes covered the streets and squares; the houses collapsed, the roofs were thrown upon the foundations, and the foundations were scattered; thirty thousand inhabitants of every age and both sexes were crushed under the ruins. Whistling and swearing, the sailor said: "There'll be something to pick up here." "What can be the sufficient reason for this phenomenon?" said Pangloss. "It is the last day!" cried Candide. The sailor immediately ran among the debris, dared death to find money, found it, seized it, got drunk, and having slept off his wine, purchased the favors of the first woman of good-will he met on the ruins of the houses and among the dead and dying. Pangloss, however, pulled him by the sleeve. "My friend," said he, "this is not well, you are disregarding universal reason, you choose the wrong time." "Blood and 'ounds!" he retorted, "I am a sailor and I was born in Batavia; four times have I stamped on the crucifix during four voyages to Japan; you have found the right man for your universal reason!" Candide had been hurt by some falling stones; he lay in the street covered with debris. He said to Pangloss: "Alas! Get me a little wine and oil; I am dying." "This earthquake is not a new thing," replied Pangloss. "The town of Lima felt the same shocks in America last year; similar causes produce similar effects; there must certainly be a train of sulphur underground from Lima to Lisbon." "Nothing is more probable," replied Candide; "but, for God's sake, a little oil and wine." "What do you mean, probable?" replied the philosopher; "I maintain that it is proved." Candide lost consciousness, and Pangloss brought him a little water from a neighboring fountain. Next day they found a little food as they wandered among the ruins and gained a little strength. Afterward they worked like others to help the inhabitants who had escaped death. Some citizens they had assisted gave them as good a dinner as could be expected in such a disaster; true, it was a dreary meal; the hosts watered their bread with their tears, but Pangloss consoled them by assuring them that things could not be otherwise. "For," said he, "all this is for the best; for, if there is a volcano at Lisbon, it cannot be anywhere else; for it is impossible that things should not be where they are; for all is well." A little, dark man, a familiar of the Inquisition, who sat beside him, politely took up the conversation, and said: "Apparently, you do not believe in original sin; for, if everything is for the best, there was neither fall nor punishment." "I most humbly beg your excellency's pardon," replied Pangloss still more politely, "for the fall of man and the curse necessarily entered into the best of all possible worlds." "Then you do not believe in free-will?" said the familiar. "Your excellency will pardon me," said Pangloss; "free-will can exist with absolute necessity; for it was necessary that we should be free; for in short, limited will . . ." Pangloss was in the middle of his phrase when the familiar nodded to his armed attendant who was pouring out port or Oporto wine for him.

After the earthquake which destroyed three-quarters of Lisbon, the wise men of that country could discover no more efficacious way of preventing a total ruin than by giving the people a splendid *auto-da-fé*. It was decided by the university of Coimbre that the sight of several persons being slowly burned in great ceremony is an infallible secret for preventing earthquakes. Consequently they had arrested a Biscayan convicted of having married his fellow-godmother, and two Portuguese who, when eating a chicken, had thrown away the bacon; after dinner they came and bound Dr. Pangloss and his disciple Candide, one because he had spoken and the other because he had listened with an air of approbation; they were both carried separately to extremely cool apartments, where there was never any discomfort from the sun; a week afterward each was dressed in a sanbenito and their heads were ornamented with paper mitres; Candide's mitre and sanbenito were painted with flames upside down and with devils who had neither tails nor claws; but Pangloss's devils had claws and tails, and his flames were upright. Dressed in this manner they marched in procession and listened to a most pathetic sermon, followed by lovely plain-song music. Candide was flogged in time to the music, while the singing went on; the Biscayan and the two men who had not wanted to eat bacon were burned, and Pangloss was hanged, although this is not the custom. The very same day, the earth shook again with a terrible clamour. Candide, terrified, dumbfounded, bewildered, covered with blood, quivering from head to foot, said to himself: "If this is the best of all possible worlds, what are the others? Let it pass that I was flogged, for I was flogged by the Bulgarians, but, O my dear Pangloss! The greatest of philosophers! Must I see you hanged without knowing why! O my dear Anabaptist! The best of men! Was it necessary that you should be drowned in port! O Mademoiselle Cunegonde! The pearl of women! Was it necessary that your belly should be slit!" He was returning, scarcely able to support himself, preached at, flogged, absolved and blessed, when an old woman accosted him and said: "Courage, my son, follow me."

The second account of the Lisbon earthquake, by James Newman—contemporary American lawyer, diplomat, mathematician, journalist and author—catches the setting of the earthquake and the mood of the continent very well. The catastrophe, which lasted only ten minutes, brought great physical damage and destruction. But in that ten minutes, "an era came to an end." Europe was never quite the same again, for that brief interval changed the course of human thought. This stability of 17th century philosophy, the ordered domestic tranquility of the prosperous, and the confidence of the common man, all tumbled before what many regarded as the wrath and the judgment of God. It was this prevailing mood of optimism that Voltaire attacked, although the extension of his own view, into a sustained cynicism, as others from Rousseau to Newman have pointed out, is, itself, too extreme. If the "all is well" brand of optimism leads to a life of shallow unconcern and indifference, it is also not clear that an "all is meaningless" brand of pessimism leads to a life which is more productive. The quest and debate, raised by the Lisbon earthquake still continue.

THE LISBON EARTHQUAKE

James R. Newman

At about 9:30 A.M. on All Saints' Day, Saturday, November 1, 1755, a massive earthquake convulsed the great city of Lisbon in Portugal. There were three distinct shocks separated by intervals of about a minute. "The first alarm"—I quote from Sir Thomas Kendrick's description—"was a rumbling noise that many people said sounded like that of exceptionally heavy traffic in an adjacent street, and this was sufficient to cause great alarm and make the buildings tremble; then there was a brief pause and a devastating shock followed, lasting over two minutes, that brought down roofs, walls, and façades of churches, palaces and houses and shops in a dreadful deafening roar of destruction. Close on this came a third trembling to complete the disaster, and then a dark cloud of suffocating dust settled foglike on the ruins of the city. It had been a clear, bright morning, but in a few moments the day turned into the frightening darkness of night." The earthquake lasted about ten minutes; but in that ten minutes an era came to an end. After the shaking of Lisbon, Europe was not the same, as after Hiroshima the world was changed.

For the bicentenary of the disaster, the director and principal librarian of the British Museum prepared a scholarly and entertaining study based upon the immense literature occasioned by the event. Since the fall of Rome in the fifth century no other happening had so shocked Western civilization. Not only in Portugal but throughout Europe there was an outpouring of diagnosis and commentary. Scientists offered explanations, preachers sermonized, philosophers speculated, poets lamented, theologians moralized and exhorted. One opinion evoked another and controversies multiplied. Sir Thomas reports the earthquake in vivid detail, but this is only the setting for his book. Primarily he is concerned with the effect of the calamity on men's minds: how they interpreted its meaning, what light it shed on the deep question of man's place in the world, of crime and punishment and the wrath of God, of duty to one's fellows, of competing faiths and creeds, of optimism and man's prospects. Physical damage is soon repaired and the loss of human life easily forgotten; but the earthquake shaped the future by altering the course of thought.

For a few minutes of that dreadful day in November, Lisbon seemed—to observers in ships on the Tagus and on the higher ground around the city—to "sway like corn in the wind." Then avalanches of falling masonry hid the ruins under a cloud of dust. But the violent tremors, which collapsed some of the finest buildings and hundreds of the smaller houses and shops, were only the first ordeal. Within a quarter of an hour of the triple shock, fires broke out in different parts of the town, and as the conflagration mounted, a third disaster occurred. The waters of the Tagus "rocked and rose menacingly, and then poured in three great towering waves over its banks, breaking with their mightiest impact on the shore between the Alcántara docks and [the palace square] the Terreiro do Paço."

Some 10,000 to 15,000 persons, it is said, lost their lives in the earthquake. Original estimates were wildly exaggerated, and the official statistics compiled with great difficulty afterward were not reliable. But the careful historian, Moreira de Mendonça, thought that not more than 5,000 of Lisbon's 275,000 persons were killed on the first of November, the casualties being doubled or trebled during the course of the month. The loss of property was severe, the fire doing the most damage. Much could have been salvaged from the shattered buildings, but the flames destroyed pictures, furniture, tapestries, jewelry and plate, and enormous stocks of merchandise. The foreign traders alone—principally British and Hamburg merchants—lost about £10,000,000 worth of goods. Among the irreplaceable treasures incinerated in a single building, the palace of the Marquês de Louriçal, were two hundred pictures including works by Titian, Correggio, and Rubens, a library of 18,000 printed books, 1000 manuscripts, and a huge collection of maps and charts relating to the Portuguese voyages of discovery and colonization in the East and in the New World. Seventy thousand books perished in the burning of the King's palace.

Measured by its death roll and damage, Lisbon was not one of the greatest disasters of its kind. India, China, and Japan have suffered much more terrible earthquakes. The Kwanto quake of 1923, which almost leveled Tokyo and Yokohama, killed 100,000 persons. Nonetheless, the earthquake of Novem-

Much had been learnt as to the diffusion of basalt in Europe, and many excellent drawings had been published of the remarkable prismatic structure of this rock. But no serious attempt seems to have been made to grapple with the problem of its origin. Some absurd notions had indeed been entertained on this subject. The long regular pillars of basalt, it was gravely suggested, were jointed bamboos of a former period, which had somehow been converted into stone. The similarity of the prisms to those of certain minerals led some mineralogists to regard basalt as a kind of schorl, which had taken its geometrical forms in the process of crystallization. Romé de Lisle is even said to have maintained that each basalt prism ought to have a pyramidal termination, like the schorls and other small crystals of the same nature.

Guettard, as we have seen, drew a distinction between basalt and lava, and this opinion was general in his time. The basalts of Central and Western Europe were usually found on hill tops, and displayed no cones or craters, or other familiar sign of volcanic action. On the contrary, they were not infrequently found to lie upon, and even to alternate with, undoubted sedimentary strata. They were, therefore, not unnaturally grouped with these strata, and the whole association of rocks was looked upon as having had one common aqueous origin. It was also a prevalent idea that a rock which had been molten must retain obvious traces of that condition in a glassy structure. There was no such conspicuous vitreous element in basalt, so that this rock, it was assumed, could never have been volcanic. As Desmarest afterwards contended, those who made such objections could have but little knowledge of volcanic products.

We may now proceed to trace how the patient and sagacious Inspector of French industries made his memorable contribution to geological theory. It was while traversing a part of Auvergne in the year 1763 that he detected for the first time columnar rocks in association with the remains of former volcanoes. On the way from Clermont to the Puy de Dôme, climbing the steep slope that leads up to the plateau of Prudelle, with its isolated outlier of a lava-stream that flowed long before the valley below it had been excavated, he came upon some loose columns of a dark compact stone which had fallen from the edge of the overlying sheet of lava. He found similar columns standing vertically all along the mural front of the lava, and observed that they were planted on a bed of scoriæ and burnt soil, beneath which lay the old granite that forms the foundation rock of the region. He noticed still more perfect prisms a little further on, belonging to the same thin cake of dark stone that covered the plain which leads up to the foot of the great central puy.

Every year geological pilgrims now make their way to Auvergne, and wander over its marvellous display of cones, craters and lava-rivers. Each one of them climbs to the plateau of Prudelle, and from its level surface gazes in admiration across the vast fertile plain of the Limagne on the one side, and up to the chain of the puys on the other. Yet how few of them connect that scene with one of the great triumphs of their science, or know that it was there that Desmarest began the observations which directly led to the fierce contest over the origin of basalt!

That cautious observer tells us that amidst the infinite variety of objects around him, he drew no inference from this first occurrence of columns, but that his attention was aroused. He was kept no long time in suspense on the subject. "On the way back from the Puy de Dôme," he tells us, "I followed the thin sheet of black stone and recognised in it the characters of a compact lava. Considering further the thinness of this crust of rock, with its underlying bed of scoriæ, and the way in which it extended from the base of hills that were obviously once volcanoes, and spread out over the granite, I saw in it a true lava-stream which had issued from one of the neighbouring volcanoes. With this idea in my mind, I traced out the limits of the lava, and found again everywhere in its thickness the faces and angles of the columns, and on the top their cross-section, quite distinct from each other. I was thus led to believe that prismatic basalt belonged to the class of volcanic products, and that its constant and regular form was the result of its ancient state of fusion. I only thought then of multiplying my observations, with the view of establishing the true nature of the phenomenon, and its conformity with what is to be found in Antrim—a conformity which would involve other points of resemblance."

He narrates the course of his discoveries as he journeyed into the Mont Dore, detecting in many places fresh confirmation of the conclusion he had formed. But not only did he convince himself that the prismatic basalts of Auvergne were old lava-streams, he carried his induction much further and felt assured that the Irish basalts must also have had a volcanic origin. "I could not doubt," he says, "after these varied and repeated observations, that the groups of prismatic columns in Auvergne belonged to the same conformation as those of Antrim, and that the constant and regular form of the columns must have resulted from the same cause in both regions. What convinced me of the truth of this opinion was the examination of the material constituting the Auvergne columns with that from the Gaint's Causeway, which I found to agree in texture, colour and hardness, and further, the sight of two engravings of the Irish locality which at once recalled the scenery of parts of Mont Dore. I draw, from this recognised resemblance and the facts that establish it, a deduction which appears to be justified by the strength of the analogy—namely, that in the Giant's Causeway, and in all the prismatic masses which present themselves along the cliffs of the Irish coast, in short even among the truncated summits of the interior, we see the operations of one or more volcanoes which are extinct, like those of Auvergne. Further, I am fully persuaded that in general these groups of polygonal columns are an infallible proof of an old volcano, wherever the stone composing them has a compact texture, spangled with brilliant points, and a black or grey tint."

Here, then, was a bold advance in theoretical as well as observational geology. Not only was the discovery of Guettard confirmed, that there had once been active volcanoes in the heart of France, but materials were obtained for explaining the origin of certain enigmatical rocks which, though they had been found over a large part of Europe, had hitherto remained a puzzle to mineralogists. This explanation, if it were confirmed, would show how widely volcanic action prevailed over countries wherein no sign of an eruption has been witnessed since the earliest ages of human history.

Desmarest was in no hurry to publish his discovery. Unlike some modern geologists, who rush in hot haste into print, and overload the literature of the science with narratives of rapid and imperfect observations, he kept his material beside him, revolving the subject in his mind, and seeking all the information that he could bring to bear upon it. He tells us that in the year following his journey in Auvergne, he spent the winter in Paris, and while there, laid before the Intendant of Auvergne the desirability of having the volcanic region mapped. His proposition was accepted, and Pasumot, one of the state surveyors, was entrusted with the task of making a topographical map of the region from Volvic to beyond Mont Dore. The whole of the summer of 1764 was taken up with this work. Desmarest accompanied the geographer, who himself had a large acquaintance with the mineralogy of his day. The final result was the production of a map which far surpassed anything of the kind that had before been attempted, in the accuracy, variety, and clearness of its delineations of volcanic phenomena.

CHAPTER 4

Philosophy of Geology

Geology, it is sometimes claimed, is unique amongst the sciences in being chiefly concerned with historical development, rather than present state, and by its lack of a distinctive methodology, such as those which are characteristic of other sciences. Unsatisfactory as both statements are as generalizations, they contain enough truth to merit our concern.

All sciences contain a historical element; it is the relative historical emphasis in which they differ. Meaningful discourse in biology or astronomy, for example, would be impossible without it. G.G. Simpson has drawn a useful distinction between the immanent processes and products which are the universal components of the earth—and indeed of the whole material universe—and the particular state of the earth or some part of it at any given moment. Such a state, though always contingent upon and the product of the interactions of immanent processes and products, is itself unique.

The goal of geology is so to describe and understand the present processes, constitution and composition of the earth as to establish relationships between them, and thus to interpret earlier historical products and sequences. The study of existing states, processes and components involves the application of all of the sciences, including especially physics, chemistry and biology, even though the particular abstractions favored by geologists may not always be those chosen by these other scientists. The uniqueness of geology lies in their combination and historical extrapolation to remote periods. It is not the events themselves of these periods that are thus studied, but rather their results and records. In this, geology infers causes from results, rather than predicts results from causes, as does the less-historical science.

Geology shares with other sciences the common goal of establishing generalizations (laws) of wide applicability which describe recurrent and invariable relationships between objects or states. Such laws, as Simpson has pointed out, are so massively abstractive in their basis, that it would be unwarranted to expect to derive them from the particular interaction of immanent variables which produce the unique sequences of events we call history. The very uniqueness of these events precludes their use as a basis for the establishment of the recurrent relationships involved in 'laws.'

The extracts that follow can do little more than introduce some aspects of these philosophical but fundamental aspects of geologic science. The interested reader can find no better introduction to the subject than Simpson's excellent paper, and especially his discussion of predictive testing and predictability.

Simplicity

The aim of science is to seek the simplest explanations of complex facts. We are apt to fall in to the error of thinking that our facts are simple, because simplicity is the goal of our quest. The guiding motto in the life of every natural philosopher should be, "seek simplicity and distrust it."

—Alfred North Whitehead

James Hutton (1726-1797) was born in Edinburgh and qualified as a doctor after studying medicine in Edinburgh and Paris. He never practiced his profession, however, but instead devoted most of his life to practical agriculture and to study and writing concerning the history of the Earth. He is regarded as the father of modern geology. His System of the Earth, *from which the present abstract is taken, was presented to the Royal Society of Edinburgh in the spring of 1785. In it he argued cogently for two processes which account for the present composition and structure of the Earth; first, that the rocks of the Earth's surface are formed from loose materials which have been subsequently consolidated; and second, that sediments formed on the sea floor can subsequently be uplifted and transformed into land, at which time igneous rocks and veins may be intruded into the uplifted area. He argued for the widespread demonstration of terrestrial weathering and of the action of circulating water and heat in leading to liquefaction, and he also explored the questions of continental uplift, being convinced that volcanoes provided visible proof of the manifestation of the power required to produce such results. It was in this context that he was also led to argue for the need for an indefinite period of time and to make his often quoted statement, "with respect to human observation, this world has neither a beginning nor an end."*

The publication of Hutton's work, supplemented after his death by the publication of his friend John Playfair's book Illustrations of the Huttonian Theory of the Earth, *was of major importance in the establishment of geology as a science, not only because they provided the deathblow for the earlier claims that rocks were the result of a single process (the Neptunist-Plutonist controversy), but also because Hutton, like Lyell after him, advocated the efficacy of existing processes and the immensity of geologic time.*

CONCERNING THE SYSTEM OF THE EARTH

James Hutton

The purpose of this Dissertation is to form some estimate with regard to the time the globe of this Earth has existed, as a world maintaining plants and animals; to reason with regard to the changes which the earth has undergone; and to see how far an end or termination to this system of things may be perceived, from the consideration of that which has already come to pass.

As it is not in human record, but in natural history, that we are to look for the means of ascertaining what has already been, it is here proposed to examine the appearances of the earth, in order to be informed of operations which have been transacted in time past. It is thus that, from principles of natural philosophy, we may arrive at some knowledge of order and system in the economy of this globe, and may form a rational opinion with regard to the course of nature, or to events which are in time to happen.

The solid parts of the present land appear, in general, to have been composed of the productions of the sea, and of other materials similar to those now found upon the shores. Hence we find reason to conclude,

1st, That the land on which we rest is not simple and original, but that it is a composition, and had been formed by the operation of second causes.

2dly, That, before the present land was made, there had subsisted a world composed of sea and land, in which were tides and currents, with such operations at the bottom of the sea as now take place. And,

Lastly, That, while the present land was forming at the bottom of the ocean, the former land maintained plants and animals; at least, the sea was then inhabited by animals, in a similar manner as it is at present.

Hence we are led to conclude, that the greater part of our land, if not the whole, had been produced by operations natural to this globe; but that, in order to make this land a permanent body, resisting the operations of the waters, two things had been required; *1st*, The * consolidation of masses formed by collections of loose or incoherent materials; *2dly*, The elevation of those consolidated masses from the bottom of the sea, the place where they were collected, to the stations in which they now remain above the level of the ocean.

Here are two different changes, which may serve mutually to throw some light upon each other; for, as the same subject has been made to undergo both these changes, and as it is from the examination of this subject that we are to learn the nature of those events, the knowledge of the one may lead us to some understanding of the other.

Thus the subject is considered as naturally divided into two branches, to be separately examined: *First*, by what natural operation strata of loose materials had been formed into solid masses; *secondly*, By what power of nature the consolidated strata at the bottom of the sea had been transformed into land.

With regard to the *first* of these, the consolidation of strata, there are two ways in which this operation may be conceived to have been performed; first, by means of the solution of bodies in water, and the after concretion of these dissolved substances, when separated from their solvent; *secondly*, the fusion of bodies by means of heat, and the subsequent congelation of those consolidating substances.

---

* There are two senses in which the term *solidity* is used; one of these is in opposition to *fluidity*, the other to *vacuity*. When the change from a fluid state to that of solidity, in the first sense, is to be expressed, we shall employ the term *concretion;* consequently, the consolidation of a mass is only to be understood as in opposition to its vacuity, or porousness.

With regard to the operation of water, it is *first* considered, how far the power of this solvent, acting in the natural situation of those strata, might be sufficient to produce the effect; and here it is found, that water alone, without any other agent, cannot be supposed capable of inducing solidity among the materials of strata in that situation. It is, 2*dly*, considered, how far, supposing water capable of consolidating the strata in that situation, it might be concluded, from examining natural appearances, that this had been actually the case? Here again, having proceeded upon this principle, that water could only consolidate strata with such substances as it has the power to dissolve, and having found strata consolidated with every species of substance, it is concluded, that strata in general have not been consolidated by means of aqueous solution.

With regard to the other probable means, heat and fusion, these are found to be perfectly competent for producing the end in view, as every kind of substance may by heat be rendered soft, or brought into fusion, and as strata are actually found consolidated with every different species of substance.

A more particular discussion is then entered into: Here, consolidating substances are considered as being classed under two different heads, viz. Siliceous and sulphureous bodies, with a view to prove, that it could not be by means of aqueous solution that strata had been consolidated with those particular substances, but that their consolidation had been accomplished by means of heat and fusion.

Sal Gem, as a substance soluble in water, is next considered, in order to show that this body had been last in a melted state; and this example is confirmed by one of fossil alkali. The case of particular septaria of iron-stone, as well as certain crystallized cavities in mineral bodies, are then given as examples of a similar fact; and as containing, in themselves, a demonstration, that all the various mineral substances had been concreted and crystallized immediately from a state of fusion.

Having thus proved the actual fusion of the substances with which strata had been consolidated, in having such fluid bodies introduced among their interstices, the case of strata, consolidated by means of the simple fusion of their proper materials, is next considered; and examples are taken from the most general strata of the globe, viz. siliceous and calcareous. Here also demonstration is given, that this consolidating operation had been performed by means of fusion.

Having come to this general conclusion, that heat and fusion, not aqueous solution, had preceded the consolidation of the loose materials collected at the bottom of the sea, those consolidated strata, in general, are next examined, in order to discover other appearances, by which the doctrine may be either confirmed or refuted. Here the changes of strata, from their natural state of continuity, by veins and fissures, are considered; and the clearest evidence is hence deduced, that the strata have been consolidated by means of fusion, and not by aqueous solution; for not only are strata in general found intersected with veins and cutters, an appearance inconsistent with their having been consolidated simply by previous solution; but, in proportion as strata are more or less consolidated, they are found with the proper corresponding appearances of veins and fissures.

With regard to the second branch, in considering by what power the consolidated strata had been transformed into land, or raised above the level of the sea, it is supposed that the same power of extreme heat, by which every different mineral substance had been brought into a melted state, might be capable of producing an expansive force, sufficient for elevating the land, from the bottom of the ocean, to the place it now occupies above the surface of the sea. Here we are again referred to nature, in examining how far the strata, formed by successive sediments or accumulations deposited at the bottom of the sea, are to be found in that regular state, which would necessarily take place in their original production; or if, on the other hand, they are actually changed in their natural situation, broken, twisted, and confounded, as might be expected, from the operation of subterranean heat, and violent expansion. But, as strata are actually found in every degree of fracture, flexure, and contortion, consistent with this supposition, and with no other, we are led to conclude, that our land had been raised above the surface of the sea, in order to become a habitable world; as well as that it had been consolidated by means of the same power of subterranean heat, in order to remain above the level of the sea, and to resist the violent efforts of the ocean.

This theory is next confirmed by the examination of mineral veins, those great fissures of the earth, which contain matter perfectly foreign to the strata they traverse; matter evidently derived from the mineral region, that is, from the place where the active power of fire, and the expansive force of heat, reside.

Such being considered as the operations of the mineral region, we are hence directed to look for the manifestation of this power and force, in the appearances of nature. It is here we find eruptions of ignited matter from the scattered volcanoes of the globe; and these we conclude to be the effects of such a power precisely as that about which we now inquire. Volcanoes are thus considered as the proper discharges of a superfluous or redundant power; not as things accidental in the course of nature, but as useful for the safety of mankind, and as forming a natural ingredient in the constitution of the globe.

The doctrine is then confirmed, by examining this earth, and by finding everywhere, beside the many marks of ancient volcanoes, abundance of subterraneous or unerupted lava, in the basaltic rocks, the Swedish trap, the toadstone, ragstone, and whinstone of Britain and Ireland, of which particular examples are cited; and a description given of the three different shapes in which that unerupted lava is found.

The peculiar nature of this subterraneous lava is then examined; and a clear distinction is formed between this basaltic rock and the common volcanic lavas.

Lastly, The extension of this theory, respecting mineral strata, to all parts of the globe, is made, by finding a perfect similarity in the solid land through all the earth, although, in particular places, it is attended with peculiar productions, with which the present inquiry is not concerned.

A theory is thus formed, with regard to a mineral system. In this system, hard and solid bodies are to be formed from soft bodies, from loose or incoherent materials, collected together at the bottom of the sea; and the bottom of the ocean is to be made to change its place with relation to the centre of the earth, to be formed into land above the level of the sea, and to become a country fertile and inhabited.

That there is nothing visionary in this theory, appears from its having been rationally deduced from natural events, from things which have already happened; things which have left, in the particular constitutions of bodies, proper traces of the manner of their production; and things which may be examined with all the accuracy, or reasoned upon with all the light, that science can afford. As it is only by employing science in this manner that philosophy enlightens man with the knowledge of that wisdom or design which is to be found in nature, the system now proposed, from unquestionable principles, will claim the attention of scientific men, and may be admitted in our speculations with regard to the works of nature, notwithstanding many steps in the progress may remain unknown.

By thus proceeding upon investigated principles, we are led to conclude, that, if this part of the earth which we now inhabit had been produced, in the course of time, from the materials of a former earth, we should, in the examination of our land, find data from which to reason, with regard to the nature of that world, which had existed during the period of time in which the present earth was forming; and thus we might be brought to understand the nature of that earth which had preceded this; how far it had been similar to the present, in producing plants and nourishing animals. But this interesting point is perfectly ascertained, by finding abundance of every manner of vegetable production, as well as the several species of marine bodies, in the strata of our earth.

Having thus ascertained a regular system, in which the present land of the globe had been first formed at the bottom of the ocean, and then raised above the surface of the sea, a question naturally occurs with regard to time; what had been the space of time necessary for accomplishing this great work?

In order to form a judgment concerning this subject, our attention is directed to another progress in the system of the globe, namely, the destruction of the land which had preceded that on which we dwell. Now, for this purpose, we have the actual decay of the present land, a thing constantly transacting in our view, by which to form an estimate. This decay is the gradual ablution of our soil, by the floods of rain; and the attrition of the shores, by the agitation of the waves.

If we could measure the progress of the present land, towards it dissolution by attrition, and its submersion in the ocean, we might discover the actual duration of a former earth; an earth which had supported plants and animals, and had supplied the ocean with those materials which the construction of the present earth required; consequently, we should have the measure of a corresponding space of time, viz. that which had been required in the production of the present land. If, on the contrary, no period can be fixed for the duration or destruction of the present earth, from our observations of those natural operations, which, though unmeasurable, admit of no dubiety, we shall be warranted in drawing the following conclusions; 1*st*, That it had required an indefinite space of time to have produced the land which now appears; 2*dly*, That an equal space had been employed upon the construction of that former land from whence the materials of the present came; *lastly*, That there is presently laying at the bottom of the ocean the foundation of future land, which is to appear after an indefinite space of time.

But, as there is not in human observation proper means for measuring the waste of land upon the globe, it is hence inferred, that we cannot estimate the duration of what we see at present, nor calculate the period at which it had begun; so that, with respect to human observation, this world has neither a beginning nor an end.

An endeavour is then made to support the theory by an argument of a moral nature, drawn from the consideration of a final cause. Here a comparison is formed between the present theory, and those by which there is necessarily implied either evil or disorder in natural things; and an argument is formed, upon the supposed wisdom of nature, for the justness of a theory in which perfect order is to be perceived. Or,

According to the theory, a soil, adapated to the growth of plants, is necessarily prepared, and carefully preserved; and, in the necessary waste of land which is inhabited, the foundation is laid for future continents, in order to support the system of this living world.

Thus, either in supposing Nature wise and good, an argument is formed in confirmation of the theory, or, in supposing the theory to be just, an argument may be established for wisdom and benevolence to be perceived in nature. In this manner, there is opened to our view a subject interesting to man who thinks; a subject on which to reason with relation to the system of nature; and one which may afford the human mind both information and entertainment.

Thomas Chrowder Chamberlin (1843-1928) attended Beloit College to study theology, but later decided to follow a career in geology, subsequently attending the University of Michigan. In the course of a long career, he served as principal of a high school in Wisconsin, professor at the State Normal School at Whitewater, Wisconsin, and professor at Beloit, as well as Director of the Wisconsin Geological Survey. He later became successively Chief of the Glacial Division of the U.S. Geological Survey—taking part in the Peary Relief Expedition—and President of the University of Wisconsin, an office which he occupied for five years. Resigning in 1892 to return to active scientific work, he was appointed the first chairman of the Department of Geology at the University of Chicago, and taught there for the next twenty-six years. Chamberlin made substantial contributions to glacial geology, the study of diastrophism, the origin of the Earth, and the philosophy of scientific investigation. It is from his writings in connection with this last category that the present extract is taken. The paper was published in Science *in 1890 and is a useful summary of the method used to such effect by Chamberlin and his contemporaries. Chamberlin was one of the founding members of the Geological Society of America and served as president of that body in 1905.*

THE METHOD OF MULTIPLE WORKING HYPOTHESES

T. C. Chamberlin

There are two fundamental classes of study. The one consists in attempting to follow by close imitation the processes of previous thinkers, or to acquire by memorizing the results of their investigations. It is merely secondary, imitative, or acquisitive study. The other class is primary or creative study. In it the effort is to think independently, or at least individually, in the endeavor to discover new truth, or to make new combinations of truth, or at least to develop an individualized aggregation of truth. The endeavor is to think for one's self, whether the thinking lies wholly in the fields of previous thought or not. It is not necessary to this habit of study that the subject-material should be new; but the process of thought and its results must be individual and independent, not the mere following of previous lines of thought ending in predetermined results. The demonstration of a problem in Euclid precisely as laid down is an illustration of the former; the demonstration of the same proposition by a method of one's own or in a manner distinctively individual is an illustration of the latter; both lying entirely within the realm of the known and the old.

Creative study, however, finds its largest application in those subjects in which, while much is known, more remains to be known. Such are the fields which we, as naturalists, cultivate; and we are gathered for the purpose of developing improved methods lying largely in the creative phase of study, though not wholly so.

Intellectual methods have taken three phases in the history of progress thus far. What may be the evolutions of the future it may not be prudent to forecast. Naturally the methods we now urge seem the highest attainable. These three methods may be designated, first, the method of the ruling theory; second, the method of the working hypothesis; and, third, the method of multiple working hypotheses.

As in the earlier days, so still, it is the habit of some to hastily conjure up an explanation for every new phenomenon that presents itself. Interpretation rushes to the forefront as the chief obligation pressing upon the putative wise man. Laudable as the effort at explanation is in itself, it is to be condemned when it runs before a serious inquiry into the phenomenon itself. A dominant disposition to find out what is, should precede and crowd aside the question, commendable at a later stage, "How came this so?" First full facts, then interpretations.

The habit of precipitate explanation leads rapidly on to the development of tentative theories. The explanation offered for a given phenomenon is naturally, under the impulse of self-consistency, offered for like phenomena as they present themselves, and there is soon developed a general theory explanatory of a large class of phenomena similar to the original one. This general theory may not be supported by any further considerations than those which were involved in the first hasty inspection. For a time it is likely to be held in a tentative way with a measure of candor. With this tentative spirit and measurable candor, the mind satisfies its moral sense, and deceives itself with the thought that it is proceeding cautiously and impartially toward the goal of ultimate truth. It fails to recognize that no amount of provisional holding of a theory, so long as the view is limited and the investigation partial, justifies an ultimate conviction.

It is in this tentative stage that the affections enter with their blinding influence. Love was long since represented as blind, and what is true in the personal realm is measurably true in the intellectual realm. Important as the intellectual affections are as stimuli and as rewards, they are nevertheless dangerous factors, which menace the integrity of the intellectual processes. The moment one has offered an original explanation for a phenomenon which seems satisfactory, that moment affection for his intellectual child springs into existence; and as the explanation grows into a definite theory, his parental affections cluster about his intellectual offspring, and it grows more and more dear to him, so that, while he holds it seemingly tentative, it is still lovingly tentative, and not impartially tentative.

Briefly summed up, the evolution is this: a premature explanation passes into a tentative theory, then into an adopted theory, and then into a ruling theory.

When the last stage has been reached, unless the theory happens, perchance, to be the true one, all hope of the best results is gone. To be sure, truth may be brought forth by an investigator dominated by a false ruling idea. His very errors may indeed stimulate investigation on the part of others. But the condition is an unfortunate one. Dust and chaff are mingled with the grain in what should be a winnowing process.

As previously implied, the method of the ruling theory occupied a chief place during the infancy of investigation. It is an expression of the natural infantile tendencies of the mind, though in this case applied to its higher activities, for in the earlier stages of development the feelings are relatively greater than in later stages.

Unfortunately it did not wholly pass away with the infancy of investigation, but has lingered along in individual instances to the present day, and finds illustration in universally learned men and pseudo-scientists of our time.

The defects of the method are obvious, and its errors great. If I were to name the central psychological fault, I should say that it was the admission of intellectual affection to the place that should be dominated by impartial intellectual rectitude. Efforts at reform followed.

The first great endeavor was repressive. The advocates of reform insisted that theorizing should be restrained, and efforts directed to the simple determination of facts. The effort was to make scientific study factitious instead of causal. Because theorizing in narrow lines had led to manifest evils, theorizing was to be condemned. The reformation urged was not the proper control and utilization of theoretical effort, but its suppression. Its weakness lay in its narrowness and its restrictiveness. There is no nobler aspiration of the human intellect than desire to compass the cause of things. The disposition to find explanations and to develop theories is laudable in itself. It is only its ill use that is reprehensible. The vitality of study quickly disappears when the object sought is a mere collocation of dead unmeaning facts.

The inefficiency of this simply repressive reformation becoming apparent, improvement was sought in the method of the working hypothesis. This is affirmed to be *the* scientific method of the day, but to this I take exception. The working hypothesis differs from the ruling theory in that it is used as a means of determining facts, and has for its chief function the suggestion of lines of inquiry; the inquiry being made not for the sake of the hypothesis, but for the sake of facts. Under the method of the ruling theory, the stimulus was directed to the finding of facts for the support of the theory. Under the working hypothesis, the facts are sought for the purpose of ultimate induction and demonstration, the hypothesis being but a means for the more ready development of facts and of their relations, and the arrangement and preservation of material for the final induction.

It will be observed that the distinction is not a sharp one, and that a working hypothesis may with the utmost ease degenerate into a ruling theory. Affection may as easily cling about an hypothesis as about a theory, and the demonstration of the one may become a ruling passion as much as of the other.

Conscientiously followed, the method of the working hypothesis is a marked improvement upon the method of the ruling theory; but it has its defects—defects which are perhaps best expressed by the ease with which the hypothesis becomes a controlling idea. To guard against this, the method of multiple working hypotheses is urged. It differs from the former method in the multiple character of its genetic conceptions and of its tentative interpretations. It is directed against the radical defect of the two other methods; namely, the partiality of intellectual parentage. The effort is to bring up into view every rational explanation of new phenomena, and to develop every tenable hypothesis respecting their cause and history. The investigator thus becomes the parent of a family of hypotheses: and, by his parental relation to all, he is forbidden to fasten his affections unduly upon any one. In the nature of the case, the danger that springs from affection is counteracted, and therein is a radical difference between this method and the two preceding. The investigator at the outset puts himself in cordial sympathy and in parental relations (of adoption, if not of authorship) with every hypothesis that is at all applicable to the case under investigation. Having thus neutralized the partialities of his emotional nature, he proceeds with a certain natural and enforced erectness of mental attitude to the investigation, knowing well that some of his intellectual children will die before maturity, yet feeling that several of them may survive the results of final investigation, since it is often the outcome of inquiry that several causes are found to be involved instead of a single one. In following a single hypothesis, the mind is presumably led to a single explanatory conception. But an adequate explanation often involves the co-ordination of several agencies, which enter into the combined result in varying proportions. The true explanation is therefore necessarily complex. Such complex explanations of phenomena are specially encouraged by the method of multiple hypotheses, and constitute one of its chief merits. We are so prone to attribute a phenomenon to a single cause, that, when we find an agency present, we are liable to rest satisfied therewith, and fail to recognize that it is but one factor, and perchance a minor factor, in the accomplishment of the total result. Take for illustration the mooted question of the origin of the Great Lake basins. We have this, that, and the other hypothesis urged by different students as the cause of these great excavations; and all of these are urged with force and with fact, urged justly to a certain degree. It is practically demonstrable that these basins were river-valleys antecedent to the glacial incursion, and that they owe their origin in part to the pre-existence of those valleys and to the blocking-up of their outlets. And so this view of their origin is urged with a certain truthfulness. So, again, it is demonstrable that they were occupied by great lobes of ice, which excavated them to a marked degree, and therefore the theory of glacial excavation finds support in fact. I think it is furthermore demonstrable that the earth's crust beneath these basins was flexed downward, and that they owe a part of their origin to crust deformation. But to my judgment neither the one nor the other, nor the third, constitutes an adequate explanation of the phenomena. All these must be taken together, and possibly they must be supplemented by other agencies. The problem, therefore, is the determination not only of the participation, but of the measure and the extent, of each of these agencies in the production of the complex result. This is not likely to be accomplished by one whose working hypothesis is pre-glacial erosion, or glacial erosion, or crust deformation, but by one whose staff of working hypotheses embraces all of these and any other agency which can be rationally conceived to have taken part in the phenomena.

Laws of Nature

The chess-board is the world; the pieces are the phenomena of the universe; the rules of the game are what we call the laws of Nature. The player on the other side is hidden from us. We know that his play is always fair, just, and patient. But also we know, to our cost, that he never overlooks a mistake, or makes the smallest allowance for ignorance.

—Thomas Henry Huxley

G. K. Gilbert (1843-1918) was one of the most respected and influential of all American geologists during the late 19th and early 20th centuries, and his influence upon subsequent students has been profound. He served as Civil Geologist for the Surveys west of the One Hundredth Meridian, and his work in the southwestern states brought him in contact with John Wesley Powell, with whom he collaborated. He became chief geologist of the U.S. Geological Survey in 1890. He is frequently spoken of as one of the foremost philosophers of geology, but his contribution to philosophy was modest, being contained chiefly in two papers published in 1886 and 1896. Where Gilbert did excel was in the rigorous application of a few simple, sound, but not original, scientific principles to the specific problems with which he was concerned. In this he was unrivaled. He himself knew and clearly stated that his scientific approach contained nothing novel. Gilbert's particular contribution has been well described in the writings of James Gilluly, distinguished contemporary geologist, longtime member of the U.S. Geological Survey and Professor Emeritus at the University of California, Santa Cruz. Gilluly has shown that this method involved the use of what we should now refer to as multiple working hypotheses a method that goes back to Hume, Carr, Locke and Bacon. Gilbert exemplified this beautifully in his papers in recording and interpreting the phenomena that he studied; by developing hypotheses to account for them, by predicting the consequences of each of the various hypotheses and then by testing these consequences against new observations. In the tracing out of these principles in practice, Gilbert was meticulous, providing an outstanding example of the application of the philosophy which he preached. There are few places where this is better shown than in his description of some of the shoreline beaches of ancient Lake Bonneville. The present extract from his monograph of the subject, published in 1890, discusses the various methods by which shoreline features may be identified and distinguished from other similar structures. Lake Bonneville, now preserved as a remnant of the once much larger Tertiary lake, has been the subject of intensive study by geologists for the last century, but it was Gilbert's pioneering work that laid the foundation for our present understanding.

THE DISCRIMINATION OF SHORE FEATURES

G. K. Gilbert

A shore is the common margin of dry land and a body of water. The elements of its peculiar topography are little liable to confusion so long as they are actually associated with land on one side and water on the other; but after the water has been withdrawn, their recognition is less easy. They consist merely of certain cliffs, terraces, and ridges; and cliffs, terraces, and ridges abound in the topography of land surfaces. In the following pages the topographic features characteristic of ancient shores will be compared and contrasted with other topographic elements likely to create confusion.

Such a discrimination as this has not before been attempted, although the principal distinctions upon which it is based have been the common property of geologists for many years. The contrast of stream terraces with shore terraces was clearly set forth by Dana in the American Journal of Science in 1849, and has been restated by Geikie in his Text-Book of Geology. It was less clearly enunciated by the elder Hitchcock in his Illustrations of Surface Geology.

A cliff is a topographic facet, in itself steep, and at the same time surrounded by facets of less inclination. The only variety belonging to the phenomena of shores is that to which the name "sea-cliff" has been applied. It will be compared with the cliff of differential degradation, the stream cliff, the coulée edge, the fault scarp, and the land-slip cliff.

It is a familiar fact that certain rocks, mainly soft, yield more rapidly to the agents of erosion than certain other rocks, mainly hard. It results from this, that in the progressive degradation of a country by subaerial erosion the minor reliefs are generally occupied by hard rocks while the minor depressions mark the positions of soft rocks. Where a hard rock overlies one much softer, the erosion of the latter proceeds so rapidly that the former is sapped, and being deprived of its support falls away in blocks, and is thus wrought at its margin into a cliff. In regions undergoing rapid degradation such cliffs are exceedingly abundant.

It is the invariable mark of a cliff of differential degradation that the rock of the lower part of its face is so constituted as to yield more rapidly to erosion than the rock of the upper part of its face. It is strictly dependent on the constitution and structure of the terrane. It may have any form, but since the majority of rocks are stratified in broad, even sheets, and since the most abrupt alternations of texture occur in connection with such stratification, a majority of cliffs of differential degradation exhibit a certain uniformity and parallelism of parts. The crest of such a cliff is a line parallel to the base, and other associated cliffs run in lines approximately parallel. The most conspicuous of the cliffs of stratified rocks occur where the strata are approximately horizontal; and these more often than any others have been mistaken for sea-cliffs.

The most powerful agent of land erosion is the running stream, and, in regions undergoing rapid degradation, corrasion by streams so far exceeds the general waste of the surface that their channels are cut down vertically, forming cliffs on either hand. These cliffs are afterward maintained by lateral corrasion, which opens out the valley of the stream after the establishment of a base level has checked the vertical corrasion. Such cliffs are in a measure independent of the nature of the rock, and are closely associated with the stream. They stand as a rule in pairs facing each other and separated only by the stream and its flood plain. The base of each is a line inclined in the direction of the stream channel and in the same degree. The crest is not parallel thereto, but is an uneven line conforming to no simple law.

The viscosity of a lava stream is so great, and this viscosity is so augmented as its motion is checked by gradual cooling, that its margin after congelation is usually marked by a cliff of some height. The distinguishing characters of such a cliff are that the rock is volcanic, with the superficial features of a subaerial flow. It has probably never been mistaken for a sea-cliff, and receives mention here only for the sake of giving generality to the classification of cliffs.

The faulting of rocks consists in the relative displacement of two masses separated by a fissure. The plane of the fissure is usually more or less vertical, and by virtue of the displacement one mass is made to project somewhat above the other. The portion of the fissure wall thus brought to view constitutes a variety of cliff or escarpment, and has been called a fault scarp. In the Great Basin such scarps are associated with a great number of mountain ranges, appearing generally at their bases, just where the solid rock of the mountain mass is adjoined by the detrital foot slope. They occasionally encroach upon the latter, and it is in such case that they are most conspicuous as well as most likely to be mistaken for sea-cliffs. Although in following the mountain bases they do not vary greatly in altitude, yet they never describe exact contours, but ascend and descend the slopes of the foot hills. The crest of such a cliff is usually closely parallel to the base for long distances, but this parallelism is not absolute. The two lines gradually converge at either end of the displacement. In exceptional instances they converge rapidly, giving the cliff a somewhat abrupt termination, and in such case a new cliff appears *en échelon,* continuing the displacement with a slight offset.

The land-slip differs from the fault chiefly in the fact that it is a purely superficial phenomenon, having its whole history upon a visible external slope. It occurs usually in unconsolidated material, masses of which break loose and move downward short distances. The cliffs produced by their separation from the general or parent mass, are never of great horizontal extent, and have no common element of form except that they are concave outward. They frequently occur in groups, and are apt to contain at their bases little basins due to the backward canting which forms part of the motion of the sliding mass.

The sea-cliff differs from all others, first, in that its base is horizontal, and, second, in that there is associated with it at one end or other a beach, a barrier, or an embankment. A third valuable diagnostic feature is its uniform association with the terrace at its base; but in this respect it is not unique, for the cliff of differential degradation often springs from a terrace. Often, too, the latter is nearly horizontal at base, and in such case the readiest comparative test is found in the fact that the sea-cliff is independent of the texture and structure of the rocks from which it is carved, while the other is closely dependent thereon.

The sea-cliff is distinguished from the stream-cliff by the fact that it faces an open valley broad enough and deep enough to permit the generation of efficient waves if occupied by a lake. It is distinguished from the coulée edge by its independence of rock structure and by its associated terrace. It differs from the fault scarp in all those peculiarities which result from the attitude of its antecedent; the water surface concerned in the formation of the sea-cliff is a horizontal plane; the fissure concerned in the formation of the fault scarp is a less regular but essentially vertical plane. The former crosses the inequalities of the preexistent topography as a contour, the latter as a traverse line.

The land-slip cliff is distinguished by the marked concavity of its face in horizontal contour. The sea-cliff is usually convex, or, if concave, its contours are long and sweeping. The former is distinguished also by its discontinuity.

Skeleton of Old

Afflicted skeleton of old, doomed to damnation
Soften, thou stone, the hearts of this wicked generation!

—**Johann J. Scheuchzer**

George Gaylord Simpson, for many years curator of mammals at the American Museum of Natural History, Professor of Geology at Columbia University, and more recently Agassiz Professor of Vertebrate Paleontology at Harvard, is one of the most influential contemporary geologic writers. His wide-ranging studies on fossil mammals led him to produce a new classification of mammals, while his books on evolution and taxonomy have altered the whole approach to those important studies. He is also a philosopher of distinction, and a present essay shows the refinement of the "simple" concept of causation which has taken place since the time of G. K. Gilbert. The present essay is not easy reading, but it is of fundamental importance to anyone interested in the philosophical implications of historical science.

HISTORICAL SCIENCE

George G. Simpson

The simplest definition of history is that it is change through time. It is, however, at once clear that the definition fails to make distinctions which are necessary if history is to be studied in a meaningful way. A chemical reaction involves change through time, but obviously it is not historical in the same sense as the first performance by Lavoisier of a certain chemical experiment. The latter was a nonrecurrent event, dependent on or caused by antecedent events in the life of Lavoisier and the lives of his predecessors, and itself causal of later activities by Lavoisier and his successors. The chemical reaction involved has no such causal relationship and has undergone no change before or after Lavoisier's experiment. It always has occurred and always will recur under the appropriate historical circumstances, but as a reaction in itself it has no history.

A similar contrast between the historical and the nonhistorical exists in geology and other sciences. The processes of weathering and erosion are unchanging and nonhistorical. The Grand Canyon or any gully is unique at any one time but is constantly changing to other unique, nonrecurrent configurations as time passes. Such changing, individual geological phenomena are historical, whereas the properties and processes producing the changes are not.

The unchanging properties of matter and energy and the likewise unchanging processes and principles arising therefrom are *immanent* in the material universe. They are nonhistorical, even though they occur and act in the course of history. The actual state of the universe or of any part of it at a given time, its configuration, is not immanent and is constantly changing. It is *contingent* in Bernal's term, or *configurational*, as I prefer to say. History may be defined as configurational change through time, i.e, a sequence of real, individual but interrelated events. These distinctions betweens the immanent and the configurational and between the nonhistorical and the historical are essential to clear analysis and comprehension of history and of science. They will be maintained, amplified, and exemplified in what follows.

Definitions of science have been proposed and debated in innumerable articles and books. Brief definitions are inevitably inadequate, but I shall here state the one I prefer: Science is an exploration of the material universe that seeks natural, orderly relationships among observed phenomena and that is self-testing. Apart from the points that science is concerned only with the material or natural and that it rests on observation, the definition involves three scientific activities: the description of phenomena, the seeking of theoretical, explanatory relationships among them, and some means for the establishment of confidence regarding observations and theories. Among other things, later sections of this essay will consider these three aspects of historical science.

Historical science may thus be defined as the determination of configurational sequences, their explanations, and the testing of such sequences and explanations. (It is already obvious and will become more so that none of the three phases is simple or thus sufficiently described.)

Geology is probably the most diverse of all the sciences, and its status as in part a historical science is correspondingly complex. For one thing, it deals with the immanent properties and processes of the physical earth and its constituents. This aspect of geology is basically nonhistorical. It can be viewed simply as a branch of physics (including mechanics) and chemistry, applying those sciences to a single (but how complex!) object: the earth. Geology also deals with the present configuration of the earth and all its parts, from core to atmosphere. This aspect of geology might be considered nonhistorical insofar as it is purely descriptive, but then it also fails to fulfill the whole definition of a science. As soon as theoretical, explanatory relationships are brought in, so necessarily are changes and sequences of configurations, which are historical. The fully scientific study of geological configurations is thus historical science. This is the *only* aspect of geology that is peculiar to this science, that is simply geology and not also something else. (Of course I do not mean that it can be studied without reference to other aspects of geology and to other sciences, both historical and nonhistorical.)

Paleontology is primarily a historical science, and it is simultaneously biological and geological. Its role as a part of historical biology is obvious. In this role, like all other aspects of biology, it involves all the immanent properties and processes of the physical sciences, but differs from them not only in being historical but also in that its configurational systems are incomparably more complex and have feedback and information storage and transmittal mechanisms unlike any found in the inorganic realm. Its involvement in geology and inclusion in that science as well as in biology are primarily due to the fact that the history of organisms runs parallel with, is environmentally contained in, and continuously interacts with the physical history of the earth. It is of less philosophical interest but of major operational importance that paleontology, when applicable, has the highest resolving power of any method yet discovered for determining the sequence of strictly geological events. (That radiometric methods may give equal or greater resolution is at present a hope and not a fact.)

In principle, the observational basis of any science is a straight description of what is there and what occurs, what Lloyd Morgan used to call "plain story." In a physical example, plain story might be the specifications of a pendulum and observations of its period. A geological plain story might describe a bed of arkose, its thickness, its attitude, and its stratigraphic and geographic position. An example of paleontological plain story would be the occurrence of a specimen of a certain species at a particular point in the bed of arkose. In general, the more extended plain stories of historical science would describe configurations and place them in time.

In fact, plain story in the strictest, most literal sense plays little part in science. Some degree of abstraction, generalization, and theorization usually enters in, even at the first observational level. The physicist has already abstracted a class of configurational systems called "pendulums" and assumes that only the length and period need be observed, regardless of other differences in individuals of the class, unless an observation happens to disagree with the assumption. Similarly, the geologist by no means describes all the characteristics of the individual bed of arkose and its parts but has already generalized a class "arkose" and adds other details, if any, only in terms of such variations within the class as are considered pertinent to his always limited purpose. The paleontologist has departed still further from true, strict plain story, for in recording a specimen as of a certain species he has not only generalized a particularly complex kind of class but has also reached a conclusion as to membership in that class that is not a matter of direct observation at all.

Every object and every event is unique if its configurational aspects are described in full. Yet, and despite the schoolteachers, it may be said that some things are more unique than others. This depends in the first place on the complexity of what is being described, for certainly the more complex it is the more ways there are in which it may differ from others of its general class. A bed of arkose is more complex than a pendulum, and an organism is to still greater degree more complex than a bed of arkose. The hierarchy of complexity and individual uniqueness from physics to geology to biology is characteristic of those sciences and essential to philosophical understanding of them. It bears on the degree and kind of generalization characteristic of and appropriate to the various sciences even at the primary observational level. The number of pertinent classes of observations distinguished in physics is much smaller than in geology, and much smaller in geology than in biology. For instance, in terms of taxonomically distinguished discrete objects, compare the numbers of species of particles and atoms in physics, of minerals and rocks in geology, and of organisms in biology. Systems and processes in these sciences have the same sequence as to number and complexity.

Another aspect of generalization and degree of uniqueness arises in comparison of nonhistorical and historical science and in the contrast between immanence and configuration. In the previous examples, the physicist was concerned with a nonhistorical and immanent phenomenon: gravitation. It was necessary to his purpose and inherent in his method to eliminate as far as possible and then to ignore any historical element and any configurational uniqueness in the particular, individual pendulum used in the experiment. He sought a changeless law that would apply to all pendulums and ultimately to all matter, regardless of time and place. The geologist and paleontologist were also interested in generalization of common properties and relationships between one occurrence of arkose and another, between one specimen and another of a fossil species, but their generalizations were of the configurational and not the immanent properties and were, or at least involved, historical and not only nonhistorical science. The arkose or the fossil had its particular as well as its general configurational properties, its significant balance of difference and resemblance, not only because of immanent properties of its constituents and immanent processes that had acted on it, but also because of its history, the configurational sequence by which these individual things arose. The latter aspect, not pertinent to the old pendulum experiment or to almost anything in the more sophisticated physics of the present day, is what primarily concerns geology and paleontology as historical sciences, or historical science in general.

A scientific law is a recurrent, repeatable relationship between variables that is itself invariable to the extent that the factors affecting the relationship are explicit in the law.

Laws, as thus defined, are generalizations, but they are generalizations of a very special kind. They are complete abstractions from the individual case. Laws are inherent, this is *immanent*, in the nature of things as abstracted entirely from contingent configurations, although always acting on those configurations.

Until recently the theoretical structure of the nonhistorical physical sciences consisted largely of a body of laws or supposed laws of this kind. The prestige of these sciences and their success in discovering such laws were such that it was commonly believed that the proper scientific goal of the historical sciences was also to discover laws. Supposed laws were proposed in all the historical sciences. By way of example in my own field, paleontology, I may mention "Dollo's law" that evolution is irreversible, "Cope's law" that animals become larger in the course of evolution, or "Williston's law" that repetitive serial structures in animals evolve so as to become less numerous but more differentiated. The majority of such supposed laws are no more than descriptive generalizations. For example, animals do not invariably become larger in time. Cope's law merely generalizes the observation that this is a frequent tendency, without establishing any fixed relationship among the variables possibly involved in this process.

The search for historical laws is, I maintain, mistaken in principle. Laws apply, in the dictionary definition "under the same conditions," or in my amendment "to the extent that factors affecting the relationship are explicit in the law," or in common parlance "other things being equal." But in history, which is a sequence of real, individual events, other things never are equal. Historical events, whether in the history of the earth, the history of life, or recorded human history, ae determined by the immanent characteristics of the universe acting on and within particular configurations, and never by either the immanent or the configurational alone. If laws thus exclude factors inextricably and significantly involved in real events, they cannot belong to historical science.

It is further true that historical events are unique, usually to a high degree, and hence cannot embody laws defined as recurrent, repeatable relationships. Apparent repetition of simple events may seem to belie this. A certain person's repeatedly picking up and dropping a certain stone may seem to be a recurrent event in all essentials, but there really is no applicable *historical* law. Abstraction of a law from such repeated events leads to a nonhistorical law of immanent relationships, perhaps in this case of gravity and acceleration or perhaps of neurophysiology, and not to a historical law of which this particular person, picking up a certain stone, at a stated moment, and dropping it a definite number of times would be a determinate instance. In less trivial and more complex events, it is evident that the extremely intricate configurations involved in and necessary, for example, as antecedents for the erosion of the Grand Canyon or the origin of *Homo sapiens* simply cannot recur and that there can be no laws of such one-of-a-kind events. (Please bear in mind that the true, immanent laws are equally necessary and involved in such events but that they remain nonhistorical; the laws would have acted differently and the historical events, the change of configuration, would have been different if the configuration had been different; this historical element is not included in the operative laws.)

It might be maintained that my definition of law is old-fashioned and is no longer accepted in the nonhistorical sciences either. Many laws of physics, considered nonhistorical, are now conceived as statistical in nature, involving not an invariable relationship but an average one. The old gas laws or the new laws of radioactive decay are examples.

The historical scientist here notes that a real gas in a real experiment has *historical* attributes that are *additional* to the laws affecting it. Every molecule of a real gas has its individual history. Its position, direction of motion, and velocity at a given moment (all parts of the total configuration) are the outcome of that history. It is, however, quite impractical and, for the purposes of physics, unnecessary to make an historical study of the gas. The gas laws apply well enough "other things being equal," which means here that the simple histories of the molecules tend, as observation shows, to produce a statistical result so nearly uniform that the historical, lawless element can be ignored for practical purposes.

The laws immanent in the material universe are not statistical in essence. They act invariably in variable historical circumstances. The pertinence of statistics to such laws as those of gases is that they provide a generalized description of usual historical circumstances in which those laws act, and not that they are inherent in the laws themselves. Use of statistical expressions, not as laws but as generalized descriptions, is common and helpful in all science and especially in historical science. For example, the statistical specifications of land forms or of grain size in sediments clearly are not laws but descriptions of configurations involved in and arising from history.

To speak of "laws of history" is either to misunderstand the nature of history or to use "laws" in an unacceptable sense, usually for generalized descriptions rather than formulations of immanent relationships.

Uniformitarianism has long been considered a basic principle of historical science and a major contribution of geology to science and philosophy. In one form or another it does permeate geologic and historical thought to such a point as often to be taken for granted. Among those who have recently given conscious attention to it, great confusion has arisen from conflicts and obscurities as to just what the concept is. To some, uniformitarianism (variously defined) is a law of history. Others, maintaining that it is not a law, have tended to deny its significance. Indeed, in any reasonable or usual formulation, it is not a law, but that does not deprive it of importance. It is commonly defined as the principle that the present is the key to the past. That definition is, however, so loose as to be virtually meaningless in application. A new, sharper, and clearer definition in modern terms is needed.

Uniformitarianism arose around the turn of the 18th to 19th centuries, and its original significance can be understood only in that context. It was a reaction against the then prevailing school of catastrophism, which had two main tenets: (1) the general belief that God has intervened in history, which therefore has included both natural and supernatural (miraculous) events; and (2) the particular proposition that earth history consists in the main of a sequence of major catastrophes, usually considered as of divine origin in accordance with the first tenet. Uniformitarianism, as then expressed, had various different aspects and did not always face these issues separately and clearly. On the whole, however, it embodied two propositions contradictory to catastrophism: (1) earth history (if not history in general) can be explained in terms of natural forces still observable as acting today; and (2) earth history has not been a series of universal or quasi-universal catastrophes but has in the main been a long, gradual development—what we would now call an evolution. (The term "evolution" was not then customarily used in this sense.) A classic example of the conflicting application of these principles is the catastrophist belief that valleys are clefts suddenly opened by a supernally ordered revolution as against the uniformitarian belief that they have been gradually formed by rivers that are still eroding the valley bottoms.

Both of the major points originally at issue are still being argued on the fringes of science or outside it. To most geologists, however, they no longer merit attention from anyone but a student of human history. It is a necessary condition and indeed part of the definition of science in the modern sense that only natural explanations of material phenomena are to be sought or can be considered scientifically tenable. It is interesting and significant that general acceptance of this principle (or limitation, if you like) came much later in the historical than in the nonhistorical sciences. In historical geology it was the most important outcome of the uniformitarian-catastrophist controversy. In historical biology it was the still later outcome of the Darwinian controversy and was hardly settled until our own day. (It is still far from settled among nonscientists.)

As to the second major point originally involved in uniformitarianism, there is no *a priori* or philosophical reason for ruling out a series of *natural* worldwide catastrophes as dominating earth history. However, this assumption is simply in such flat disagreement with everything we now know of geological history as to be completely incredible. The only issues still valid involve the way in which natural processes still observable have acted in the past and the sense in which the present is a key to the past.

Then what uniformity principle, if any, is valid and important? The distinction between immanence and configuration (or contingency) clearly points to one: the postulate that immanent characteristics of the material universe have not changed in the course of time. By this postulate all the immanent characteristics exist today and so can, in principle, be observed or, more precisely, inferred as generalizations and laws from observations. It is in this sense that the present is the key to the past. Present immanent properties and relationships permit the interpretation and explanation of history precisely because they are *not* historical. They have remained unchanged, and it is the configurations that have changed. Past configurations were never quite the same as they are now and were often quite different. Within those different configurations, the immanent characteristics have worked at different scales and rates at different times, sometimes combining into complex processes different from those in action today. The uniformity of the immanent characteristics helps to explain the fact that history is not uniform. (It could even be said that uniformitarianism entails catastrophes, but the paradox would be misleading if taken out of context.) Only to the extent that past configurations resembled the present in essential features can past processes have worked in a similar way.

That immanent characteristics are unchanging may seem at first sight either a matter of definition or an obvious conclusion, but it is neither. Gravity would be immanent (an inherent characteristic of matter *now*) even if the law of gravity had changed, and it is impossible to prove that it has not changed. Uniformity, in this sense, is an unprovable postulate justified, or indeed required, on two grounds. First, nothing in our incomplete but extensive knowledge of history disagrees with it. Second, only with this postulate is a rational interpretation of history possible, and we are justified in seeking—as scientists we *must* seek—such a rational interpretation. It is on this basis that I have assumed on previous pages that the immanent is unchanging.

Scientific Truth

Every great scientific truth goes through three stages. First, people say it conflicts with the Bible. Next, they say it has been discovered before. Lastly, they say they always believed it.

—Louis Agassiz

CHAPTER 5

Life's Biography

Earth science is concerned not only with the physical processes and development of the Earth, but also with the history of the myriad groups of organisms for which the planet provides a home. Indeed, each is in some respects the key to the other. The stratified rocks of the Earth's crust and the fossils they contain provide the pages of the record of the history of life; but the fossils that represent that record in turn provide the methods of arranging and numbering the pages of Earth's later history and grouping them into chapters. The fossil record, imperfect and incomplete as it is, is of unique importance to our understanding of life; for it also furnishes the possibility of studying something of the processes that gave rise to the first living things, and which lie behind the sweeping transformations they have undergone. In one sense, it is only because of the perspective the fossil record provides, that we can study evolution on anything but the most trivial scale. Living organisms, such as the fruitfly *Drosophila*, which have provided the experimental evidence on which evolutionary theory is based, have been studied only over a limited span of time and a relatively small number of generations. In the fossil record alone there lies the opportunity to study the great cavalcade of living things with whose teeming descendants we now share the planet Earth.

The extract by Frank H. T. Rhodes describes human history in evolutionary perspective.

EARTH AND MAN

Frank H. T. Rhodes

We share an Inhabited Planet. It is a remarkable thing that within our particular solar system, and the very limited area of which we have any direct knowledge, this planet alone is known to be inhabited. Of course, beyond our planetary system, many other such planets may well exist. One of the dramatic things as one reads the accounts of the astronauts returning to the Earth from the Moon is the incredible beauty of the planet Earth, in comparison with the darkness of space and the aridity and bleakness of the lunar landscape. As they approach nearer and nearer to our own planet, the bleakness and the grayness recede as they marvel at the colorful exuberance of the Earth. The thing that makes the difference is that ours is a living planet. The discovery that we share the Earth with other creatures is not a new one. Man has always known that he was part of the animal creation. He has always known that there were many different kinds of animals, and it requires no particular scientific skill to distinguish one from another. It has always been known that animals existed in great abundance. A swarm of locusts would convince anyone that this was the case. Yet, the question that constantly nagged our ancestors was the question, "Whence came this astonishing diversity of animals and plants?" There are over a million different species of animals, and over a third of a million different species of plants. How did they originate? Three kinds of answers were given by our great-grandparents.

The first suggestion was that living things arose by spontaneous generation and have existed since the creation of the planet. But the fossil record came to light, and somehow that had to be related to what was known of living creatures themselves. So, a second viewpoint developed to account for this record and for the origin of kinds of living things. It was supposed that there had been a series of great catastrophes within the history of the Earth, of which the flood of Noah was the most recent. New life came into existence spontaneously by divine command. But then a catastrophe overwhelmed the world, destroying it and producing massive extinction of living things, so that ultimately, there was a new creation, another episode of renewal. This, it was suggested, happened thirty or thirty-five times. Then, you remember, came those who in the 19th Century, and even earlier, offered another explanation, first of all hesitatingly, and later with more conviction. Lamarck, Alfred Russell Wallace, and Charles Darwin insisted that there was a continuity and an orderliness in the world of living things, as real as the orderliness in the world of non-living things. Darwin argued that reproduction produced variation, which was acted on by natural selection, so that those better fitted to the environment in which they lived gradually left more offspring than those that were less well-adapted. That was the secret behind the great diversity of living things, it was suggested. A hundred years of reflection has tended to confirm that particular hypothesis.

There are two implications of that evolutionary point of view. The first is that we are related to everything else in the universe, and our net effect as an agent of human action has scarcely been benevolent. There is an interdependence in which we are not only our brother's keeper, but, literally, the keeper of every other species on the plant. The second implication was even more profound, because the writings of Darwin threw the contemporary world into turmoil. You remember that the year after the publication of Darwin's book the British Associate for the Advancement of Science met at Oxford. You can still go to Oxford into the Natural History Museum to an upstairs room, now used to house collections, where a memorable debate took place. Four hundred people crowded inside a lecture theatre on a June evening in 1860 where the great debate of the century took place.

So it was that this debate was hostile from the start. The battle lines were drawn, and the nature of man became for fifty years the central theme of a furious debate between the emergent biological and geological sciences and established religion. You remember Tennyson having read "Chamber's Vestiges," crying out in anguish,

> Are God and Nature then at strife
> That Nature lends such evil dreams?
> So careful of the type she seems
> So careless of the single life.

Even a man such as George Bernard Shaw could remark, "If it could be proved that the whole universe has been produced by such natural selection, only fools and rascals could bear to live." Even Marxism, which was then dawning upon the world, had its own particular teleological view of history. In fact, Karl Marx offered to dedicate *Das Kapital* to Darwin, but Darwin politely declined the honor. It did seem, however, that the growth and understanding concerning the Inhabited Planet had overthrown all the security of man. The most remarkable thing about the publication of the *Origin of Species*, was not the reception which it received from the scientific public, but this worldwide confusion and outburst which it produced amongst men of all interests and persuasions. Philosophers, politicians, theologians, literary critics, historians, classical scholars, and the man in the street—all alike took it upon themselves to assess its worth. As so varied a group studied it, so their verdicts also varied—some accepted and respected Darwin's conclusions, others viewed them with suspicion, but most rejected them out of hand, and denounced both Darwinism and all its supposed implications. "As for the book, some treasured it, some burnt it, and some, undecided, like the Master of Trinity College, Cambridge, merely hid it!" Scientific theories, philosophies, political systems, ethical standards, revolutionary movements and social reforms, economic *laissez faire*—all these and more were established, modified, or justified upon Darwin's premises. Indeed, Darwinism soon became all things to all men.

In our own day, a century later, Darwin's theory, having found its place and left its mark in a host of different fields of inquiry, is now seen in its true perspective, and most of the clamor has subsided around it. The uninformed condemnation and hostility on the one hand, and the extravagances of many popular science writers on the other, have both been largely forgotten. There is, perhaps, only one area of human activity in which misunderstanding persists, for in the realm of religious belief, evolutionary theory still poses an "unsolved conflict," as Dr. David Lack called his recent book.

And yet, the truth is that the theory of evolution is neither anti-theistic nor theistic. So far as religion is concerned, it is strictly neutral, for it is a theory of the mechanism of descent with modification which seeks to explain *how* new species arise. It does indeed correct certain former ideas of the manner of creation, it does suggest that natural selection proceeds by natural laws, but like all other scientific theories, it provides no interpretation of the natural laws themselves, for it no more proves them to be the result of "pure" chance, than it proves them to be the servant and expression of purpose.

This conclusion should not surprise us, for it is true of science as a whole. And the neutrality arises, not because of some inadequacy in evolutionary theory, but because the student of evolution deliberately excludes from this explanation all reference to final cause. Unlike his Greek predecessor, he concerns himself not with the question "Why did life develop?" but with the question, "How did life develop?" To this latter question, we now begin to understand the answer. But the mind of man is such that, even as we understand the "How?" of life, we gaze towards the ultimate "Why?" To that question, evolutionary theory gives no answer.

This need not and does not imply that the question "Why?" is meaningless or irrelevant. It means only that we must look elsewhere for the answer, for the questions "How?" and "Why?" are not alternative and competitive, but rather complementary to one another. And this complementary relationship, which it has taken a century for the world to assimilate, brings us back to Darwin, for it is brilliantly summarized by the three short quotations from Whewell, Butler, and Bacon with which he prefaced the *Origin of Species*.

So, painfully, and slowly, Man has discovered himself as part of the inhabited planet. It is no accident that it was the young Charles Darwin, the geologist rather than the biologist, who provided us with the clue which was finally to solve the riddle of human relationships. As we look back in geological perspective today, we see a grand continuity, for biological evolution is but the last phase of a greater, threefold evolution, a continuous majestic process. First, there was a cosmic evolution, perhaps ten billion years ago in which the universe and all its elements came into existence. This was an evolution involving a low molecular order, but involving gigantic bodies of low structural complexity, such as stars, and galaxies. Then, perhaps five billion years ago, came the aggregation of cold planetary material around a star in this particular corner of our own local group galaxy, This was Earth's birthday. But there was a watershed around 3.5 billion years ago wherein that cosmic evolution which produced the Earth and stars provided a series of inorganically produced organic compounds, which came together for the first time to initiate a new kind of evolution: the organic evolution with natural selection about which we have been talking. We now understand at least some of the phases by which that change may have come. And so, for over three billion years, that particular evolutionary process complemented, but did not displace, the cosmic evolution that we have just described. But then a few thousand years ago, there emerged a third and new kind of evolution: a kind that Julian Huxley called psycho-social evolution, in which organic evolutionary change was partly displaced by tradition, and later by language and writing, so that each new generation does not have to learn by instinct or by firsthand experience, but where knowledge and precepts can be handed on from one generation to another. So the last generation stands on the shoulders of the accumulated wisdom of all its ancestors.

Geology

Geology is the science which investigates the successive changes that have taken place in the organic and inorganic kingdoms of nature...

—**Sir Charles Lyell**

"The Flowering Earth" by Donald C. Peattie is taken from a book by the same name. The short extract which we include is an elegant example of the author's sense of beauty and integrity. Though a botanist by training, Peattie's writing has a magical quality, comparable to that of more widely read authors. The extract gives a graphic account of the towering sequoia forests of the High Sierras and links them with their prehistoric forebears.

THE FLOWERING EARTH

Donald C. Peattie

What we love, when on a summer day we step into the coolness of a wood, is that its boughs close up behind us. We are escaped, into another room of life. The wood does not live as we live, restless and running, panting after flesh, and even in sleep tossing with fears. It is aloof from thoughts and instincts; it responds, but only to the sun and wind, the rock and the stream—never, though you shout yourself hoarse, to propaganda, temptation, reproach, or promises. You cannot mount a rock and preach to a tree how it shall attain the kingdom of heaven. It is already closer to it, up there, than you will grow to be. And you cannot make it see the light, since in the tree's sense you are blind. You have nothing to bring it, for all the forest is self-sufficient; if you burn it, cut, hack through it with a blade, it angrily repairs the swathe with thorns and weeds and fierce suckers. Later there are good green leaves again, toiling, adjusting, breathing—forgetting you.

For this green living is the world's primal industry; yet it makes no roar. Waving its banners, it marches across the earth and the ages, without dust around its columns. I do not hold that all of that life is pretty; it is not, in purpose, sprung for us, and moves under no compulsion to please. If ever you fought with thistles, or tried to pull up a cattail's matted rootstocks, you will know how plants cling to their own lives and defy you. The pond-scums gather in the cistern, frothing and buoyed with their own gases; the storm waves fling at your feet upon the beach the limp sea-lettuce wrenched from its submarine hold—reminder that there too, where the light is filtered and refracted, there is life still to intercept and net and by it proliferate. Inland from the shore I look and see the coastal ranges clothed in chaparral—dense shrubbery and scrubbery, close-fisted, intricately branched, suffocating the rash rambler in the noon heat with its pungency. Beyond, on the deserts, under a fierce sky, between the harsh lunar ranges of unweathered rock, life still, somehow, fights its way through the year, with thorn and succulent cell and indomitable root.

Between such embattled life and the Forest of Arden, with its ancient beeches and enchanter's nightshade, there is not great biologic difference. Each lives by the cool and cleanly and most commendable virtue of being green. And though that is not biological language, it is the whole story in two words. So that we ought not speak of getting at the root of a matter, but of going back to the leaf of things. The orator who knows the way to the country's salvation and does not know that the breath of life he draws was blown into his nostrils by green leaves, had better spare his breath. And before anyone builds a new state upon the industrial proletariat, he will be wisely cautioned to discover that the source of all wealth is the peasantry of grass.

The reason for these assertions—which I do not make for metaphorical effect but maintain quite literally—is that the green leaf pigment, called chlorophyll, is the one link between the sun and life; it is the conduit of perpetual energy to our own frail organisms.

For inert and inorganic elements—water and carbon dioxide of the air, the same that we breathe out as a waste—chlorophyll can synthesize with the energy of sunlight. Every day, every hour of all the ages, as each continent and, equally important, each ocean rolls into sunlight, chlorophyll ceaselessly creates. Not figuratively, but literally, in the grand First Chapter Genesis style. One instant there are a gas and water, as lifeless as the core of earth or the chill of space; and the next they are become living tissue—mortal yet genitive, progenitive, resilient with all the dewy adaptability of flesh, ever changing in order to stabilize some unchanging ideal of form. Life, in short, synthesized, plant-synthesized, light-synthesized. Botanists say photosynthesized. So that the post-Biblical synthesis of life is already a fact. Only when man has done as much, may he call himself the equal of a weed.

Plant life sustains the living world; more precisely, chlorophyll does so, and where, in the vegetable kingdom, there is not chlorophyll or something closely like it, then that plant or cell is a parasite—no better, in vital economy, than a mere animal or man. Blood, bone and sinew, all flesh is grass. Grass to mutton, mutton to wool, wool to the coat on my back—it runs like one of those cumulative nursery rhymes, the wealth and diversity of our material life accumulating from the primal fact of chlorophyll's activity. The roof of my house, the snapping logs upon the hearth, the desk where I write, are my imports from the plant kingdom. But the whole of modern civilization is based upon a whirlwind spending of the plant wealth long ago and very slowly accumulated. For, fundamentally, and away back, coal and oil, gasoline and illuminating gas had green origins too. With the exception of a small amount of water [and atomic] power, a still smaller of wind and tidal mills, the vast machinery of our complex living is driven only by these stores of plant energy.

We, then, the animals, consume those stores in our restless living. Serenely the plants amass them. They turn light's active energy to food, which is potential energy stored for their own benefit. Only if the daisy is browsed by the cow, the maple leaf sucked of its juices by an insect, will that green leaf become of our kind. So we get the song of a bird at dawn, the speed in the hoofs of the fleeing deer, the noble thought in the philosopher's mind. So Plato's Republic was builded on leeks and cabbages.

Animal life lives always in the red; the favorable balance is written on the other side of life's page, and it is written in chlorophyll. All else obeys the thermodynamic law that energy forever runs down hill, is lost and degraded. In economic language, this is the law of diminishing returns, and it is obeyed by the cooling stars as by man and all the animals. They float down its Lethe stream. Only chlorophyll fights up against the current. It is the stuff in life that rebels at death, that has never surrendered to entropy, final icy stagnation. It is the mere cobweb on which we are all suspended over the abyss.

So we made a prompt start that morning, with little ceremony about it but with some reverence preparing in us, for we went to visit giants in the earth. Of all that has survived from the Mesozoic, which began two hundred million years ago and ended about 55,000,000 B.C., Sequoia is the king. It is so much a king that, deposed today from all but two corners of its empire, superseded, outmoded, exiled and all but exterminated, it still stands without rival. And from all over the world, those who can make the pilgrimage come sooner or later to its feet, and do it homage.

Of Sequoia there are two species left, though once they were as various and abundant as are today the pines, their lesser brothers. One is the coastal redwood of California, which is the tallest tree in the world, and the other is the Big Tree of the Sierra Nevada, which is the mightiest in bulk. These two surviving species were here before the last glacial period. But as a genus or clan of species Sequoia has its roots in a day of fabulous eld. This noble line knew the tyrant lizards; through its branches swept the pterodactyls on great batty wings. As they saw the coming of the first birds, crawling up out of lizard shapes, so the forebears of our Sequoia witnessed the evolution of the first mammals when these still laid eggs, when they were low-skulled opossum-like things, when they became scuttling rodents that perhaps, gnawing and sucking at dinosaur eggs, brought down that giant dynasty from its very base.

Sequoia as a tribe saw the rise of all the most clever and lovely types of modern insects—the butterflies and moths, the beetles and bees and ants. Yet since there were then none of the intricate inter-relationships that have developed between modern flower and modern bee, Sequoia sowed the wind. It had flowers of an antique sort, flowers by technical definition, at least; petals and scent they had none. But their pollen must have been golden upon that ancient sunlight, and the communicable spark of futurity was in it. For Sequoia towers still upon its mountain top, and I was going there. . . .

It is a long climb still through the foothills of the Sierra. But now I sit up, with a lifted face. Beyond, higher in the east, portent is gathering. It takes shape, cloud-colored, gleaming with a stern reality where the sun smites a rocky forehead. Then appears that eternally moving miracle—snow in the summer sky. Sierra Nevada. . . .

The forests march upon the car; the ruddy soaring trunks of the sugar pines close around in escort. One hundred and two hundred feet overhead, their foliage is not even visible, screened by the lower canopy spread by western yellow pines which are giants in themselves. Groves of white fir, smelling like Christmas morning, troop between the yellow pines. Aisles of incense cedar with gracious down-sweeping boughs and flat sprays of gleaming foliage invite the eye down colonnaded avenues, fragrance drifting from their censers that appear to smoke with the long afternoon light. It grows darker with every mile, darker and deeper in moss and lichen, dim with the dimness of a vanished era. We have got back into earliest spring, at this altitude, and the blossoming dogwood troops along, illuminating the dusky places with a white laughter.

. . . Now, as the land of sunny levels has fallen remotely out of sight, there is a prescience in the cold air, of grandeur. We have climbed into the shadows; the drifts of snow are thicker between great roots, and richer grows the livid green mantle of staghorn lichen that clothes all Sierra wood in green old age. The boles of the sugar pines, which are kings, give place before the coming of an emperor. The sea sound of the forest deepens a tone in pitch. The road is twisting to find some way between columns so vast they block the view. They are not in the scale of living things, but geologic in structure, fluted and buttressed like colossal stone work, weathered to the color of old sandstone. They are not the pillars that hold up the mountains. They are Sequoia. The car has stopped, and I am standing in the presence.

Centuries of fallen needles make silence of my step, and the command upon the air, very soft, eternal, is to be still. I am at the knees of gods. I believe because I see, and to believe in these unimaginable titans strengthens the heart. Five thousand years of living, twelve million pounds of growth out of a tiny seed. Three hundred vertical feet of growth, up which the water travels every day dead against gravity from deep in the great root system. Every ounce, every inch was built upward from the earth by the thin invisible stream of protoplasm that has been handed down by the touch of pollen from generation to generation, for a hundred million years. Ancestral Sequoias grew here before the Sierra was uplifted. Today they look down upon the plains of men. No one has every known a Sequoia to die a natural death. Neither insects nor fungi can corrupt them. Lightning may smite them at the crown and break it; no fire gets to the heart of them. They simply have no old age, and the only down trees are felled trees.

In their uplifted hands they permit the little modern birds, the passerine song birds, vireos and warblers, tanagers and thrushes, to nest and call. I heard, very high above me in the luminous glooms, voices of such as these. I saw, between the huge roots that kept a winter drift, the snowplant thrust through earth its crimson fist. A doe—so long had I stood still—stepped from behind the enormous bole and, after a long dark liquid look, ventured with inquiring muzzle to touch my outheld hand. Bright passing things, these nestle for an hour in the sanctuary of the strong and dark, the vast and incalculably old.

That day I stood upon a height in time that let me glimpse the Mesozoic. It followed the Coal Age, the age of the fern forests, and it was itself the age of Gymnosperms. Sequoias are Gymnosperms. So are the pines, the larches, spruces, fir, yew, cypress, cedar—all that we call conifers, though there are other Gymnosperms that do not bear cones.

The Gymnosperms are, literally translating, "the naked-seeded" plants. For their seed is not completely enclosed in any fruit or husk, as it is in the higher modern plants that truly fruit and flower. Neither is the Gymnosperm egg or ovule completely enclosed in an ovary, as in the true flowers. To make an analogy, you could say that the Gymnosperms are plants without wombs, while the Angiosperms, the true flowering plants with genuine fruits, are endowed with that engendering sanctuary.

But though the seeds of the Gymnosperms are naked, they are seeds, and the seed is mightier than the spore. For the seed contains an embryo. Spores are very many and very small; they blow lightly about the world and find a lodging easily. But the seed is weighted with a great thing. Within even the tiniest lies the germ of a fetal plantlet, its fat cotyledons or first baby leaves till crumpled in darkness, its primary rootlet ready to thrust and suckle at the breast of earth.

This vital secret was inherited from the seed-ferns, back in misty days when the ferns were paramount. The conifers bore it forward; the true flowering plants were to carry it on and spread it in blossoming glory. Of that there was no sign in the Mesozoic forests. They must have been dark with an evergreen darkness, upright with a stern colonnaded strength. For they developed the power of building wood out of earth, not the punky wood of the tree ferns, but timber as we know it.

And we know no timber like the conifers'. No other trees are cut on such a scale. Where they grow, wooden cities swiftly rise, railroads are bent to them, mushroom fortunes arise from them, great fleets are built to export them. Scandinavia is one vast lumber camp, supplying western Europe; Port Oxford cedar of Oregon crosses the ocean in a perpetual stream of logs, supplying Japan and China; Kauri pines of New Zealand feed the wood hunger of barren Australia. The world's books and newspapers are printed on coniferous pulp; it is driving silk and cotton to the wall, as sources of cellulose and textile fibre. For beautiful grains, for capacity to take stains, the evergreen woods are incomparable. The living conifers are to us what the dead coal forests are.

But they can be replenished. They can be grown and cut as crops, and they yield a profit on poor sandy and rocky soil, or in swampy lands where no other crop could be hopefully tilled. Thrifty, fertile, tough, industrial, they are of all trees the most practical. Ancient in lineage beyond all others, they rise tall and straight in the pride of their aristocracy. Sea-voiced, solemn, penciled against the sky, their groves are poetic as no leafier places. Conifers stand in the sacred temple yards of Japan, where, with venerating care, their old limbs are supported by pillars. They line the solemn approaches to the tombs of the Chinese emperors at Jehol. Solomon sought them in the peaks of Lebanon for his temple. But in all the world there are none like those in our western states.

And it was in the Black Hills of Wyoming that a fragment of the Mesozoic lay hidden till the days when the West came to be called new country. Miners on their way to Deadwood, cowboys riding herd, found strange stone shapes, and broke off fragments. What lay in those calloused brown fingers, turned over curiously, ignorantly, was once sprung in the Gothic glooms of the Mesozoic forests. These were cycads, a kind of Gymnosperm which must have formed the undergrowth of those prehistoric coniferous woods, hundreds and hundreds of species of them. A few linger today, scattered thinly over the tropics of the world. Some call them sago-palms; they have an antique look, stiff, sparse and heavy; crossed in pairs upon a coffin, they impart a funebrial dignity. Cretin of stature, for the most part, growing sometimes only six feet in a thousand years, they are beloved in the Japanese dwarf horticulture, cherished in family pride there, since a cycad of even moderate size may represent a long domestic continuity.

What pride, then, and what a ring of age was there in the first set of fossil cycads from the Black Hills rim to reach the men of science at the National Museum in 1893! Professor Lester Ward hastened to the field, and what he found there, besides the bones of a great dinosaur and the petrified logs of old conifers, were not imprints but complete petrifactions. Atom by atom the living tissue had been replaced by stone. Here were hundreds of fruits, all the leaves a gloating paleobotanist could desire, perfect trunks, every detail of wood structure preserved, and dozens of species, some dwarf, some colossal.

Ward took back with him what he could. Other students hurried to the find; Yale and the Universities of Iowa and Wyoming have great collections from Deadwood, and the government museums too. Tourists carted away entire specimens, and what remained might have been utterly scattered and destroyed, had not Professor G. R. Wieland saved the last rich tract in the Black Hills. Close to the mountain where Borglum carved his heroic profiles, the scientist filed on the area under the homestead laws, and then presented his claim to his country. It has since been made Cycad National Monument.

These cycads, when the world was young and they were flourishing, must have brought into the dark monotony of the evergreen forests the first bright splashes of color. For the seeds of cycads are gorgeous scarlet or yellow or orange, borne on the edge of the leaf or commonly in great cones. They are sweet and starchy to the taste, and perhaps Archaeopteryx, that first feathered bird in all time, crunched them in the teeth that he still kept, reminder of his lizard ancestry. So, it may be, the earliest animals came to aid in the dissemination of plants, as squirrels do today, and birds. Somehow, at least, the cycads over-ran the world. Their reign had grandeur, but its limits narrowed. There is evidence that some of the Mesozoic cycads flowered only at the end of their immensely long lives—a thousand years, perhaps. Then, after one huge cone of fruit was set, the plant died to the very root. A hero's death, but a plan ill fit to breed a race of heroes. In the cupped hand of the future lay other seeds, with a fairer promise.

A conversation with Gertrude Stein

...Niagara has power and it has form and it is beautiful for thirty seconds, but the water at the bottom that has been Niagara is no better and no different from the water at the top that will be Niagara. Something wonderful and terrible has happened to it, but it is the same water and nothing at all would have happened if it had not been for an aberration in one of nature's forms. The river is the true form and it is a very satisfactory form for the water and Niagara is altogether wrong.

(Quoted by John H. Preston)

The life of prehistoric "man" has a unique interest for us, who are "his"
descendants. R. Claiborne, gives a lively, though simplified, account of
the life style of some early hominids in an extract from his book Climate,
Man, and History.

HABITS AND HABITATS

R. Claiborne

We pick up man's fossil record at just about the time that the Pliocene was
giving way to the Glacial or Pleistocene epoch. What the fossils reveal is
a creature—several creatures, in fact—unequivocally adapted to ground life.
The long drying out of the Pliocene had done its work well. When we
examine the skulls of these animals, we find that the foramen magnum—the
hole through which the nerves of the spinal cord enter the skull—lies almost
at the bottom, rather than toward the rear as in tree-dwelling apes, meaning
that the creatures' heads were balanced on the tops of their spines, like our
own, rather than hunched forward. Leg, foot, and pelvis bones tell the same
story: the animals stood and moved erect, or nearly so, though their gait was
apparently a trot or waddle rather than the striding walk of our own species.
Their feet were no longer capable of gripping a tree-branch, as their ancestors'
feet presumably had been; doubtless they could and did climb trees for food
or safety, even as men do today, but they did so no more efficiently than
we do.

I have said "animals"—but were these creatures merely animals? Here we
must drag in that old cliché of academic dialogue: Define your terms. What
is a mere animal? Or, putting it in a more meaningful way, what is a man?

We have already noted that the Pliocene ground apes almost certainly
used tools. Chimpanzees do—and not simply as missiles. They have been
seen to use sticks for digging and rocks for cracking nuts; they poke thin
twigs into termite nests and then withdraw them covered with delicious
termites; on occasion they even carry the tools around with them. These
observations force us to define man as an animal that makes tools *to a pattern.*
It does not simply modify objects, but does so in a uniform, customary, one
might almost say habitual way.

Judged by this criterion, the apelike creatures that inhabited southern and
eastern Africa in the early Pleistocene stood on the very threshold of humanity, for they fashioned tools out of stones and, in some cases, fashioned
them to a pattern. The tools are terribly primitive: pebbles ranging from ping-
pong to billiard ball size, with a few chips knocked off one end to make a
crude cutting or chopping edge. Some of them, indeed, were apparently
chipped by nature—e.g., being rolled about in the bed of a stream—and are
identifiable as tools only through being dug up on "living floors" miles from

where similar pebbles are found today. Others are so similar to the "nature-made" group as to recall the classic remark of a French prehistorian: "Man made one, God made ten thousand. God help the man who has to distinguish the one from the ten thousand!" Yet by carefully examining the appearance and angle of the chipping, we can be reasonably sure that man, not God, made them.

And here we get into one of those hassles over terminology which reminds us again that science is not quite as rational as some of its practiners would have us believe. When the first skull of one of these man-apes was discovered in South Africa, nobody, not even its discoverer, was expecting to find a manlike creature in that region. Accordingly it was christened merely *Australopithecus*—the "southern ape." Subsequently, other specimens were discovered with tools which made it clear that they were more than apes, and were christened *Paranthropus* ("very like man"), *Plesianthropus* ("nearly man") and *Zinjanthropus* ("man of Zinj"—an old word for East Africa). Obviously the splitters have been at work here. Indeed, they have pushed one group of fossils into still another genus, our own—*Homo*, meaning "man" without trimmings. This step Elwyn Simons, who might be called the lumper's lumper, calls "a conjecture derived from a hypothesis based on an assumption." Simons and the other lumpers have tried to bring some order out of this terminological chaos by reducing all the genera to one. But they were stuck with the original generic name coined in 1924, so that the earliest creature that was definitely not an ape is still known as the southern ape—Australopithecus.

Much of what we know about Australopithecus we owe to the famous diggings at Olduvai Gorge in Tanzania, East Africa. The gorge, a sort of miniature Grand Canyon, was cut out by a now-vanished river which exposed fossil deposits covering, with only one important gap, something like a million and a half years of prehistory. Moreover, by a singular bit of geological luck, the gorge lies close to two now-extinct volcanoes, Ngorongoro and Lemagrut, whose periodic eruptions during the Pleistocene laid down numerous layers of ash by which, using the potassium-argon technique, the Olduvai deposits can be dated. Due largely to the skillful and devoted labors of Louis Leakey and his wife Mary, Olduvai has yielded up a singularly detailed and dated record of the ground apes' emergence into humanity. Their findings have also helped to make chronological sense out of man-ape fossils from South Africa's Transvaal, which, though quite as abundant as those in Olduvai, did not have the luck to be laid down in a volcanic area.

Perhaps the most interesting fact about the Australopithecuses of Olduvai and the Transvaal is that they came in at least two (Leakey would say three or four) species, the Australopithecus africanus, was a slender four footer, no bigger than a modern pygmy. Its cousin, Australopithecus robustus (Paranthropus, Zinjanthropus) was, as its name suggests, bigger—about five feet—and burlier. From all the evidence, the two species seem to have been more or less contemporary (at Olduvai, much more than less) during the earlier part of their careers on earth. The simultaneous existence of two closely related man-ape species is puzzling, because it is exceptional. There is today only one species of gorilla, one species of chimpanzee, of orangutan, and, for that matter, of man. The reason seems to lie in a general principle of ecology: Where two closely related species occupy the same territory (and at Olduvai we are pretty certain they did), they must be adapted to different ways of life. If both occupy the same "ecological niche," either one species will displace the other in fairly short order or, much more likely, they will not evolve into two species in the first place.

One clue to the explanation of the two Australopithecine species comes from the bones themselves. I have on my living room table a plaster cast of a robustus skull which I have doctored with furniture stain so that it makes a fairly plausible-looking fossil. It does not, needless to say, look much like a human skull; perhaps the most striking difference emerges when I turn it over and examine the upper teeth (the lower jaw was missing in the specimen from which the cast was made). The biting teeth (incisors) are not very remarkable; they are, if anything, smaller than my own. But the molars are enormous—massive, and so broad that they look as if they had been set in the jaw sideways.

The top and back of the skull are also missing, but from descriptions of it I know that robustus had a bony ridge along the top of his cranium, similar to, but smaller than, that on the skull of a modern gorilla. Mechanically, the function of such a ridge is to give a strong anchorage to the massive thick muscles which move the lower jaw. Thus the ridge with its vanished muscular attachments, tells the same story as the massive molars: robustus lived on a bulky diet which required lots of chewing. It seems very likely, then, that like the gorilla he was a vegetarian, or close to it, living largely or entirely on the traditional primate diet of fruits and shoots during the rainy season and on roots and seeds during the dryer parts of the year. This conjecture is strengthened by miscroscopic studies of robustus teeth; these reveal scratches apparently made by sand grains, which presumably were carried into his mouth with the roots.

Australopithecus africanus, with neither skull ridge nor massive molars, seems to have adapted to a considerably more catholic diet. From bones found on his living floors he ate lizards, tortoises, various kinds of small mammals, and the young of some medium-sized species such as antelopes—along with, we can be sure, such vegetable foods as he could get.

This hypothesis (and I should stress that it is still only a hypothesis) explains a number of other apparent facts about the two species. From what I have said about their contrasting life patterns, it will be obvious that while both could survive in a fairly moist savannah (which the Olduvai area seems to have been some two million years ago), only africanus could have managed to make a living on a dry savanna or steppe, since the vegetable foods to which robustus' anatomy—and perhaps other things—adapted him would have been too sparse for survival.

The South African deposits seem to confirm this, notably in the sand found in the various layers. Where we find africanus fossils, we find sand whose grains are rounded—evidently by being blown along for considerable distances, perhaps from the Kalahari some hundreds of miles to the west. And wind-blown sand, obviously, means a dry climate. The robustus fossils, on the other hand, are found with sand having sharp, angular grains—evidence that the climate was wetter. Still further confirmation comes from another curious fact. Robustus did not evolve significantly between his first appearance in the fossil record, some two million years ago, and his final bow, perhaps as much as a million and a half years later. During the same period, however, africanus changed into quite a different creature—larger, brainier, and in other respects more human—to the point where he had become a full-fledged member of the genus *Homo*.

It seems at least plausible that the difference stemmed from diet. I have already noted the important mental differences between herbivores and predators. Vegetable food, provided it is tolerably plentiful, offers little mental challenge, since all you have to do is find it and eat it. Animals, however, must not only be found but also caught—which, unless you are as superbly equipped by nature as the cats and dogs, means living by your wits. Africanus, it seems, must have been smarter than robustus. It is quite possible, indeed, that of the two species only he was a toolmaker. At any rate, no fabricated tools have ever been found with robustus bones—unless africanus bones were also present. Africanus fossils, on the other hand, often occur without robustus remains—but with tools.

Thus it appears that not all the ground apes learned the lesson of the savanna equally well. Robustus, sticking to a vegetable diet and adapting himself to it anatomically, remained tied to a particular climate, one that was neither too dry nor too wet. Africanus, on the other hand, evolved in the direction of less anatomical specialization, not more. From an evolutionary standpoint, he cultivated his mind, and reaped his reward by ultimately becoming the one animal that can survive amid rain forest, desert, or polar ice. Robustus remained a manlike ape, ultimately driven to the evolutionary wall (and quite possibly eaten) by his more adaptable cousin. Africanus became man.

CHAPTER 6

Modern Problems in Geology

We have attempted in this book to put our present knowledge of the Earth into a more general perspective. This clearly has its benefits, for it emphasizes the provisional nature of all knowledge, and the continuing refinement to which an explanation is subject. But it has its dangers too, for it is temptingly easy to see the errors of the past only as a prelude to the truth of the present. In fact, our present truths and explanations share the inherent limitations and weaknesses of all earlier and now sometimes obsolete formulations of truth. There is, however, one safeguard against the cynicism and despair that this discovery tends to produce, and that is the fact that all truth, and especially scientific truth, has to be grounded in the world of common experience. To the extent that our theories become more generalized, more inclusive, and more powerful in their predictive power on the Earth we inhabit, to that extent we may claim some progress in our thinking.

Such progress is of no little concern to those who are our fellow passengers on spaceship Earth. We select a single topic to show the implications of our continuing quest for understanding: earthquakes, which are, of all earth's features, the most dangerous and terrifying to man. Our brief extracts deal with earthquake prediction and control, and the understanding of their origin in relation to a more general theory of the earth.

J. Tuzo Wilson, a leading geophysicist, and one of the pioneers of current plate tectonic theory, is principal of Erindale College, University of Toronto. He wrote the following in 1972.

MAO'S ALMANAC—3,000 YEARS OF KILLER EARTHQUAKES

J. Tuzo Wilson

Earthquakes have always been one of China's great scourges. In the last three millenniums, there has been a major quake every six years on the average, including what is probably the most destructive temblor ever recorded anywhere. It shook Kansu Province so severely in A.D. 1556 that falling buildings and walls killed more than 820,000 persons recorded by name. Nobody knows how many were killed whose names were not recorded. Small wonder then that the regime of Mao Tse-tung has instructed the Institute of Geophysics and the Institute of Geology of China's Academy of Sciences to concentrate on the predicting of earthquakes.

Of greatest interest are the Chinese discoveries about the distribution of earthquakes and the cycles of seismic activity. These are based upon records that go back more than 3,000 years and are considered reliable over the past 2,752 years. From them Chinese seismologists have put together a four-stage model of the process that first accumulates and then releases the energy that causes earthquakes. The only other study that is at all comparable covers the seismic history of the eastern Mediterranean over the past 2,000 years, and it does not reveal the clear cycles of activity shown in the Chinese records.

If the rate at which earthquakes release energy could be controlled, the damage they inflict might be considerably reduced. Before control can be seriously thought of, however, the nature of quakes must be better understood. This understanding requires international cooperation.

My wife, Isabel, and I entered China via Rangoon, flying along the great Irrawaddy River to Mandalay, then over the Burma "hump" and across the plains of China. Monsoon clouds hid most of the landscape until we neared Shanghai at dusk, but I was surprised by the extent of water—rivers, lakes, canals, and flooded paddy fields—that reflected the bright moonlight off the vast delta. I was even more surprised by the twinkles of tiny electric lights that marked the countryside: none of the brilliance of Western cities, but proof of extensive electrification.

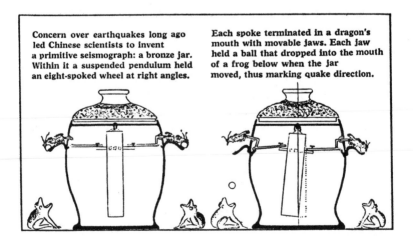

Concern over earthquakes long ago led Chinese scientists to invent a primitive seismograph: a bronze jar. Within it a suspended pendulum held an eight-spoked wheel at right angles. Each spoke terminated in a dragon's mouth with movable jaws. Each jaw held a ball that dropped into the mouth of a frog below when the jar moved, thus marking quake direction.

Shanghai's airport resembled that of a city in the central plains of North America except for one thing: It was virtually deserted. The absence of planes, automobiles, and bustle focused our attention on a long double line of cheerful young Chinese dressed in blue track suits, enthusiastically clapping. They formed a welcoming corridor for our traveling companions, a group of young Africans and Arabs bound for the great Afro-Asian Table Tennis Friendship Invitational Tournament soon to begin in Peking. Loudspeakers boomed martial music and exhilarating choruses.

The sound was more Russian than Chinese. It combined with the architecture of the airport's imposing new terminal and the four Ilyushin planes on the tarmac to suggest to us that we might have flown over the wrong mountains and landed in the Soviet Union. Frequently during the following weeks we had reason to recall the close connections that had existed between China and the Soviet Union up to 1960. Our hosts often reminded us, however, that those connections had been completely broken.

Isabel and I were met by half a dozen charming and attentive scientists, officials, and interpreters. They expressed sympathy for our fatiguing, long journey and led us into the terminal, where we sat in easy chairs and were offered Chinese tea and cigarettes. I had scarcely been asked for the baggage checks and my passport when the bags appeared. Isabel's passport, our health certificates, and all questions of foreign currency were waived. We were foreign friends, guests of the academy. It was the simplest and most gracious entry I recall making into any country.

We spent the night at a wonderful old hotel in Shanghai. It was identified only by a street number, but I have surmised, since leaving China, that it must originally have been Victor Sassoon's Cathay Hotel. It overlooked the Whampo river near a clock tower that chimed "The East Is Red."

The next morning we boarded a crack steam train, one of the more appealing aspects of a bygone age. The engine had four scarlet drive wheels, three red flags between its eyebrows, and gold Chinese characters on its tender. As we rolled past the fields and villages of the Yangtze plain on our way to Peking, we could see people working everywhere. A young interpreter who pointed out improvements on every side apologized rather unnecessarily for his lack of fluency in English. He mentioned his greater familiarity with Russian, from which he had recently switched because he could foresee no future for Russian interpreters in China.

We were met at the Peking station by the acting director of the Institute of Geophysics, the deputy director of the Institute of Geology, a senior scientist named Mrs. Woo, several officials from the foreign relations section of the Academy of Sciences, two interpreters, and a Canadian Embassy counselor whose black-felt hat towered above the blue Mao caps. We were greeted cordially and then driven to our hotel, where we were told to "rest yourselves a little and then we shall return to discuss the program."

I had hoped to be able to return to the places I had visitied in 1958, in order to assess the changes wrought by the Cultural Revolution. It had swept through the academic life of China like a typhoon, shutting off all scientific publications, essentially closing universities for four years, and sending the professors out to plant rice, work in factories, or otherwise gather firsthand knowledge of workers and peasants so as to be able to apply academic knowledge to the problems of the countryside while undergoing re-education in the thought of Chairman Mao.

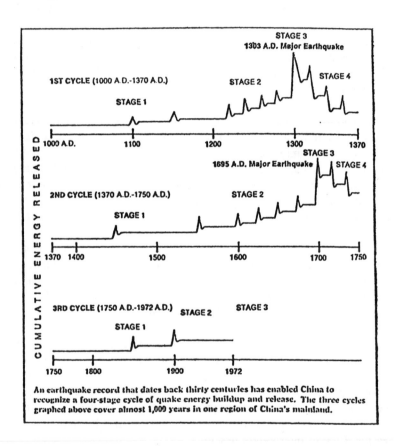

An earthquake record that dates back thirty centuries has enabled China to recognize a four-stage cycle of quake energy buildup and release. The three cycles graphed above cover almost 1,000 years in one region of China's mainland.

My 1958 visit had taken me to Peking, Siam, Lanchow, and Canton. However, I was told that Lanchow was out of the question for 1971, apparently because cities as far inland as it is are unprepared to receive visitors with traditional Chinese graciousness. The substitute itinerary we agreed upon included all the other cities I had seen in 1958 plus Yenan (famous as the point where Mao's monumental "long march" in the early 1930s ended), Nanking, Shanghai, and Hangchow. Relatively, this would expose me to the

Chinese equivalent of a chunk of North America bounded by the cities of New York, Atlanta, New Orleans, Chicago, and Winnipeg.

The historical study of large earthquakes has long been the specialty of Dr. Lee Shan-pang, a former California Institute of Technology student of Professor Beno Gutenberg and author of a three-volume classic on the subject. The earliest records he had found were from 1198 B.C. These he did not consider reliable. But he plotted the location and magnitude of 530 major earthquakes that had occurred in China since 780 B.C. By his estimation, all of these had reached a magnitude of six or greater on the Richter scale of measurement. Dr. Lee, Dr. Ku, and the revolutionary committee of the Institute of Geophysics allowed me to photograph a manuscript map of those 530 quakes. Seventeen of them were graded as greater than magnitude eight, and about eighty were between magnitudes seven and eight.

Anyone with a grasp of modern geophysics could have seen at a glance that the earthquakes marked on the maps were concentrated along definite lines of tectonic weakness. One of these lines goes straight north from Kunming and crosses the Yellow River. Another extends down the Red River to Hanoi. A third cuts across the Yellow River near Sian to the neighborhood of Peking. Other quake epicenter lines pass along the Altai and Tien Shan mountains, encircle Tibet, underlie Taiwan, and skirt the Russian border at Vladivostok.

With instruments derived from those designed by Caltech professor Hugo Benioff, Chinese have plotted the rate of energy release along the great faults revealed by their maps. Thanks to their long historical record, they have fixed the dates and magnitudes of earthquakes along each line over a range of time unequaled anywhere else in the world.

They have found that they can divide the earthquake history of each tectonically active area into cycles. Each cycle has four stages. During the first stage, which may last a century or more, the strain on the rocks accumulates. Only a few small earthquakes occur in this period. The second stage is one of transition, during which earthquakes become larger and more frequent. In the third stage the main energy is released. There is one huge earthquake, possibly followed by a few other large earthquakes within a short span of years. The fourth stage of the cycle is characterized by lesser aftershocks, completing the cycle.

In Shansi Province, west of Peking, for example, one cycle extended from A.D. 1000 to 1370. A second cycle ran from 1370 to 1750, culminating in a big shock near Peking that killed 100,000 people. A third cycle started in 1750. It is now in stage two and may be approaching stage three. If stage three is at hand, great shocks are expectable in the near future in the vicinity of Peking. The Chinese did not quite admit this to me, but such an anticipation would be an obvious explanation for the stress that is being laid on the predicting of earthquakes in China today.

The second article on earthquake prediction is by Peter J. Wyllie of the University of Chicago.

EARTHQUAKES AND CONTINENTAL DRIFT

Peter J. Wyllie

California's worst earthquake in 38 years shook the residents of Los Angeles from their sleep at 5:59 a.m. on February 9th, 1971. Widespread damage occurred with many patients killed beneath the rubble of two collapsed hospitals. Freeways were closed by fallen overpasses, damaged bridges, or buckled roadways. Many fires were started by broken gas lines and electrical cables. Thousands of people and homes were menaced by a cracked dam in the San Fernando valley. This was the tenth major earthquake in recorded California history since 1812.

Nerve-wracking as this earthquake was, with aftershocks continuing for months afterwards, the event was insignificant compared with predictions that had been made a few years earlier. In the latter half of the 1960's, clairvoyants and soothsayers predicted that the State of California would soon split asunder, and one half would drift away and founder in the Pacific Ocean. This separation would be accompanied by devastating earthquakes and tidal waves. A number of religious sects announced that Doomsday was imminent, and they prepared themselves for the end of the world. The population of California has often been warned that sooner or later the state will be shaken by another earthquake, but the normal educated citizen knew perfectly well that California could not sail off into the Pacific Ocean, and the topic for a while provided entertaining material for cocktail party banter. And yet . . . what about that theory of continental drift and sea-floor spreading? Many articles in the popular press informed readers that most geologists now believe in the reality of continental drift.

In 1966 and 1967 various lines of evidence led to the formulation and widespread acceptance of a modern version of the old hypothesis of continental drift. This was proclaimed as a revolution in the earth sciences. The concept known as sea-floor spreading serves as a master plan which appears to accommodate and link together many topics in the earth sciences which previously had been treated as subjects for independent research. A theory for the whole earth provides a framework for the explanation of localized geological phenomena. In particular, earthquakes are caused by the slow movement of one part of California independently of the main American continent. This was grist for the clairvoyants and some rather distorted versions of the theory were circulated. It began to appear that the prophesied schism of the state had the respectable scientific backing of a highly touted new global theory of earth sciences. By 1969, when the prophets of doom specified dates, the telephone switchboards into university departments of geology and geophysics began to hum with enquires from the anxious citizenry. The growing concern of the California public made newspaper headlines across the country. For example, an issue of the *Chicago Tribune* on March 30th, 1969, reported: "Disaster talk runs rampant in California: quake will destroy state, seers say."

Monday, April 14th, 1969, the day most frequently cited for the great event, was awaited with interest, with concern, or with disdain by various groups of people. The day came and went. Another Monday had started another working week and the state remained in one piece. No earthquakes rumbled from the faults to test the stability of the fine high-rise buildings in the urban centers. The citizens laughed and told each other that they had known it was a hoax all along. It was reported that a number of prominent politicians whose business had forced them to be out of state on April 14th returned to their offices in California. Some disappointed religious sects prayed that next time Armageddon was announced it would really come.

The predictions of the seers and soothsayers actually erred only in one respect: they used the wrong time scale. According to scientifically-based predictions, one section of California is indeed destined to slide away from the rest of the state, at the rate of inches per year, and to disappear 60 million years hence beneath Alaska. The seers, in their far-sighted visions, compressed the anticipated events of a few million years into a few hours.

Let us now examine deduced history and anticipated events with the proper time scale and see to what extent the revolutionary new global theory is capable of explaining and predicting the occurrence of earthquakes.

An earthquake is a minor consequence of the global process of sea-floor spreading, but the magnitude of the devastation that it can wreak in human terms is almost beyond belief. About 500,000 earthquakes are recorded by instruments every year, but only about 50,000 can be felt without instruments. Of these, about 100 are strong earthquakes, and about two are very strong, strong enough to cause widespread devastation. It has been estimated that during the period 1900-1968 earthquakes caused the death of about 800,000 people, with property damage on the order of 10 billion dollars. In the United States more than 1,500 people were killed, and property damage was about 1.3 billion dollars. Most deaths are due to secondary causes such as building collapse, fires, landslides and giant waves called *tsunamis*. Underdeveloped areas with poorly constructed homes, often on loosely compacted earth, suffer most in loss of life.

Plate tectonics explains the observation that earthquakes are concentrated in narrow belts, and distinction among the three types of plate boundaries can be made on the basis of the type of earthquakes occurring. For example, the boundary between the Pacific and South American plates is of the compressive or collision type, where the Pacific plate slides underneath the other.

In this environment the epicenters of the earthquakes occur close to a plane that starts at the floor of the oceanic trench and dips at an angle of about 45° beneath the continent. The cause of the deep earthquakes lies in the slippage of material in the direction of this plane to depths of 700 kms. In contrast, in California the boundary between the Pacific and North American plates is a fracture zone where the two plates are slipping past each other. Shallow earthquakes occur in this environment following the build-up of strain caused by horizontal stresses.

Peru is situated in the belt of compression between the Pacific and South American plates. The Andes Cordillera are slowly rising in response to the tremendous compressive pressure below, producing magnificent mountain peaks but great instability. The deadliest earthquake in the recorded history of Latin America occurred in Peru on May 31st, 1970. At least 50,000 people died, another 100,000 were injured, and nearly 1 million out of the population of 13 million were left homeless. Estimates of property damage exceeded 250 million dollars.

The epicenter of the shock for the earthquake was located beneath the ocean trench, about 16 miles west of the coast, and 27 miles below the surface. As the energy vibrations were transmitted to the continent, towns and villages collapsed. In an area covering 25,000 square miles, eighty percent of the adobe houses were destroyed or made uninhabitable. Many of the villages are situated on poorly consolidated alluvial soils. Additional damage was caused by secondary effects, including a massive avalanche of ice, water, mud, rocks and other debris that swept down a valley in the Cordillera Blanca. This type of avalanche, a *huayco*, killed 41,000 in Ecuador and Peru in 1797, and another in 1939 killed 40,000 in Chile.

The *huayco* had its source about 9 miles east of the resort town of Yungay and about 12,000 feet higher in the mountains. Apparently, the earthquake broke loose a mass of glacial ice and rock about 3,000 feet wide and a mile long. This moved downslope, picking up earth and rocks, and melting during descent. Enormous boulders were hurled across the valley with calculated velocities as high as 248 miles per hour. Survivors of the earthquake in Yungay were praying in the streets amid the wrecked buildings when they heard the thunderous roar of the *huayco* as it swept down the valley. Within a few minutes, the wreckage was covered by earth, mud, and boulders with thickness up to 45 feet in places.

Many nations rushed to aid the stricken country, but it will probably take 10 to 15 years to repair the damage. Meanwhile, more strain builds up beneath the ocean trench and Andes mountains as the Pacific plate is forced beneath the South American continent.

As the Pacific plate moves northwards relative to the North American plate, intermittent slippage along the San Andreas fault system causes shallow-focus earthquakes in California. Effects of the 1971 Los Angles shock were outlined at the beginning of this article. In the 1906 San Francisco earthquake about 390 people were killed, with property damage estimated at 400 million dollars. Much of this damage was caused by the great fire which raged out of control for three days because broken water mains left no water for fire fighting.

A report released in 1970 described the possible effects if an earthquake equivalent to that of 1906 were to recur. According to the report, many buildings would collapse. The alluvial soils around the Bay area would turn into quicksand and buildings would sink. Communications would be disrupted as the freeways buckled and railway tracks twisted. At least one of the many dams would fail, releasing flood water from the reservoir to inundate large areas. Water mains would break. Estimated property damage would amount to about 30 billion dollars, and estimated deaths number tens or even hundreds of thousands. The total of all previous losses caused by earthquakes in the United States are insignificant in comparison.

According to the theory of plate tectonics earthquakes are concentrated in belts, and they are infrequent in the large stable plates. The most violent earthquakes known to have occurred in the United States, however, were located well within the North American plate.

The New Madrid earthquakes shook the central part of the Mississippi valley in 1811 and 1812 with such intensity that they were felt over two-thirds of the United States. Eight hundred miles away, in Washington, D.C., sleepers were awakened, dishes and windows were rattled, and walls were cracked. The vibrations rang church bells in Boston. The earthquakes caused major changes in topography over an area of 50,000 square miles. Large areas sank from 3 to 9 feet, and depressions 20 to 30 miles long and 4 to 5 miles wide were produced. The course of the Mississippi River was changed. The area was sparsely populated at the time, and the loss of life was small but not precisely known.

It is the power of explanation and prediction that makes the new global theory so different from previous concepts in the earth sciences. The fact that data from the study of earthquakes gives strong support for the sea-floor spreading theory is closely related to increased understanding of the distribution of, and the characteristics of earthquakes. With the new theory and greatly improved geophysical instruments, some scientists were optimistic in 1970 that accurate prediction of earthquakes would be possible within three to five years, and that their control may follow very soon after this. Considerable research effort is directed to the subject of earthquake control, especially by federal and state agencies and universities in California, where attention is concentrated on the San Andreas Fault system.

It was discovered only recently that along parts of some faults in the San Andreas system deformation is accomplished by fault slippage without accompanying earthquakes. This process is called fault creep. While parts of a fault may creep in distinct episodes at rates up to 10 cms/year, other parts of the same fault may not creep at all. In these locked portions the deformation causes accumulation of elastic strain energy with the prospect that later energy release will produce a large earthquake. Hopes for earthquake control lie in development of methods for releasing locked portions of faults so that the crustal blocks slide past each other with only small energy-dissipating tremors.

Before prediction becomes possible, it is necessary to learn the patterns of movement, and energy storage and release. There are now more than 100 seismic stations deployed to locate the boundaries of moving blocks in California from the distribution of small earth tremors, with the help of new, ultrasensitive instruments.

Fault creep is accompanied by tilting of the ground. The changes are very small, but careful measurement of ground tilt and rates of fault creep can be correlated with subsequent earthquake tremors.

Clear sky lightning has traditionally been regarded as a precursor of earthquakes in Japan. This has some scientific basis. There is evidence that the electrical and magnetic properties of rocks can be changed by strain, and it has been calculated that in rocks where the mineral quartz is arranged in suitable crystalline order and texture, the effects of strain could generate enough voltage to produce lightning. There is also evidence that fluctuations in the earth's magnetic field are associated with earthquakes.

The measurement of earthquake activity, fault creep, accumulation of strain in rocks, ground tilt, and continuous monitoring of the magnetic field provides an assemblage of clues which should soon permit short-range prediction of earthquakes with a warning interval of hours or days. Present knowledge of the processes involved in the generation of earthquakes is not sufficient to guide an engineering program for earthquake control, but there is good evidence that earthquakes can be triggered by several means.

The occurrence of earthquakes appears to be linked with the construction of dams and the impounding of large masses of water in reservoirs. Earthquakes also appear to be associated with the injection of fluids into the ground. Underground explosions of nuclear devices have also been associated with increased earthquake activity, apparently by releasing natural strain energy in the region. Earthquake hazards can possibly be reduced if fluid injections and perhaps explosions can be employed to free locked portions of fault systems to release the accumulating strain before it builds up to dangerous levels.

The role of water or other interstitial fluid in earthquake activity is receiving considerable attention. The first serious suggestion that earthquakes could be controlled arose from study of the series of earthquakes triggered near Denver by injection of waste fluids into a deep disposal well at the Army's

Rocky Mountain Arsenal. The case here is strong for a cause-and-effect relationship between the injection of fluids and triggering of earthquakes. Small amounts of water may produce the effects of a lubricated surface on which fault displacements can occur. The injection of water may free locked fault systems.

Earthquake control may thus become feasible eventually, but only in regions where the locked fault zones can be reached by drilling. This approach will not work for the earthquakes originating beneath ocean trenches, like the 1970 Peru earthquake. Many engineers emphasize that it is not the earthquakes that kill people, but the failure of buildings that people construct. They maintain that the best approach is not to predict and control earthquakes, but to erect sound buildings, bridges and dams on relatively safe sites. Then comes the problem of judging the safety of development sites and the effectiveness of building codes.

In California and Japan, where earthquake activity is accepted as a fact of life, building codes include provisions for earthquake-resistant construction. No such provisions are deemed necessary in the central U.S.A. where the land is considered to be stable. If the New Madrid earthquakes had occurred this year the wreckage could have been catastrophic. At the present time, we have no way of predicting the likelihood of such an occurrence in the supposedly stable plates, but we do know that their frequency is low.

There are social, economic and political problems associated with the prediction of earthquakes. Suppose that a scientific team predicted the occurrence of an earthquake of significant magnitude centered in a densely populated urban region two days hence. Would the city authorities order evacuation? Imagine the cost of such an evacuation, the social upheaval, and the economic loss through interruption of normal business operations. If the earthquake failed to materialize on schedule how long would the population stay away, waiting for the shock? Perhaps the authorities would decide that the consequences of a possible earthquake might be less harmful to the city than the definite social and economic disruption of an evacuation. Perhaps they would then issue warnings and advice about taking shelter beneath strong tables or door jambs, and mobilize the civil defense. What political recriminations would then result if the prediction was followed by a major earthquake that levelled the city and killed thousands!

If faced with this kind of decision, city leaders would probably regret scientific progress and wish that our ideas on the occurrence of earthquakes had remained in the state of mystery illustrated by the following quotation from Shakespeare's *King Henry IV:*

> Oft the teeming earth
> Is with a kind of colic pinch'd and vex'd

By the imprisoning of unruly wind
Within her womb; which for enlargement striving
Shakes the old beldam earth and topples down
Steeples and moss-grown towers.

No individual could be blamed for not anticipating an accident arising from Mother Nature's discomfort.

Nature

Only Supernaturalists really see Nature.... To treat her as God, or as Everything, is to lose the whole pith and pleasure of her. Come out, look back, and then you will see.... this astonishing cataract of bears, babies, and bananas: this immoderate deluge of atoms, orchids, oranges, cancers, canaries, fleas, gases, tornadoes and toads. How could you ever have thought this was the ultimate reality? How could you ever have thought that it was merely a stage-set for the moral drama of men and women? She is herself. Offer her neither worship nor contempt. Meet her and know her.

—C.S. Lewis

Part II

GEOLOGY AND THE INDIVIDUAL

All the varied inhabitants of the earth, plants and animals alike, are affected by geologic processes and phenomena, but none so much as the group classified as *Homo sapiens*. We humans cannot escape geology—the study of the earth—even if we wish. For the foreseeable future, we are irrevocably tied to the earth. Thus, the individual is created and constrained by the materials of which the earth is made and the various forces and processes that act upon it. Individuals have applied geology from the time a primitive man consciously selected a particular type of rock for a tool or weapon, because he was aware from experience that it was harder and denser than others, or kept in his possession another mineral or rock because of its clarity or pleasing color. With but a moment of reflection, each reader can quickly recall many ways in which geology relates to the individual: some as fundamental as where he obtains his food and fuel, with what materials he constructs his home, where he builds it, and, often, how he earns his living and where he spends his leisure.

One of the satisfying aspects of geology is that it is a science in which much of the research, the bulk of the field work and many of the solutions to problems can still be accomplished by individuals. Be it a solitary geochemist working in a laboratory with isotopes, a structural geologist mapping in the solitude of the Alps, or a paleontologist collecting and classifying fossils, there is continued individual effort. With this, an individuality of approach and the personal touch is not unusual, either in the way the study is made or in the written record of such work.

From the many possibilities that can be used to relate geology to the individual, three very different themes have been chosen as a sampler for the reader. The first group of selections illustrates the humor of individual geologists and some humorous aspects of geology. This is followed by a section which shows the relationship of geology to the world of literature and the arts. To close the section, individual flair and its place in the science is suggested by the work of two geological investigators in attacking and solving thorny problems.

Doubts and Certainties

If a man will begin with certainties, he shall end in doubts; but if he will be content to begin with doubts, he shall end in certainties.

—**Francis Bacon**

CHAPTER 7

Humor in Geology

In the stereotyped view of a scientist, two of the more popular adjectives used are "dour" and "humorless." Selections from this section suggest that as a group, geologists and other scientists are as varied and diverse in appearance, personality and outlook as any other more or less random group of people. Some even have a well-developed sense of humor, as the following extracts reveal.

At one time or another, almost every geologist has been presented with a peculiarly-formed rock mass which is interpreted by the finder to be almost anything a well-rounded imagination permits. Loren Eiseley, curator of Early Man at the University of Pennsylvania Museum, with whimsy and a touch of pathos, relates how he reacted to a specimen of unusual proportions and contours.

THE PETRIFIED WOMAN

Loren Eiseley

In the proper books, you understand, there is no such thing as a petrified woman. I knew that, because I was a professional bone hunter. But bone hunters, like the men of other professions, have bad seasons. We had made queries in a score of western towns and tramped as many canyons. We had sent the institution for which we worked a total of one Oligocene turtle and a bag of rhinoceros bones. Our luck had to change. Somewhere there had to be fossils.

I was cogitating on the problem under a coating of lather in a barber shop with an 1890 chair when I became aware of a voice. You can hear a lot of odd conversation in barber shops, particularly in the back country. What caught my ear was something about "petrified." "I'm a-tellin' ya," the man's voice boomed. "A petrified woman, right out in that canyon. But he won't show it, not to nobody."

171

I managed to push an ear up through the lather. "Mister," I said, "I'm reckoned a kind of specialist in these matters. Where is this woman, and how do you know she's petrified?"

I knew perfectly well she wasn't, of course. Flesh doesn't petrify like wood or bone. Just the same, you can never tell what will turn up. Once I had a mammoth vertebra handed to me with the explanation that it was a petrified griddle cake.

Yes, he told me, the woman was petrified, all right. Old Man Buzby wasn't a feller to say it if 'tweren't so. And it weren't no part of a woman; it was a *whole* woman. Buzby had said that, too. But Buzby was a queer one. An old bachelor, you know. And when the boys had wanted to see it, the old man had clammed up on where it was. A keepin' it all to hisself, he was. But seein' as I was interested in these things and a stranger, he might talk to me and no harm done. It was the trail to the right and out and up to the overhang of the hills. A little tar-papered shack there.

At the back of the house we found the skull of a big, long-horned, extinct bison hung up under the eaves. It was a nice find, and we coveted it.

Buzby invited us into the neat two-room shack to see his collection of arrowheads. He was precise about his Indian relics as he was precise about everything. But I sensed, after a while, a touch of pathos—the pathos of a man clinging to order in a world where the wind changed the landscape before morning and not even a dog could help you contain the loneliness of your days.

"Someone told me in town you might have a wonderful fossil up here," I finally ventured, poking in a box of arrowheads and watching his shy, tense face.

"That would be Ned Burner," he said. "He talks too much."

"I'd like to see it," I said, carefully avoiding the word *woman*. "It might be of great value to science."

I could see him hesitating. It was plain that he wanted to show us, but the prospect was half-frightening. As he talked on, I began to see what he wanted. He intended to show it to us in the hope we would confirm his belief that it was a petrified woman. At last he said, "Why don't you camp here tonight, Doctor? Maybe in the morning—"

I remembered the sound of the wind the next morning. In that country the wind never stopped. I think everyone there was a little mad because of it. It starts on the flats and goes down into those canyons and through them with weird noises, flaking and blasting at every loose stone or leaning pinnacle. It scrapes the sand away from pipy concretions till they stand out like strange, distorted sculptures. It leaves great stones teetering on wineglass stems. I began to suspect what we'd find.

Once he had given his consent and started, Buzby hurried on ahead, eager and panting. Up. Down. Up. Over boulders and splintered deadfalls of timber. Higher and higher into the back country. Toward the last he outran us, and I couldn't hear what he was saying. The wind whipped it away.

But there he stood, finally, at a niche under the canyon wall. He had his hat off, and for a moment was oblivious to us. He might almost have been praying. "This must be it," I said to Mack. "Watch yourself." Then we stepped forward.

It was a concretion, of course—an oddly shaped lump of mineral matter— just as I had figured after seeing the wind at work in those miles of canyon. It wasn't a bad job, at that. There were some bumps in the right places, and marks that might be the face, if your imagination was strong. Mine wasn't just then.

Buzby didn't wait for me to speak. He blurted with a terrible intensity that embarrassed me, "She—she's beautiful, isn't she?"

"It's remarkable," I said. "Quite remarkable." And then I just stood there not knowing what to do.

He seized on my words with such painful hope that Mack backed off and started looking for fossils in places where he knew perfectly well there weren't any.

I didn't catch it all; I couldn't possibly. The words came out in a long, aching torrent—the torrent dammed up for years in the heart of a man not meant for this place, nor for the wind at night by the windows, nor the empty bed, nor the neighbors 20 miles away. You're tough at first. He must have been to stick there. And then suddenly you're old. You're old and you're beaten, and there must be something to talk to and to love. And if you haven't got it, you'll make it in your head or out of a stone in a canyon wall.

He had found her, and he had a myth of how she came there, and now he came up and talked to her in the long afternoon heat while the dust devils danced in his failing corn. It was progressive. I saw the symptoms. In another year, she would be talking to him.

"It's true, isn't it, Doctor?" he asked me, looking up with that rapt face, after kneeling beside the niche. "You can see it's her. You can see it plain as day." For the life of me I couldn't see anything except a red scar writhing on the brain of a living man who must have loved somebody once, beyond words and reason.

"Now, Mr. Buzby," I started to say then, and Mack came up and looked at me. This, in general, is when you launch into a careful explanation of how concretions are made, so that the layman will not make the same mistake again. Mack just stood there looking at me in that stolid way of his. I couldn't go on with it.

But I saw where this was going to end. I saw it suddenly and too late. I opened my mouth while Mr. Buzby clasped his hands and tried to regain his composure. I opened my mouth, and I lied in a way to damn me forever in the halls of science. "Mr. Buzby," I said, "that—er—figure is astonishing.

It is a remarkable case of preservation. We must have it for the museum."

The light in his face was beautiful. He believed me now. He believed himself. He came up to the niche again, and touched her lovingly.

"It's okay," I whispered to Mack. "We won't have to pack the thing out. He'll never give her up."

That's where I was a fool. He came up to me, his eyes troubled and unsure, but very patient.

"I think you're right, Doctor," he said. "It's selfish of me. She'll be safer with you. If she stays here, somebody will smash her. I'm not well." He sat down on a rock and wiped his forehead. "I'm sure I'm not well. I'm sure she'll be safer with you. Only I don't want her in a glass case where people can stare at her. If you can promise that, I—"

"I can promise that," I said, meeting Mack's eyes across Buzby's shoulder.

"And if I come there, I can see her?"

I knew I would never meet him again in this life.

"Yes," I said. "You can see her there." I waited, and then I said, "We'll get the picks and plaster ready. Now that bison skull at your house . . ."

It was two days later, in the truck, that Mack spoke to me. "Doc."

"Yeah."

"You know what the Old Man is going to say about shipping that concretion. It's heavy. Must be three hundred pounds with the plaster."

"Yes, I know."

Mack was pulling up slow along the abutment of a bridge. It was the canyon of the big Piney. He got out and went to the rear of the truck. "Doc, give me a hand with this, will you?"

I took one end, and we heaved together. It's a long drop into the big Piney. I didn't look, but I heard the thing break on the stones.

"I wish I hadn't done that," I said.

"It was only a concretion," Mack answered. "The old geezer won't know."

"I don't like it," I said. "Another week in that wind and I'd have believed in her myself—maybe I do anyhow. Let's get out of here. I tell you I don't like it. I don't like it at all."

"It's a hundred more to Valentine," Mack said.

He put the map away and slid over and gave me the wheel.

Geotimes is the monthly publication of the American Geological Institute which has a membership of 15,000 geologists and earth scientists. Widely read and anticipated by many of the members is a monthly column of general observation by Robert L. Bates, professor emeritus of geology at Ohio State University.

THE GEOLOGIC COLUMN

Robert L. Bates

The high cost of filling the old bookcase is borne out by further studies of the shelving index (SI) since the March column. For example, *Studies of Appalachian geology: northern and maritime* (Wiley/Interscience) will set you back $29.50 for 1.2 inches of book, giving an SI of $24.58, the highest so far e..countered. The SI of *Basalts* (same publisher) is $22; of *Sedimentology and ore genesis* (Elsevier), $18.33; and of *Handbook of world salt resources* (Plenum), $17.90. To line your shelves with books of this kind would cost you an average of $20.70 per front inch, or $745 per yard. This is obviously carrying a joke too far.

At this point I'm pleased to retract an injustice to my old friend GSA Special Paper 88, *Saline deposits*. Clearly its SI of $15.83 is not 'astronomical', as formerly stated. Indeed, if you're a GSA member you can get one copy at a discount of 20 per cent, for an SI of only $12.67. The new *Ore deposits of the U.S., 1933–1967*, has an SI to AIME members of a mere $6.

Robert Metz of Newark State College says that as recently as five years ago he bought 9 volumes of James Hall's *Paleontology of New York* from the State Survey at the original price, getting 17.375 inches of classical paleontology for $23.50. This gives an almost incredible SI of $1.35. Imagine being able to one-up your office mates at a bargain rate like that!

As a 'non-attending member' of the International Congress at Prague, Laing Ferguson of Mt Allison U. paid $28 and has already got 7 inches of publications, with more presumably to come. When all are in, he expects to end up with an SI of only about $3.50. This is by no means a record (as we just saw), and besides the Congress meets only once every four years. But consider the long-term prestige! 'The presence of numerous sets on one's shelves certainly lends more charisma than merely North American journals, as they suggest that the occupant of the office might have actually *been* to Algiers, London, Copenhagen, New Delhi, and Prague.' True indeed.

If you are really desperate to fill shelves, you can join the AAAS for $12 and receive *Science* every week; in 1968 this came to 9.75 inches, for a record low SI of $1.23. OK, but *Science* is mostly nongeological, and charismawise it's nowhere.

Eugene Swick of Lafayette, La., says a friend calculated that to fill the four empty shelves in his bookcase with AAPG *Bulletins* would take 24 years and cost $600 (assuming no increase in dues). After thinking it over, he (a) resigned from AAPG, (b) removed the shelves, (c) invested $300 in assorted old guns and rifles, and (d) mounted these in the bookcase. 'He has now enveloped himself in an aura of manliness, and furthermore has saved $300, which he plans to spend on booze. How's that for scientific ingenuity?'

When a fellow writes a geologic report, he's got to name it something. Since the title is the banner under which the paper will march and be identified, obviously it should be clear, informative, and reasonably brief. Unfortunately this simple goal is frequently missed—whether through author's whim, editorial caprice, or bureaucratic decree is often hard to say.

These reflections are prompted by a title recently sent in by Louis Moyd, with the comment that it's a classic example of not giving anything away ahead of time, or close-to-the-chestmanship. This is *Geology of townships 149 and 150* (J. A. Robertson, 1968, Ont. dept. mines geol. rept. 57). There's no problem here if you happen to know that townships 149 and 150 include Elliott Lake, one of the world's major uranium-producing districts; if you don't, this ranks near the top in the least-informative category.

A title that's no more informative, but has the undeniable virtue of brevity, is *Q* (L. Knopoff, 1964, *Reviews of geophysics* 4: 625–60). Just *Q*. According to a colleague who's a member of the In crowd, this is a seismic-wave attenuation factor. Despite the omission of this fact from the title, it is hard to fault anything so gloriously brief.

A remarkable title indeed is promised for the October GSA *Bulletin*, namely *Submersible observations in the Straits of Florida*. Interpretation here is fraught with uncertainty. A friend claims that the first word is a noun and that they've finally trained submersibles to take notes, whereas I contend that it's an adjective and the title refers to notes taken on land and later submerged, for reasons undoubtedly explained in the report. Or of course the expression could mean observations recorded with one of those pens that write under water. Ah well—by the time you read this, you can relieve the suspense by checking the report itself.

I'll be happy to entertain nominations for other categories: nouniest, longest, and most pretentious are possibilities that suggest themselves. (Let's put a 1945-or-later limit on this.) Any nominations from the floor?

*   *   *

The other day I read that the late Senator Wherry of Nebraska was famous for making oral bloopers, such as "Indigo China' and the 'chief joints of staff'. He is reported to have said that a witness would have 'opple amportunity' to reply to an accusation. Such booboos are easy to laugh at, but unfortunately they're also easy to make, as everyone knows who has ever addressed an audience. I once heard an anxious speaker at an AAPG session, short of time and beset by projection troubles, just barely avoid coining the admirable phrase 'a doubly anti plunging cline'. Don't snicker. It could happen to you.

*   *   *

Show me a rock-knocker in the basement and I'll show you a subsurface geologist.

Discovery

The outcome of any serious research can only be to make two questions grow where only one grew before.

—**Thorstein Veblen**

This tongue-in-cheek article reviews compressional velocities of lunar rocks with those of earth rocks and other "terrestrial materials."

PROPERTIES AND COMPOSITION OF LUNAR MATERIALS: EARTH ANALOGIES

Edward Schreiber and Orson L. Anderson

Abstract. *The sound velocity data for the lunar rocks were compared to numerous terrestrial rock types and were found to deviate widely from them. A group of terrestrial materials were found which have velocities comparable to those of the lunar rocks, but they do obey velocity-density relations proposed for earth rocks.*

Certain data from Apollo 11 and Apollo 12 missions present some difficulties in that they require explanations for the signals received by the lunar seismograph as a result of the impact of the lunar module (LEM) on the lunar surface (*1*). In particular, the observed signal does not resemble one due to an impulsive source, but exhibits a generally slow build-up of energy with time. In spite of the appearance of the returned lunar samples, the lunar seismic signal continued to ring for a remarkably long time—a characteristic of very high Q material. The lunar rocks, when studied in the laboratory, exhibited a low Q (*2*). Perhaps most startling of all, however, was the very low sound velocity indicated for the outer lunar layer deduced from the LEM impact signal. The data obtained on the lunar rocks and finds agree well with the results of the Apollo 12 seismic experiment (*2,3*). These rock velocities are startlingly low. The measured velocities on a vesicular medium grained, igneous rock (10017) having a bulk density of 3.2 g/cm³ were v_p = 1.84, and v_s = 1.05 km/sec. The results for a microbreccia (10046) with a bulk density of 2.2 g/cm³ were v_p = 1.25 and v_s = 0.74 km/sec for the compressional (v_p) and shear (v_s) velocities.

It was of some interest to consider the behavior of these lunar rocks in terms of the expected behavior based on measurements of earth materials. Birch (*4*) first proposed a simple linear relation between compressional velocity and density for rocks. This relation was examined further by Anderson (*5*) who showed that this was a first approximation to a more general relation, derivable from a dependence of the elastic moduli with the density through a power function. Comparison of the results obtained from the returned lunar rocks with the predictions of these relationships expresses graphically the manner they deviate from the behavior of rocks found on earth. The velocities are remarkably lower than what would be predicted from either the Birch or Anderson relationships.

To account for this very low velocity, we decided to consider materials other than those listed initially by Birch (*4*) or more detailed compilation of

Table 1. Comparison of compressional velocities of lunar
rocks and various earth materials.

Lunar rocks and cheeses	v_p (km/sec)
Sapsego (Swiss)	2.12
Lunar Rock 10017	1.84
Gjetost (Norway)	1.83
Provolone (Italy)	1.75
Romano (Italy)	1.75
Cheddar (Vermont)	1.72
Emmenthal (Swiss)	1.65
Muenster (Wisconsin)	1.57
Lunar Rock 10046	1.25
Sedimentary rocks	v_p (km/sec)
Dolomite	5.6
Dolomite	4.69
Limestone	5.06
Limestone	5.97
Greywacke	5.4
Greywacke	6.06
Sandstone	4.90
Metamorphic rocks	v_p (km/sec)
Schist	5.1
Slate	5.39
Charnockite	6.15
Gneiss	4.9
Marble	6.02
Quartzite	5.6
Amphibolite	6.70
Eclogite	6.89
Igneous rocks	v_p (km/sec)
Granite	5.9
Syenite	5.7
Diorite	5.78
Oligoclase	6.40
Andesite	5.23
Gabbro	5.8
Gabbro	6.8
Norite	6.50
Diabase	6.33
Minerals	v_p (km/sec)
Corundum	10.8
Periclase	9.69
Spinel	9.91
Garnet	8.53
Quartz	6.05
Hematite	7.90
Olivine	8.42
Trevorite	7.23
Lime	7.95

Anderson and Liebermann (6). The search was aided by considerations of much earlier speculations concerning the nature of the moon (7), and a significant group of materials was found which have velocities that cluster about those actually observed for lunar rocks.

These materials are summarized in Table 1, where, for emphasis, common rock types found on earth are listed for comparison. The materials studied were chosen so as to represent a broad geographic distribution in order to preclude any bias that might be introduced by regional sampling. It is seen that these materials exhibit compressional velocities that are in consonance with those measured for the lunar rocks—which leads us to suspect that perhaps old hypotheses are best, after all, and should not be lightly discarded.

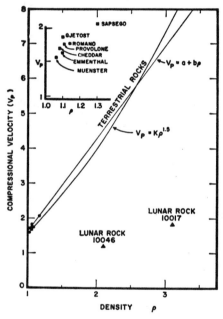

Fig. 1. Comparison of the velocity of sound for rocks with that of earth materials.

A comparison of these low velocity materials with the predictions of Birch and of Anderson is shown in Fig. 1. It is at once apparent that these materials do yield values of velocity that are predicted by these relations for their densities. Thus the curve of Birch for the rock types diabase, gabbros, and eclogites fit the cheeses surprisingly well. This apparent inconsistency, in that the cheeses do obey these relationships by having a velocity appropriate to their density, in contrast to the lunar rocks with which they compare so well, may readily be accounted for when one considers how much better aged the lunar materials are.

In Harper's Magazine about a decade ago an article was published entitled, "Oxford's Magnificent Oddballs," in which William Buckland (1784–1856), a geologist, was honored by being named the most lurid and fabled of the Oxford eccentrics. Yet Buckland was a geologist of ability, author of 53 articles, twice President of the Geological Society, Fellow of the Royal Society, first President of the British Association for the Advancement of Science and the recipient of the Wollaston Medal, the highest geological honor that could be bestowed.

Buckland was a truly eloquent and enthusiastic lecturer and highly regarded in contemporary circles. In fact, a readership at Oxford was created specifically for him and was financed by a treasury stipend advanced by the Prince Regent.

Mrs. Elizabeth O. B. Gordon, his daughter, prepared a book on his life and correspondence which shows that William Buckland had a well-developed sense of humor and considerable stage-presence. Extracted from his daughter's book is an account of a student's reaction to a Buckland lecture and a poem about Bucklandian antics in the classroom by a fellow professor. Humor, eccentricity, and ability all characterize this unusual Oxford don.

WILLIAM BUCKLAND

Elizabeth O. B. Gordon

"He lectured on the Cavern of Torquay, the now famous Kent's Cavern. He paced like a Franciscan Preacher up and down behind a long show-case, up two steps, in a room in the old Clarendon. He had in his hand a huge hyena's skull. He suddenly dashed down the steps—rushed, skull in hand, at the first undergraduate on the front bench—and shouted, 'What rules the world?' The youth, terrified, threw himself against the next back seat, and answered not a word. He rushed then on me, pointing the hyena full in my face—'What rules the world?' 'Haven't an idea,' I said. 'The stomach, sir,' he cried (again mounting his rostrum), 'rules the world. The great ones eat the less, and the less the lesser still.'"

The Professor's *forte* as a lecturer in these early days excited the rhyming propensities of his College friend Shuttleworth, afterwards Warden of New College, and subsequently Bishop of Chichester. The lecture which suggested the following lines was probably delivered early in 1822.

> In Ashmole's ample dome, with look sedate,
> 'Midst heads of mammoths, heads of houses sate;
> And tutors, close with undergraduates jammed,
> Released from cramming, waiting to be crammed.

Above, around, in order due displayed,
The garniture of former worlds was laid:
Sponges and shells in lias moulds immersed,
From Deluge fiftieth, back to Deluge first;
And wedged by boys in artificial stones,
Huge bones of horses, now called mammoths' bones;
Lichens and ferns which schistose beds enwrap,
And—understood by most professors—trap.
Before the rest, in contemplative mood,
With sidelong glance, the inventive Master stood,
And numbering o'er his class with still delight,
Longed to possess them cased in stalactite.
Then thus with smile suppressed: "In days of yore
One dreary face Earth's infant planet bore;
Nor land was there, nor ocean's lucid flood,
But, mixed of both, one dark abyss of mud;
Till each repelled, repelling by degrees,
This shrunk to rock, that filtered into seas;
Then slow upheaved by subterranean fires,
Earth's ponderous crystals shot their prismy spires;
Then granite rose from out the trackless sea,
And slate, for boys to scrawl—when boys should be—
But earth, as yet, lay desolate and bare;
Man was not then,—but Paramoudras were.
'Twas silence all, and solitude; the sun,
If sun there were, yet rose and set to none,
Till fiercer grown the elemental strife,
Astonished tadpoles wriggled into life;
Young encrini their quivering tendrils spread,
And tails of lizards felt the sprouting head.
(The specimen I hand about is rare.
And very brittle; bless me, sir, take care!)
And high upraised from ocean's inmost caves,
Protruded corals broke the indignant waves.
These tribes extinct, a nobler race succeeds:
Now sea-fowl scream amid the plashing reeds;
Now mammoths range, where yet in silence deep
Unborn Ohio's hoarded waters sleep,
Now ponerous whales...
                            (Here, by the way, a tale
I'll tell of something, very like a whale.
An odd experiment of late I tried,
Placing a snake and hedgehog side by side;
Awhile the snake his neighbour tried t' assail,
When the sly hedgehog caught him by the tail,
And gravely munched him upwards joint by joint,—
The story's somewhat shocking, but in point.)

Now to proceed:—
The earth, what is it? Mark its scanty bound—
'Tis but a larger football's narrow round;
Its mightiest tracts of ocean—what are these?
At best but breakfast tea-cups, full of seas.
O'er these a thousand deluges have burst,
And quasi-deluges have done their worst."

The lecture ends with a couplet which the facetious writer observes of its own accord "slides into verse, and hitches in a rhyme":—

Of this enough. On Secondary Rock,
To-morrow, gentlemen, at two o'clock.

Buckland was greatly pleased, on his return from this long sojourn on the Continent, to be greeted with the following epitaph written by his friend Whately, afterwards the famous Archbishop of Dublin. He had the verses lithographed, and gave copies to his friends, so that they are more known than many of the clever verses written by Dr. Shuttleworth and Mr. Duncan.

ELEGY

Intended for Professor Buckland. December 1st, 1820.

BY RICHARD WHATELY

Mourn, Ammonites, mourn o'er his funeral urn,
  Whose neck ye must grace no more,
Gneiss, granite, and slate, he settled your date,
  And his ye must now deplore.
Weep, caverns, weep with unfiltering drip,
  Your recesses he'll cease to explore;
For mineral veins and organic remains
  No stratum again will he bore.

Oh, his wit shone like crystal; his knowledge profound
  From gravel to granite descended,
No trap could deceive him, no slip could confound,
  Nor specimen, true or pretended;
He knew the birth-rock of each pebble so round,
  And how far its tour had extended.

His eloquence rolled like the Deluge retiring,
  Where mastodon carcases floated;
To a subject obscure he gave charms so inspiring,
  Young and old on geology doated.
He stood out like an Outlier; his hearers, admiring,
  In pencil each anecdote noted.

Where shall we our great Professor inter,
　　That in peace may rest his bones?
If we hew him a rocky sepulchre,
　　He'll rise and break the stones,
And examine each stratum that lies around,
For he's quite in his element underground.

If with mattock and spade his body we lay
　　In the common alluvial soil,
He'll start up and snatch those tools away
　　Of his own geological toil;
In a stratum so young the Professor disdains
That embedded should lie his organic remains.

Then exposed to the drip of some case-hardening spring
　　His carcase let stalactite cover,
And to Oxford the petrified sage let us bring
　　When he is encrusted all over;
There, 'mid mammoths and crocodiles, high on a shelf,
Let him stand as a monument raised to himself."

Buckland's championship of the glacial theory was the subject of a poetic "Dialogue between Dr. Buckland and a Rocky Boulder," written by his friend Philip Duncan. The following are the lines:—

Buckland, *loquitur*.

Say when, and whence, and how, huge Mister Boulder,
And by what wondrous force hast thou been rolled here?
Has some strong torrent driven thee from afar,
Or hast thou ridden on an icy car?
Which, from its native rock once torn like thee,
Has floundered many a mile throughout the sea,
And stranded thee at last upon this earth,
So distant from thy primal place of birth;
And having done its office with due care,
Was changed to vapour, and was mixed in air.

Boulder, *respondit*.

Thou great idolater of stocks and stones,
Of fossil shells and plants and buried bones;
Thou wise Professor, who wert ever curious
To learn the true, and to reject the spurious,
Know that in ancient days an icy band
Encompassed around the frozen land,

Until a red-hot comet, wandering near
The strong attraction of this rolling sphere,
Struck on the mountain summit, from whence torn
Was many a vast and massive iceberg borne,
And many a rock, indented with sharp force
And still-seen striae, shows my ancient course;
And if you doubt it, go with friend Agassiz
And view the signs in Scotland and Swiss' passes.

Scientific Power

The means by which we live have outdistanced the ends for which we live.
Our scientific power has outrun our spiritual power. We have guided
missiles and misguided men.

—Martin Luther King, Jr.

CHAPTER 8

Geology and Poetry

This chapter concerns geology and poetry. What is poetry? "Poetry," Wordsworth declared, "is the image of man and nature . . . [It] takes its origin from emotion recollected in tranquility." To Johnson poetry was "the art of uniting pleasure with truth by calling imagination to the help of reason." "Musical thought" responded Carlyle. "The art of doing by means of words what the painter does by means of colours" affirmed Macaulay. To Coleridge, poetry was "that species of composition which is opposed to the works of science, by proposing for its immediate object, pleasure, not truth."

But poetry in one sense, is complementary to science, rather than competitive. C. A. Coulson has illustrated this well by asking (though in a quite different context) "What is a primrose?" To the casual observer

"A primrose by the river's brim
A yellow primrose was to him
And it was nothing more!"

To the classical botanist, a primrose is a member of the species *Primula vulgaris*, short plants with large, wrinkled, light green leaves, slightly notched corolla lobes, and radical peduncles, each bearing a single large flower. But to the saint, the primrose is "God's promise of spring." All descriptions apply to the same small yellow flower, yet they have little in common, and the validity of one in no way invalidates the others. Even another botanist may not use the same definitions of our classical plant taxonomist, but may emphasize the delicate biochemical mechanisms and interchanges "requiring potash, phosphates, nitrogen and water in definite proportions."

What characterizes the poet's description is its insight, its imaginative artistry, and the emotive meter of its language.

Because our experience is earthbound, it is scarcely surprising that the earth, in all its moods, features prominently in poetry, whether narrative, lyric, or dramatic. Our extracts provide a varied selection.

187

W. H. Auden (1907–1973), Anglo-American dramatist and poet, is represented here by an extract from In Praise of Limestone, *which extols the virtues of limestone terrain. It is 'the one landscape that we, the inconstant ones, are consistently homesick for . . .' because*

> 'The best and worst never stayed here long but sought
> Immoderate soils where beauty was not so external'.

Auden was born in Yorkshire, a limestone land. Yet the paradox, as he looks back at the limestone land, is that though a 'backward and dilapidated province . . .' it has the power, too, to call 'into question All the Great Powers assume; it disturbs our rights . . .' and evokes images of faultless love and life to come.

IN PRAISE OF LIMESTONE

W. H. Auden

If it form the one landscape that we, the inconstant ones,
    Are consistently homesick for, this is chiefly
Because it dissolves in water. Mark these rounded slopes
    With their surface fragrance of thyme and, beneath,
A secret system of caves and conduits; hear the springs
    That spurt out everywhere with a chuckle,
Each filling a private pool for its fish and carving
    Its own little ravine whose cliffs entertain
The butterfly and the lizard; examine this region
    Of short distances and definite places:
What could be more like Mother or a fitter background
    For her son, the flirtatious male who lounges
Against a rock in the sunlight, never doubting
    That for all his faults he is loved; whose works are but
Extensions of his power to charm? From weathered outcrop
    To hill-top temple, from appearing waters to
Conspicuous fountains, from a wild to a formal vineyard,
    Are ingenious but short steps that a child's wish
To receive more attention than his brothers, whether
    By pleasing or teasing, can easily take.
Watch, then, the band of rivals as they climb up and down
    Their steep stone gennels in twos and threes, at times
Arm in arm, but never, thank God, in step; or engaged
    On the shady side of a square at midday in
Voluble discourse, knowing each other too well to think
    There are any important secrets, unable
To conceive a god whose temper-tantrums are moral
    And not to be pacified by a clever line

Or a good lay: for, accustomed to a stone that responds,
  They have never had to veil their faces in awe
Of a crater whose blazing fury could not be fixed;
  Adjusted to the local needs of valleys
Where everything can be touched or reached by walking
  Their eyes have never looked into infinite space
Through the lattice-work of a nomad's comb; born lucky,
  Their legs have never encountered the fungi
And insects of the jungle, the monstrous forms and lives
  With which we have nothing, we like to hope, in common.
So, when one of them goes to the bad, the way his mind works
  Remains comprehensible: to become a pimp
Or deal in fake jewellery or ruin a fine tenor voice
  For effects that bring down the house, could happen to all
But the best and the worst of us...
                                That is why, I suppose,
  The best and worst never stayed here long but sought
Immoderate soils where the beauty was not so external,
  The light less public and the meaning of life
Something more than a mad camp. 'Come!' cried the granite wastes,
  'How evasive is your humour, how accidental
Your kindest kiss, how permanent is death.' (Saints-to-be
  Slipped away sighing.) 'Come!' purred the clays and gravels.
'On our plains there is room for armies to drill; rivers
  Wait to be tamed and slaves to construct you a tomb
In the grand manner: soft as the earth is mankind and both
  Need to be altered.' (Intendant Caesars rose and
Left, slamming the door.) But the really reckless were fetched
  By an older colder voice, the oceanic whisper:
'I am the solitude that asks and promises nothing;
  That is how I shall set you free. There is no love;
There are only the various envies, all of them sad.'
  They were right, my dear, all those voices were right
And still are; this land is not the sweet home that it looks,
  Nor its peace the historical calm of a site
Where something was settled once and for all: A backward
  And dilapidated province, connected
To the big busy world by a tunnel, with a certain
  Seedy appeal, is that all it is now? Not quite:
It has a worldly duty which in spite of itself
  It does not neglect, but calls into question
All the Great Powers assume; it disturbs our rights. The poet,
  Admired for his earnest habit of calling
The sun the sun, his mind Puzzle, is made uneasy
  By these marble statues which so obviously doubt
His antimythological myth; and these gamins,
  Pursuing the scientist down the tiled colonnade

With such lively offers, rebuke his concern for Nature's
  Remotest aspects: I, too, am reproached, for what
And how much you know. Not to lose time, not to get caught,
  Not to be left behind, not, please! to resemble
The beasts who repeat themselves, or a thing like water
  Or stone whose conduct can be predicted, these
Are our Common Prayer, whose greatest comfort is music
  Which can be made anywhere, is invisible,
And does not smell. In so far as we have to look forward
  To death as a fact, no doubt we are right: But if
Sins can be forgiven, if bodies rise from the dead,
  These modifications of matter into
Innocent athletes and gesticulating fountains,
  Made solely for pleasure, make a further point:
The blessed will not care what angle they are regarded from,
  Having nothing to hide. Dear, I know nothing of
Either, but when I try to imagine a faultless love
  Or the life to come, what I hear is the murmur
Of underground streams, what I see is a limestone landscape.

*John Stuart Blackie (1809–1895) was Professor of Humanities at Mar-
ischall College and later Professor of Greek at the University of Edinburgh, where he taught for 30 years. Friend of a number of contemporary
Scottish geologists, the following extracts are taken from his poem* Stra-
tigraphical Palaeontology. *This and other lighthearted examples of Paleontology in Literature are included in Archie Lamont's paper by that
name in* The Quarry Manager's Journal, *Vol. XXX, January 1947, pp.
432–41. Our extracts concern the creatures of Silurian, Devonian, Penn-
sylvanian, Triassic, Jurassic, and Pleistocene times.*

John Stuart Blackie

STRATIGRAPHICAL PALAEONTOLOGY

The waters, now big with a novel sensation,
Brought corals and buckies and bivalves to view,
Who dwell in shell houses, a softbodied nation;
But fishes and fins were yet none, or few.
Buckies and bivalves, a numberless nation!
Buckies, and bivalves, and trilobites too!
These you will find in Silurian station,
When Ramsay and Murchison sharpen your view.

COAL MEASURES AND TRIASSIC SYSTEM

God bless the fishes!—but now on the dry land,
In days when the sun shone benign on the poles,
Forests of ferns in the low and the high land
Spread their huge fans, soon to change into coals!
Forests of ferns—a wonderful verity!
Rising like palm trees beneath the North Pole;
And all to prepare for the golden prosperity
Of John Bull reposing on iron and coal!
"Now Nature the eye of the gazer entrances
With wonder on wonder from teaming abodes;
From the gills of the fish to true lungs she advances,
And bursts into blossoms of tadpoles and toads.
People Batrachian, strange, Triassic all,
Like Hippopotamus huge on the roads!
Call them ungainly, uncouth, and unclassical,
But great in the reign of the Trias were Toads!

ICHTHYOSAURUS

Behold a strange monster our wonder engages,
If dolphin or lizard your wit may defy,
Some thirty feet long on the shore of Lyme-Regis
With saw for a jaw, and a big-staring eye.
A fish or a lizard? An ichthyosaurus,
With big goggle-eyes, and a very small brain,
And paddles like mill-wheels in clattering chorus
Smiting tremendous the dreadsounding main!

FROZEN MAMMOTHS

Mammoth, Mammoth! mighty old Mammoth!
Strike with your hatchet and cut a good slice;
The bones you will find, and the hide of the mammoth,
Packed in stiff cakes of Siberian ice.

*Percy Bysshe Shelley (1792–1822), idealistic, romantic, impulsive, re-
bellious, one of the greatest of the English Romantic poets, was expelled
from Oxford for his pamphlet* The Necessity of Atheism. *Two elopements,
two marriages, and strained family relationships led to a series of short
spells of residence in Wales, England, and Switzerland, which preceded
his settling in Italy, where most of his best known poetry was written.
Shelley, a close companion of Byron, was drowned in a storm while sailing
in the Bay of Spezia. He was 30 years old.*

Prometheus Unbound, *a drama composed partly on the ruins of the* *baths of Caracaha in Rome, is his masterpiece. Prometheus, the human* *spirit, is chained and tormented by Jupiter, the God of religion. Man thus* *loses his humanity, and the poem is the record of Prometheus's (mankind's)* *liberation, and reunion with his spouse, Nature. Act 1, from which the* *present extract is taken, is set in "a Ravine of Icy Rocks in the Indian* *Caucasus. Prometheus is discovered bound to the Precipice. Panthea and* *Love are seated at his feet. Time, Night." The extract involves a conver-* *sation between Prometheus and the Earth. Prometheus 'Nailed to this wall* *of eagle-babbling mountain' cries in anguish that even the elements must* *have felt his pain.*

I ask the Earth, have not the mountains felt?
I ask yon Heaven, the all-beholding Sun,
Has it not seen?

Voices from the mountains, the springs, the air, the whirlwinds, and the *earth then declare that they have witnessed his anguish.*

PROMETHEUS UNBOUND

Percy B. Shelley

I ask the Earth, have not the mountains felt?
I ask yon Heaven, the all-beholding Sun,
Has it not seen? The Sea, in storm or calm,
Heaven's ever-changing Shadow, spread below,
Have its deaf waves not heard my agony?
Ah me! alas, pain, pain ever, for ever!
The crawling glaciers pierce me with the spears
Of their moon-freezing chrystals; the bright chains
Eat with their burning cold into my bones.
Heaven's winged hound, polluting from thy lips
His beak in poison not his own, tears up
My heart; and shapeless sights come wandering by,
The ghastly people of the realm of dream,
Mocking me: and the Earthquake-fiends are charged
To wrench the rivets from my quivering wounds
When the rocks split and close again behind:
While from their loud abysses howling throng
The genii of the storm, urging the rage
Of whirlwind, and afflict me with keen hail.
And yet to me welcome is day and night,
Whether one breaks the hoar frost of the morn,
Or starry, dim, and slow, the other climbs
The leaden-coloured cast; for then they lead
Their wingless, crawling hours, one among whom

—As some dark Priest hales the reluctant victim—
Shall drag thee, cruel King, to kiss the blood
From these pale feet, which then might trample thee
If they disdained not such a prostrate slave.
Disdain? Ah no! I pity thee. What ruin
Will hunt thee undefended thro wide Heaven!
How will thy soul, cloven to its depth with terror,
Gape like a hell within! I speak in grief,
Not exultation, for I hate no more,
As then, ere misery made me wise. The curse
Once breathed on thee I would recall. Ye Mountains,
Whose many-voiced Echoes, thro the mist
Of cataracts, flung the thunder of that spell!
Ye icy Springs, stagnant with wrinkling frost,
Which vibrated to hear me, and then crept
Shuddering through India! Thou serenest Air,
Through which the Sun walks burning without beams!
And ye swift Whirlwinds, who on poised wings
Hung mute and moveless o'er yon hushed abyss,
As thunder, louder than your own, made rock
The orbed world! If then my words had power,
Though I am changed so that aught evil wish
Is dead within! although no memory be
Of what is hate, let them not lose it now!
What was that curse? for ye all heard me speak.

FIRST VOICE: from the mountains.

Thrice three hundred thousand years
    O'er the Earthquake's couch we stood:
Oft, as men convulsed with fears,
    We trembled in our multitude.

SECOND VOICE: from the springs.

Thunder-bolts had parched our water,
    We had been stained with bitter blood,
And had run mute, 'mid shrieks of slaughter,
    Thro' a city and a solitude.

THIRD VOICE: from the air.

I had clothed, since Earth uprose,
    Its wastes in colours not their own;
And oft had my serene repose
    Been cloven by many a rending groan.

FOURTH VOICE: from the whirlwinds.

We had soared beneath these mountains
    Unresting ages; nor had thunder,

Nor yon volcano's flaming fountains,
Nor any power above or under
Ever made us mute with wonder.

FIRST VOICE.

But never bowed our snowy crest
As at the voice of thine unrest.

SECOND VOICE.

Never such a sound before
To the Indian waves we bore.
A pilot asleep on the howling sea
Leaped up from the deck in agony,
And heard, and cried, "Ah, woe is me!"
And died as mad as the wild waves be.

THIRD VOICE.

By such dread words from Earth to Heaven
My still realm was never riven:
When its wound was closed, there stood
Darkness o'er the day like blood.

FOURTH VOICE.

And we shrank back: for dreams of ruin
To frozen caves our flight pursuing
Made us keep silence—thus—and thus—
Though silence is as hell to us.

THE EARTH.

The tongueless Caverns of the craggy hills
Cried, 'Misery!' then; the hollow Heaven replied,
'Misery!' And the Ocean's purple waves,
Climbing the land, howled to the lashing winds,
And the pale nations heard it, 'Misery!'

The second extract from Shelley contains some of his finest scenic descriptions. "Lines Written in the Vale of Chamouny" includes a majestic description of Alpine glaciers, and of the towering Mont Blanc.

LINES WRITTEN IN THE VALE OF CHAMOUNY

Percy B. Shelley

The glaciers creep
Like snakes that watch their prey, from their far fountains,

Slow rolling on; there, many a precipice,
Frost and the Sun in scorn of mortal power
Have piled: dome, pyramid, and pinnacle,
A city of death, distinct with many a tower
And wall impregnable of beaming ice.
Yet not a city, but a flood of ruin
Is there, that from the boundaries of the sky
Rolls its perpetual stream; vast pines are strewing
Its destined path, or in the mangled soil
Branchless and shattered stand; the rocks, drawn down
From yon remotest waste, have overthrown
The limits of the dead and living world,
Never to be reclaimed. The dwelling-place
Of insects, beasts, and birds, becomes its spoil
Their food and their retreat for ever gone,
So much of life and joy is lost. The race
Of man flies far in dread; his work and dwelling
Vanish, like smoke before the tempest's stream,
And their place is not known. Below, vast caves
Shine in the rushing torrents' restless gleam,
Which from those secret chasms in tumult welling
Meet in the vale, and one majestic River,
The breath and blood of distant lands, for ever
Rolls its loud waters to the ocean-waves,
Breathes its swift vapours to the circling air.
Mont Blanc yet gleams on high:—the power is there,
The still and solemn power of many sights,
And many sounds, and much of life and death.
In the calm darkness of the moonless nights,
In the lone glare of day, the snows descend
Upon that Mountain; none beholds them there,
Nor when the flakes burn in the sinking sun,
Or the star-beams dart through them:—Winds contend
Silently there, and heap the snow with breath
Rapid and strong, but silently! Its home
The voiceless lightning in these solitudes
Keeps innocently, and like vapour broods
Over the snow. The secret Strength of things
Which governs thought, and to the infinite dome
Of Heaven is as a law, inhabits thee!
And what were thou, and earth, and stars, and sea,
If to the human mind's imaginings
Silence and solitude were vacancy?

That same setting that inspired Shelley also gave us Coleridge's Hymn
Before Sun-Rise in the Vale of Chamouni. *The Arve and Arveiran, rivers
that rise from the slopes of Mont Blanc, and five other torrents sweep
down its precipitous slopes. Here and there, nestled against the gleaming
ice, grow meadows of* Gentiana major, *the Alpine gentian, with its 'flowers
of loveliest blue.'*

*Samuel Taylor Coleridge (1772–1834), English critic, romantic poet
and philosopher, was a friend of Wordsworth and Robert Southey, with
whom, in the afterglow of the French Revolution, he planned an American
utopian community (a paritisocracy) which was to be founded on the banks
of the Susquehanna, although funds were not forthcoming.* The Ancient
Mariner, Christabel, *and* Kubla Khan *are amongst his best-known poems.
Addicted to opium in later life, he continued to write and lecture. Coleridge
was converted in middle years from Unitarianism to the more conventional
fold of the Church of England, and his posthumously published* Confes-
sions of an Inquiring Spirit, *and other writings did much to deepen and
liberalize contemporary Christian thought.*

HYMN BEFORE SUN-RISE, IN THE VALE OF CHAMOUNI

Samuel T. Coleridge

And you, ye five wild torrents fiercely glad!
Who called you forth from night and utter death,
From dark and icy caverns called you forth,
Down those precipitous, black, jagged rocks,
For ever shattered and the same for ever?
Who gave you your invulnerable life,
Your strength, your speed, your fury, and your joy,
Unceasing thunder and eternal foam?
And who commanded (and the silence came),
Here let the billows stiffen, and have rest?
Ye Ice-falls! ye that from the mountain's brow
Adown enormous ravines slope amain—
Torrents, methinks, that heard a mighty voice,
And stopped at once amid their maddest plunge!
Motionless torrents! silent cataracts!
Who made you glorious as the Gates of Heaven
Beneath the keen full moon? Who bade the sun
Clothe you with rainbows? Who, with living flowers
Of loveliest blue, spread garlands at your feet?—
GOD! let the torrents, like a shout of nations,
Answer! and let the ice-plains echo, GOD!
GOD! sing ye meadow-streams with gladsome voice!
Ye pine-groves, with your soft and soullike sounds!

And they too have a voice, yon piles of snow,
And in their perilous fall shall thunder, GOD!

T. A. Conrad (1803–1877) was a member of a Philadelphia Quaker family devoted to scientific studies. A superb artist, he illustrated many of his own publications on Cretaceous and Cenozoic paleontology and stratigraphy. Conrad was associated with James Hall for a time, while serving as geologist to the State of New York. Although bitterly opposed to Darwin's evolutionary theory, Conrad was a prolific author. He was, however, an archetype absent-minded paleontologist, and was even found to have described some of his own new species twice, using different names. His ode To A Trilobite *is included here.*

TO A TRILOBITE

Timothy A. Conrad

Methinks I see thee gazing from the stone
With those great eyes, and smiling as in scorn
Of notions and of systems which have grown
From relics of the time when thou wert born.

The object of his poem, The Lisbon Earthquake, *Voltaire declared in a preface, was to enquire into the prevailing mid-18th century maxim 'Whatever is, is right.'*

Voltaire "maintains that ancient and sad truth acknowledged by all men that there is evil upon earth" . . . but 'owns that no philosopher has ever been able to explain the nature of moral and physical evil . . . Revelation alone can untie the great knot . . .' In our brief extract Voltaire confronts the question

*'Say what advantage can result to all
From wretched Lisbon's lamentable fall?'*

THE LISBON EARTHQUAKE

Voltaire

Say what advantage can result to all,
From wretched Lisbon's lamentable fall?

Are you then sure, the power which could create
The universe and fix the laws of fate,
Could not have found for man a proper place,
But earthquakes must destroy the human race?
Will you thus limit the eternal mind?
Should not our God to mercy be inclined?
Cannot then God direct all nature's course?
Can power almighty be without resource?
Humbly the great Creator I entreat,
This gulf with sulphur and with fire replete,
Might on the deserts spend its raging flame,
God my respect, my love weak mortals claim;
When man groans under such a load of woe,
He is not proud, he only feels the blow.
Would words like these to peace of mind restore
The natives sad of that disastrous shore?
Grieve not, that others' bliss may overflow,
Your sumptuous palaces are laid thus low;
Your toppled towers shall other hands rebuild;
With multitudes your walls one day be filled;
Your ruin on the North shall wealth bestow,
For general good from partial ills must flow;
You seem as abject to the sovereign power,
As worms which shall your carcasses devour.
No comfort could such shocking words impart,
But deeper wound the sad, afflicted heart.
When I lament my present wretched state,
Allege not the unchanging laws of fate;
Urge not the links of the eternal chain,
'Tis false philosophy and wisdom vain.
The God who holds the chain can't be enchained;
By His blest will are all events ordained:
He's just, nor easily to wrath gives way,
Why suffer we beneath so mild a sway:
This is the fatal knot you should untie,
Our evils do you cure when you deny?
Men ever strove into the source to pry,
Of evil, whose existence you deny.
If he whose hand the elements can wield,
To the winds' force makes rocky mountains yield;
If thunder lays oaks level with the plain,
From the bolts' strokes they never suffer pain.
But I can feel, my heart oppressed demands
Aid of that God who formed me with His hands.
Sons of the God supreme to suffer all
Fated alike; we on our Father call.

After leaving Oxford, A. E. Housman (1859–1936) worked for eleven years in the Civil Service in London, before being appointed Professor of Latin at University College, London, in 1892. He became Professor of Latin at Cambridge in 1911, where he was a Fellow of Trinity. Apart from his scholarly publications in classics—one biographer described him, at his death in 1936, as "perhaps the most learned Latin scholar in the world"—he is remembered also as a poet. Housman was reserved, passionate, complex, sensitive, writing his poetry, he claimed, "as a morbid secretion." Written chiefly in times of illness or depression, his poetry is reflective on the loss of youth, the pains of friendship, and the loss of rustic simplicity. The present extract from A Shropshire Lad *contains a description of Wenlock Edge, known to countless geologists as the type area of the Middle Silurian, which is here seen through the eyes of a poet. The Wrekin is a nearby hill of Precambrian rhyolitic lava, Uriconium was an adjacent Roman city, and the Severn, the gentle river that threads its way through that enfolding landscape.*

A SHROPSHIRE LAD

A. E. Housman

On Wenlock Edge the wood's in trouble,
His forest fleece the Wrekin heaves;
The gale, it plies the saplings double,
And thick on Severn snow the leaves.
'Twould blow like this through holt and hanger
When Uricon the city stood:
'Tis the old wind in the old anger,
But then it threshed another wood.

Then, 'twas before my time, the Roman
At yonder heaving hill would stare:
The blood that warms an English yeoman,
The thoughts that hurt him, they were there.

There, like the wind through woods in riot,
Through him the gale of life blew high;
The tree of man was never quiet:
Then 'twas the Roman, now 'tis I.

The gale, it plies the saplings double,
It blows so hard, 'twill soon be gone:
To-day the Roman and his trouble
Are ashes under Uricon.

The Book of Job, written probably about twenty-five hundred years ago, contains the thoughts and conversations of a saintly man in the face of adversity. The book is a treasure of mature experience and perspective and is also a gem of literature. Job is an archetype of modern man; baffled by death, destruction, disease and loss, a good man in a tragic universe, agonizing over the justice of God and the search for wisdom and understanding.

Chapter 28, which we include, contains a moving account of man's success in mining the earth and its resources but his frustration in the search for wisdom, until he recognizes it in a knowledge of his Creator. In Chapter 38, the Lord answers Job by showing the limitations of man's experience and understanding.

WHERE SHALL WISDOM BE FOUND?
BOOK OF JOB

Chapter 28

"Surely there is a mine for silver,
and a place for gold which they refine.
² Iron is taken out of the earth,
and copper is smelted from the ore.
³ Men put an end to darkness,
and search out to the farthest bound
the ore in gloom and deep darkness.
⁴ They open shafts in a valley away from where men live;
they are forgotten by travellers,
they hang afar from men, they swing to and fro.
⁵ As for the earth, out of it comes bread;
but underneath it is turned up as by fire.
⁶ Its stones are the place of sapphires,
and it has dust of gold.

⁷ "That path no bird of prey knows,
and the falcon's eye has not seen it.
⁸ The proud beasts have not trodden it;
the lion has not passed over it.

⁹ "Man puts his hand to the flinty rock,
and overturns mountains by the roots.
¹⁰ He cuts out channels in the rocks,
and his eye sees every precious thing.
¹¹ He binds up the streams so that they do not trickle,
and the thing that is hid he brings forth to light.

¹² "But where shall wisdom be found?
  And where is the place of understanding?
¹³ Man does not know the way to it,
  and it is not found in the land of the living.
¹⁴ The deep says, 'It is not in me,'
  and the sea says, 'It is not with me.'
¹⁵ It cannot be gotten for gold,
  and silver cannot be weighed as its price.
¹⁶ It cannot be valued in the gold of Ophir,
  in precious onyx or sapphire.
¹⁷ Gold and glass cannot equal it,
  nor can it be exchanged for jewels of fine gold
¹⁸ No mention shall be made of coral or of crystal;
  the price of wisdom is above pearls.
¹⁹ The topaz of Ethiopia cannot compare with it,
  nor can it be valued in pure gold.

²⁰ "Whence then comes wisdom?
  And where is the place of understanding?
²¹ It is hid from the eyes of all living,
  and concealed from the birds of the air.
²² Abaddon and Death say,
  'We have heard a rumour of it with our ears.'

²³ "God understands the way to it,
  and he knows its place.
²⁴ For he looks to the ends of the earth,
  and sees everything under the heavens.
²⁵ When he gave to the wind its weight,
  and meted out the waters by measure;
²⁶ when he made a decree for the rain,
  and a way for the lightning of the thunder;
²⁷ then he saw it and declared it;
  he established it, and searched it out.
²⁸ And he said to man,
  'Behold, the fear of the Lord, that is wisdom;
  and to depart from evil is understanding.'"

BOOK OF JOB

Chapter 38

Then the Lord answered Job out
of the whirlwind:
² "Who is this that darkens counsel by words without
  knowledge?

³ Gird up your loins like a man,
    I will question you, and you shall declare to me.

⁴ "Where were you when I laid the foundation of the
       earth?
    Tell me, if you have understanding.
⁵ Who determined its measurements—surely you know!
    Or who stretched the line upon it?
⁶ On what were its bases sunk,
    or who laid its cornerstone,
⁷ when the morning stars sang together,
    and all the sons of God shouted for joy?

⁸ "Or who shut in the sea with doors,
    when it burst forth from the womb;
⁹ when I made clouds its garment,
    and thick darkness its swaddling band,
¹⁰ and prescribed bounds for it,
    and set bars and doors,
¹¹ and said, 'Thus far shall you come, and no farther,
    and here shall your proud waves be stayed'?

¹² "Have you commanded the morning since your days
       began,
    and caused the dawn to know its place,
¹³ that it might take hold of the skirts of the earth,
    and the wicked be shaken out of it?
¹⁴ It is changed like clay under the seal,
    and it is dyed like a garment.
¹⁵ From the wicked their light is withheld,
    and their uplifted arm is broken.

¹⁶ "Have you entered into the springs of the sea,
    or walked in the recesses of the deep.
¹⁷ Have the gates of death been revealed to you,
    or have you seen the gates of deep darkness?
¹⁸ Have you comprehended the expanse of the earth?
    Declare, if you know all this.

¹⁹ "Where is the way to the dwelling of light,
    and where is the place of darkness,
²⁰ that you may take it to its territory
    and that you may discern the paths to its home?
²¹ You know, for you were born then,
    and the number of your days is great!

²² "Have you entered the storehouses of the snow,
    or have you seen the storehouses of the hail,

²³ which I have reserved for the time of trouble,
    for the day of battle and war?
²⁴ What is the way to the place where the
      light is distributed,
    or where the east wind is scattered upon the earth?

²⁵ "Who has cleft a channel for the torrents of rain,
    and a way for the thunderbolt,
²⁶ to bring rain on a land where no man is,
    on the desert in which there is no man;
²⁷ to satisfy the waste and desolate land,
    and to make the ground put forth grass?

²⁸ "Has the rain a father,
    or who has begotten the drops of dew?
²⁹ From whose womb did the ice come forth,
    and who has given birth to the hoarfrost of heaven?
³⁰ The waters become hard like stone,
    and the face of the deep is frozen.

³¹ "Can you bind the chains of the Pleíades,
    or loose the cords of Orion?
³² Can you lead forth the Mazzaroth in their season,
    or can you guide the Bear with its children?
³³ Do you know the ordinances of the heavens?
    Can you establish their rule on the earth?

³⁴ "Can you lift up your voice to the clouds,
    that a flood of waters may cover you?
³⁵ Can you send forth lightnings, that they may go
    and say to you, 'Here we are'?
³⁶ Who has put wisdom in the clouds,
    or given understanding to the mists?
³⁷ Who can number the clouds by wisdom.
    Or who can tilt the waterskins of the heavens,
³⁸ when the dust runs into a mass
    and the clods cleave fast together?

Andrew Lawson (1861–1952) was born in Scotland, but his family moved to Canada while he was still a child. Pioneer worker on the geology of the Canadian shield, strong and contradictory in his character, he spent 55 years as a professor at Berkeley, and was a revered teacher of generations of students, who referred to him as "King." Amateur carpenter, legendary raconteur, and art connoisseur (he donated Rembrandts and Gainsboroughs in his $250,000 collection to the University of California), *he was also something of a poet.* Mente et Malleo, *written in 1888, was one of his earliest, though not one of his better, efforts, and was dedicated to the Logan Club, made up largely of members of the Geological Survey of Canada.*

MENTE ET MALLEO

Andrew C. Lawson

By thought and dint of hammering
Is the good work done whereof I sing,
And a jollier lot you'll rarely find,
Than the men who chip at earth's old rind,
And often wear a patched behind,
By thought and dint of hammering.

All summer through we're on the wing,
Kept moving by the skeeter's sting;
From Alaska unto Halifax,
With our compass and our little axe,
We make our way and pay our tax,
By thought and dint of hammering.

We crack the rocks and make them ring,
And many a heavy pack we sling;
We run our lines and tie them in,
We measure strata thick and thin,
And Sunday work is never sin,
By thought and dint of hammering.

Across the waters our paddles swing,
O'er wind and rapids triumphing;
Thro' mountain passes our slow mules trudge
As if they owed us a heavy grudge,
And often can't be got to budge
By thought and dint of hammering.

To the stars at night our thoughts we bring
But no maiden fair to our arm doth cling;
She, at Ottawa, with smiling lips,
The other fellow's ice cream sips;
You can't prevent these feminine slips
By thought and dint of hammering.

To array the "chiels that waunna ding"
Is our winter's work far into spring;
Some people think us wondrous wise;
Some maintain we're otherwise;
We're simply piercing Nature's guise
By thought and dint of hammering.

*John Milton (1608–1674), English poet and man of affairs, was educated
at Cambridge. He was born into a Catholic family, but converted to
Protestantism. Prolific as a tractarian (his* Areopagitica *is a powerful
defense of the freedom of the press, and he also wrote extensively on such
topics as divorce, Church government, and reform), he served as Latin
secretary in Cromwell's revolutionary government, being responsible for
the drafting and translation into Latin of diplomatic documents. The restoration of the monarchy led to his political downfall. He was arrested,
but escaped the execution which befell many of his colleagues. It was in
his later blindness and retirement that he wrote some of his greatest work.
His poem* Paradise Lost, *which dates from that period, is regarded as one
of the masterpieces of English epic poetry. The present extract describes,
in sublime symbolic language, the creation of the Earth.*

PARADISE LOST, BOOK VII

John Milton

Meanwhile the Son
On His great expedition now appeared,
Girt with omnipotence, with radiance crowned
Of Majesty divine, sapience and love
Immense, and all His Father in Him shone.
About His chariot numberless were poured
Cherub and Seraph, Potentates and Thrones,
And Virtues, wingèd Spirits, and Chariots winged
From the armoury of GOD, where stand of old
Myriads, between two brazen mountains lodged
Against a solemn day, harnessed at hand,
Celestial equipage; and now came forth
Spontaneous, for within them spirit lived,
Attendant on their Lord: heav'n opened wide
Her ever-during gates, harmonious sound
On golden hinges moving, to let forth
The King of glory, in His powerful Word
And Spirit coming to create new worlds.
On heav'nly ground they stood, and from the shore
They viewed the vast immeasurable abyss,
Outrageous as a sea, dark, wasteful, wild,
Up from the bottom turned by furious winds
And surging waves, as mountains, to assault
Heav'n's highth, and with the centre mix the pole.
   "'Silence, ye troubled waves, and, thou Deep, peace,'
Said then th' omnific Word; 'your discord end.'
   "Nor stayed; but, on the wings of Cherubim
Uplifted, in Paternal Glory rode

Far into Chaos and the world unborn;
For Chaos heard His voice. Him all His train
Followed in bright procession to behold
Creation, and the wonders of His might.
Then stayed the fervid wheels, and in His hand
He took the golden compasses, prepared
In GOD's eternal store, to circumscribe
This universe, and all created things.
One foot he centred, and the other turned
Round through the vast profundity obscure,
And said, 'Thus far extend, thus far thy bounds,
This be thy just circumference, O world.'
   "Thus GOD the heav'n created, thus the earth,
Matter unformed and void. Darkness profound
Covered th' Abyss; but on the watery calm
His brooding wings the Spirit of GOD outspread,
And vital virtue infused and vital warmth
Throughout the fluid mass, but downward purged
The black, tartareous, cold, infernal dregs,
Adverse to life: then founded, then conglobed
Like things to like; the rest to several place
Disparted, and between spun out the air,
And earth self-balanced on her centre hung.
   "'Let there be light,' said GOD, and forthwith light
Ethereal, first of things, quintessence pure,
Sprung from the deep, and from her native east
To journey through the aery gloom began,
Sphered in a radiant cloud, for yet the sun
Was not; she in a cloudy tabernacle
Sojourned the while. GOD saw the light was good;
And light from darkness by the hemisphere
Divided: light the Day, and darkness Night,
He named. Thus was the first day ev'n and morn:
Nor past uncelebrated, nor unsung
By the celestial quires, when orient light
Exhaling first from darkness they beheld,
Birth-day of heav'n and earth; with joy and shout
The hollow universal orb they filled,
And touched their golden harps, and hymning praised
GOD and His works, Creator Him they sung,
Both when first evening was, and when first morn.
   "Again GOD said, 'Let there be firmament
Amid the waters, and let it divide
The waters from the waters.' And GOD made
The firmament, expanse of liquid, pure,
Transparent, elemental air, diffused
In circuit to the uttermost convex

Of this great round; partition firm and sure,
The waters underneath from those above
Dividing: for as earth, so He the world
Built on circumfluous waters calm, in wide
Crystalline ocean, and the loud misrule
Of Chaos far removed, lest fierce extremes
Contiguous might distemper the whole frame:
And heav'n He named the firmament: so ev'n
And morning chorus sung the second day.
   "The earth was formed, but, in the womb as yet
Of waters embryon immature involved,
Appeared not: over all the face of earth
Main ocean flowed, not idle, but with warm
Prolific humour soft'ning all her globe
Fermented the great mother to conceive,
Satiate with genial moisture; when GOD said,
'Be gathered now, ye waters under heav'n,
Into one place, and let dry land appear.'
Immediately the mountains huge appear
Emergent, and their broad bare backs upheave
Into the clouds, their tops ascend the sky.
So high as heaved the tumid hills, so low
Down sank a hollow bottom broad and deep,
Capacious bed of waters: thither they
Hasted with glad precipitance, uprolled
As drops on dust conglobing from the dry:
Part rise in crystal wall, or ridge direct,
For haste; such flight the great command impressed
On the swift floods: as armies at the call
Of trumpet, for of armies thou hast heard,
Troop to their standard, so the watery throng,
Wave rolling after wave, where way they found;
If steep, with torrent rapture, if through plain,
Soft-ebbing: nor withstood them rock or hill,
But they, or under ground, or circuit wide
With serpent error wandering, found their way,
And on the washy oose deep channels wore,
Easy, ere GOD had bid the ground be dry,
All but within those banks, where rivers now
Stream, and perpetual draw their humid train.
The dry land, Earth; and the great receptacle
Of congregated waters He called Seas;
And saw that it was good, and said, 'Let the earth
Put forth the verdant grass, herb yielding seed,
And fruit-tree yielding fruit after her kind;
Whose seed is in herself upon the earth.'
He scarce had said, when the bare earth, till then

Desert and bare, unsightly, unadorned,
Brought forth the tender grass, whose verdure clad
Her universal face with pleasant green;
Then herbs of every leaf, that sudden flowered,
Opening their various colours, and made gay
Her bosom smelling sweet: and these scarce blown,
Forth flourished thick the clustering vine, forth crept
The swelling gourd, up stood the corny reed
Embattled in her field; and the humble shrub,
And bush with frizzled hair implicit: last
Rose, as in dance, the stately trees, and spread
Their branches hung with copious fruit, or gemmed
Their blossoms: with high woods the hills were crowned,
With tufts the valleys and each fountain side:
With borders long the rivers: that earth now
Seemed like to heav'n, a seat where gods might dwell,
Or wander with delight, and love to haunt
Her sacred shades: though GOD had yet not rained
Upon the earth, and man to till the ground
None was; but from the earth a dewy mist
Went up and watered all the ground, and each
Plant of the field; which, ere it was in the earth,
GOD made, and every herb, before it grew
On the green stem: GOD saw that it was good:
So ev'n and morn recorded the third day.

Kenneth Rexroth, self-taught, contemporary American poet, was born in 1905, and played a major role in the San Francisco literary movement of the last thirty years. Widely read and admired, he is represented here by a short poem based on the subtitle of Lyell's Principles. *The poem reflects on the supremacy of a present relationship between two people, insulated for a moment from the constraints of the "vast impersonal vindictiveness of the ruined and ruining world." Fossils, lava flows, waterfalls, and ice ages—all converge here in an intimate relationship, and "....ideograms printed on the immortal hydrocarbons of flesh and stone".*

LYELL'S HYPOTHESIS AGAIN
Kenneth Rexroth

The mountain road ends here,
Broken away in the chasm where
The bridge washed out years ago.

The first scarlet larkspur glitters
In the first patch of April
Morning sunlight. The engorged creek
Roars and rustles like a military
Ball. Here by the waterfall,
Insuperable life, flushed
With the equinox, sentient
And sentimental, falls away
To the sea and death. The tissue
Of sympathy and agony
That binds the flesh in its Nessus' shirt;
The clotted cobweb of unself
And self; sheds itself and flecks
The sun's bed with darts of blossom
Like flagellant blood above
The water bursting in the vibrant
Air. This ego, bound by personal
Tragedy and the vast
Impersonal vindictiveness
Of the ruined and ruining world,
Pauses in this immortality,
As passionate, as apathetic,
As the lava flow that burned here once;
And stopped here; and said, "This far
And no further." And spoke thereafter
In the simple diction of stone.

Naked in the warm April air,
We lie under the redwoods,
In the sunny lee of a cliff.
As you kneel above me I see
Tiny red marks on your flanks
Like bites, where the redwood cones
Have pressed into your flesh.
You can find just the same marks
In the lignite in the cliff
Over our heads. *Sequoia
Langsdorfii* before the ice,
And *sempervirens* afterwards,
There is little difference,
Except for all those years.

Here in the sweet, moribund
Fetor of spring flowers, washed,
Flotsam and jetsam together,
Cool and naked together,
Under this tree for a moment,
We have escaped the bitterness

Of love, and love lost, and love
Betrayed. And what might have been,
And what might be, fall equally
Away with what is, and leave
Only these ideograms
Printed on the immortal
Hydrocarbons of flesh and stone.

The Earth

Touch the earth, love the earth, honour the earth, her plains, her valleys, her hills, and her seas; rest your spirit in her solitary places.

—Henry Beston

CHAPTER 9

Geology and Prose

The dictionary tells us that prose is the ordinary language of writing or speaking, as opposed to verse or poetry. Samuel Coleridge (1772–1834) in his lectures on Shakespeare and Milton distinguished between prose and poetry when he observed, "I wish our clever young poets would remember my homely definitions of prose and poetry; that is, prose—words in their best order; poetry—the best words in their best order." In this section we include a group of prose selections, all concerned in one way or another with the Earth, by writers who show how to use words in their best order.

The examples given are diverse, ranging from the humor of Samuel Clemens to the taut writing of John Steinbeck, who tells of adventures in the Sea of Cortez. We also include excerpts from St. Exupéry, author of *The Little Prince*, who describes spectacular scenery viewed from the air, and T. E. Lawrence who takes the reader across the rugged terrain of Middle Eastern deserts and imparts a feeling of mysticism for these arid lands. The roll call of prose writers who write of geology and geologists is a lengthy one. In our brief collection of samples, Nobel and Pulitzer Prize-winning novelists are included, together with best-selling authors of the 1970s and giants of 200 years ago.

On June 1, 1898, Sir Archibald Geikie delivered the Romanes Lecture in the Sheldonian Theatre at Oxford University. His subject was "Types of Scenery and Their Influence on Literature." He classified the scenery of Britain into Lowlands, Uplands, and Highlands, and showed how each of these landscapes had influenced the work of a number of poets, including John Milton, William Cowper, James Thompson, Robert Burns, James Macpherson, Sir Walter Scott, and William Wordsworth. The complete essay is recommended to anyone interested in poetry or geology. Unfortunately, there is only space in our book for a single writer described by Geikie. Our selection is of the poet who was born, worked, and lived in the Lake District and who wrote of it with affection, William Wordsworth.

LANDSCAPE AND LITERATURE

Sir Archibald Geikie

The several races of mankind are marked off from each other by certain bodily and mental differences which, there can be no doubt, have been largely determined by the diverse geographical conditions of the surface of the globe. We may be disposed to put aside the first origin of these racial differences, as a problem which may for ever remain insoluble. Yet we cannot refuse to admit that in the disposition of land and sea, in the form and trend of coast-lines, in the grouping of mountains, valleys, and plains, in the disposition and flow of rivers, in the arrangement of climates, and in the distribution of vegetation and animals, a series of influences must be recognized which have unquestionably played a large part in the successive stages of human development.

Though no record of the earliest of these stages has probably survived, some of the later steps in the progress of advancement may not be beyond the reach of investigation. The connection, for example, between the circumscribing topography and geology of a country, and the mythological creed of its inhabitants offers a tempting field of inquiry, in which much may yet be gleaned. Who can doubt that the legends and superstitions of ancient Greece took their form and colour in no small measure from the mingled climates, varied scenery, and rocky structure of that mountainous land, or that the grim titanic mythology of Scandinavia bears witness to its birth in a region of rugged snowy uplands, under gloomy and tempestuous skies?

If the primeval efforts of the human imagination were thus stimulated by the more impressive features of the outer world, it is natural to believe that the same external influences would continue to exert their power during the later mental development of a people. In particular, it seems reasonable to anticipate that such potent causes would more or less make themselves felt in the growth of a national literature. The songs and ballads of the plains might be expected to present some marked diversities from those of the mountains. There may, of course, be risks of error in generalisations of this kind, especially where the writings of distinct races are compared with each other. But the risks may be reduced if we confine ourselves to the consideration of a single country and a single literature. Such at least is the task which I have undertaken on the present occasion.

One mountainous district in Britain—that of the English Lakes—claims our attention for its influence on the progress of the national literature. Of all the isolated tracts of higher ground in these islands, that of the Lake District is the most eminently highland in character. It is divisible into two entirely distinct portions by a line drawn in a north-easterly direction from Duddon Sands to Shap Fells. South of that line the hills are comparatively low and featureless, though they enclose the largest of the lakes. They are there built up of ancient sedimentary strata, like those that form so much of the similar scenery in the uplands of Wales and the South of Scotland. But to the north of the line, most of the rocks are of a different nature, and have given rise to a totally distinct character of landscape. They consist of various volcanic materials which in early Palaeozoic time were piled up around submarine vents, and accumulated over the sea-floor to a thickness of many thousand feet. They were subsequently buried under the sediments that lie to the south, but, in after ages uplifted into land, their now diversified topography has been carved out of them by the meteoric agents of denudation. Thus pike and fell, crag and scar, mere and dale, owe their several forms to the varied degrees of resistance to the general waste offered by the ancient lavas and ashes. The upheaval of the district seems to have produced a dome-shaped elevation, culminating in a summit that lay somewhere between Helvellyn and Grasmere. At least from that centre the several dales diverge, like the ribs from the top of a half-opened umbrella.

The mountainous tract of the Lakes, though it measures only some thirty-two miles from west to east by twenty-three from north to south, rises to heights of more than 3,000 feet, and as it springs almost directly from the margin of the Irish Sea, it loses none of the full effect of its elevation. Its fells present a thoroughly highland type of scenery, and have much of the dignity of far loftier mountains. Their sky-line often displays notched crests and rocky peaks, while their craggy sides have been carved into dark cliff-girt recesses, often filled with tarns, and into precipitous scars, which send long trails of purple scree down the grassy slopes.

Moreover, a mild climate and copious rainfall have tempered this natural asperity of surface by spreading over the lower parts of the fells and the bottoms of the dales a greener mantle than is to be seen among the mountains further north. Though the naked rock abundantly shows itself, it has been so widely draped with herbage and woodland as to combine the luxuriance of the lowlands with the near neighbourhood of bare cliff and craggy scar.

Such was the scenery amidst which William Wordsworth was born and spent most of his long life. Thence he drew the inspiration which did so much to quicken the English poetry of the nineteenth century, and which has given to his dales and hills so cherished a place in our literature. The scenes familiar to him from infancy were loved by him to the end with an ardent and grateful affection which he never wearied of publishing to the world. No mountain-landscapes had ever before been drawn so fully, so accurately, and in such felicitous language. Every lineament of his hills and dales is depicted as luminously and faithfully in his verse as it is reflected on the placid surface of his beloved meres, but suffused by him with an ethereal glow of human sympathy. He drew from his mountain-landscape everything that

> 'Can give an inward help, can purify
> And elevate, and harmonize and soothe.'

It brought to him 'authentic tidings of invisible things'; and filled him with

> 'The sense
> Of majesty and beauty and repose,
> A blended holiness of earth and sky.'

For his obligations to that native scenery he found continual expression.

> 'Ye mountains and ye lakes,
> And sounding cataracts, ye mists and winds
> That dwell among the hills where I was born,
> If in my youth I have been pure in heart,
> If, mingling with the world, I am content
> With my own modest pleasures, and have lived
> With God and Nature communing, removed
> From little enmities and low desires—
> The gift is yours.'

Not only did his observant eye catch each variety of form, each passing tint of colour on his hills and valleys, he felt, as no poet before his time had done, the might and majesty of the forces by which, in the mountain-world, we are shown how the surface of the world is continually modified.

> 'To him was given
> Full many a glimpse of Nature's processes
> Upon the exalted hills.'

The thought of these glimpses led to one of the noblest outbursts in the whole range of his poetry, where he gives way to the exuberance of his delight in feeling himself, to use Byron's expression, 'a portion of the tempest'—

> 'To roam at large among unpeopled glens
> And mountainous retirements, only trod

By devious footsteps; regions consecrate
To oldest time; and reckless of the storm,
. . . . . . while the mists
Flying, and rainy vapours, call out shapes
And phantoms from the crags and solid earth,
. . . . . . and while the streams
Descending from the region of the clouds,
And starting from the hollows of the earth,
More multitudinous every moment, rend
Their way before them—what a joy to roam
An equal among mightiest energies!'

In this passage Wordsworth seems to have had what he would have called 'a foretaste, a dim earnest' of that marvellous enlargement of the charm and interest of scenery due to the progress of modern science. When he speaks of 'regions consecrate to oldest time,' he has a vague feeling that somehow his glens and mountains belonged to a hoary antiquity, such as could be claimed by none of the verdant plains around. Had he written half a century later he would have enjoyed a clearer perception of the vastness of that antiquity and of the long succession of events with which it was crowded.

Limestone

I had three pieces of limestone on my desk, but I was terrified to find that they required to be dusted daily, when the furniture of my mind was all undusted still, and I threw them out the window in disgust.

—Henry Thoreau

Early in his career, the controversial and prominent humorist and satirist, Mark Twain (1835–1910) traveled extensively in the Western states and worked as a miner and journalist. From 1861 to 1864 he was assistant to his brother who was the territorial secretary of Nevada, and he also worked as a staff writer and reporter for the rip-roaring newspaper of Virginia City, Nevada, the Virginia Enterprise.

His first book, The Jumping Frog of Calaveras County, *was published in 1867. In one of his most famous books,* Roughing It, *a chapter or two is devoted to a visit to Mono Lake, California. In the excerpt chosen, Twain considers with humor and aplomb what he termed "this lonely tenant of the loneliest spot on earth."*

ROUGHING IT

Mark Twain

Mono Lake lies in a lifeless, treeless, hideous desert, eight thousand feet above the level of the sea, and is guarded by mountains two thousand feet higher, whose summits are always clothed in clouds. This solemn, silent, sailless sea—this lonely tenant of the loneliest spot on earth—is little graced with the picturesque. It is an unpretending expanse of grayish water, about a hundred miles in circumference, with two islands in its center, mere upheavals of rent and scorched and blistered lava, snowed over with gray banks and drifts of pumice-stone and ashes, the winding sheet of the dead volcano, whose vast crater the lake has seized upon and occupied.

The lake is two hundred feet deep, and its sluggish waters are so strong with alkali that if you only dip the most hopelessly soiled garment into them once or twice, and wring it out, it will be found as clean as if it had been through the ablest of washer-women's hands. While we camped there our laundry work was easy. We tied the week's washing astern of our boat, and sailed a quarter of a mile, and the job was complete, all to the wringing out.

Mono Lake is a hundred miles in a straight line from the ocean—and between it and the ocean are one or two ranges of mountains—yet thousands of sea-gulls go there every season to lay their eggs and rear their young. One would as soon expect to find sea-gulls in Kansas. And in this connection let us observe another instance of Nature's wisdom. The islands in the lake being merely huge masses of lava, coated over with ashes and pumice-stone, and utterly innocent of vegetation or anything that would burn; and sea-gulls' eggs being entirely useless to anybody unless they be cooked, Nature has provided an unfailing spring of boiling water on the largest island, and you can put your eggs in there, and in four minutes you can boil them as hard as any statement I have made during the past fifteen years. Within ten feet of the boiling spring is a spring of pure cold water, sweet and wholesome.

So, in that island you get your board and washing free of charge—and if nature had gone further and furnished a nice American hotel clerk who was crusty and disobliging, and didn't know anything about the time tables, or the railroad routes—or—anything—and was proud of it—I would not wish for a more desirable boarding-house.

Half a dozen little mountain brooks flow into Mono Lake, but *not a stream of any kind flows out of it.* It neither rises nor falls, apparently, and what it does with its surplus water is a dark and bloody mystery.

There are only two seasons in the region round about Mono Lake—and these are, the breaking up of one winter and the beginning of the next.

About seven o'clock one blistering hot morning—for it was now dead summer time—Higbie and I took the boat and started on a voyage of discovery to the two islands. We had often longed to do this, but had been deterred by the fear of storms; for they were frequent, and severe enough to capsize an ordinary rowboat like ours without great difficulty—and once capsized, death would ensue in spite of the bravest swimming, for that venomous water would eat a man's eyes out like fire, and burn him out inside, too, if he shipped a sea. It was called twelve miles, straight out to the islands—a long pull and a warm one—but the morning was so quiet and sunny, and the lake so smooth and glassy and dead, that we could not resist the temptation. So we filled two large tin canteens with water (since we were not acquainted with the locality of the spring said to exist on the large island), and started. Higbie's brawny muscles gave the boat good speed, but by the time we reached our destination we judged that we had pulled nearer fifteen miles than twelve.

We landed on the big island and went ashore. We tried the water in the canteens, now, and found that the sun had spoiled it; it was so brackish that we could not drink it; so we poured it out and began a search for the spring—for thirst augments fast as soon as it is apparent that one has no means at hand of quenching it. The island was a long, moderately high hill of ashes—nothing but gray ashes and pumice-stone, in which we sunk to our knees at every step—and all around the top was a forbidding wall of scorched and blasted rocks. When we reached the top and got within the wall, we found simply a shallow, far-reaching basin, carpeted with ashes, and here and there a patch of fine sand. In places, picturesque jets of steam shot up out of crevices, giving evidence that although this ancient crater had gone out of active business, there was still some fire left in its furnaces. Close to one of these jets of steam stood the only tree on the island—a small pine of most graceful shape and most faultless symmetry; its color was a brilliant green, for the steam drifted unceasingly through its branches and kept them always moist. It contrasted strangely enough, did this vigorous and beautiful outcast, with its dead and dismal surroundings. It was like a cheerful spirit in a mourning household.

We hunted for the spring everywhere, traversing the full length of the island (two or three miles), and crossing it twice—climbing ash-hills patiently, and then sliding down the other side in a sitting posture, plowing up smothering volumes of gray dust. But we found nothing but solitude, ashes, and a heart-breaking silence. Finally we noticed that the wind had risen, and we forgot our thirst in a solicitude of greater importance; for, the lake being quiet, we had not taken pains about securing the boat. We hurried back to a point overlooking our landing place, and then—but mere words cannot describe our dismay—the boat was gone! The chances were that there was not another boat on the entire lake. The situation was not comfortable—in truth, to speak plainly, it was frightful. We were prisoners on a desolate island, in aggravating proximity to friends who were for the present helpless to aid us; and what was still more uncomfortable was the reflection that we had neither food nor water. But presently we sighted the boat. It was drifting along, leisurely, about fifty yards from shore, tossing in a foamy sea. It drifted, and continued to drift, but at the same safe distance from land, and we walked along abreast it and waited for fortune to favor us. At the end of an hour it approached a jutting cape, and Higbie ran ahead and posted himself on the utmost verge and prepared for the assault. If we failed there, there was no hope for us. It was driving gradually shoreward all the time, now; but whether it was driving fast enough to make the connection or not was the momentous question. When it got within thirty steps of Higbie I was so excited that I fancied I could hear my own heart beat. When, a little later, it dragged slowly along and seemed about to go by, only one little yard out of reach, it seemed as if my heart stood still; and when it was exactly abreast him and began to widen away, and he still standing like a watching statue, I knew my heart did stop. But when he gave a great spring, the next instant, and lit fairly in the stern, I discharged a war-whoop that awoke the solitudes!

But it dulled my enthusiasm, presently, when he told me he had not been caring whether the boat came within jumping distance or not, so that it passed within eight or ten yards of him, for he had made up his mind to shut his eyes and mouth and swim that trifling distance. Imbecile that I was, I had not thought of that. It was only a long swim that could be fatal.

Tbe sea was running high and the storm increasing. It was growing late, too—three or four in the afternoon. Whether to venture toward the mainland or not, was a question of some moment. But we were so distressed by thirst that we decided to try it, and so Higbie fell to work and I took the steering-oar. When we had pulled a mile, laboriously, we were evidently in serious peril, for the storm had greatly augmented; the billows ran very high and were capped with foaming crests, the heavens were hung with black, and the wind blew with great fury. We would have gone back, now, but we did not dare to turn the boat around, because as soon as she got in the trough of the sea she would upset, of course. Our only hope lay in keeping her head-on to the seas. It was hard work to do this, she plunged so, and so beat and belabored the billows with her rising and falling bows. Now and then one of Higbie's oars would trip on the top of a wave, and the other one would snatch the boat half around in spite of my cumbersome steering apparatus. We were drenched by the sprays constantly, and the boat occasionally shipped water. By and by, powerful as my comrade was, his great exertions began to tell on him, and he was anxious that I should change places with him till he could rest a little. But I told him this was impossible; for if the steering oar were dropped a moment while we changed, the boat would slue around into the trough of the sea, capsize, and in less than five minutes we would have a hundred gallons of soapsuds in us and be eaten up so quickly that we could not even be present at our own inquest.

But things cannot last always. Just as the darkness shut down we came booming into port, head on. Higbie dropped his oars to hurrah—I dropped mine to help—the sea gave the boat a twist, and over she went!

The agony that alkali water inflicts on bruises, chafes, and blistered hands, is unspeakable, and nothing but greasing all over will modify it—but we ate, drank, and slept well, that night, notwithstanding.

In speaking of the peculiarities of Mono Lake, I ought to have mentioned that at intervals all around its shores stand picturesque turret-looking masses and clusters of a whitish, coarse-grained rock that resembles inferior mortar dried hard; and if one breaks off fragments of this rock he will find perfectly shaped and thoroughly petrified gulls' eggs deeply imbedded in the mass. How did they get there? I simply state the fact—for it is a fact—and leave the geological reader to crack the nut at his leisure and solve the problem after his own fashion.

At the end of a week we adjourned to the Sierras on a fishing excursion, and spent several days in camp under snowy Castle Peak, and fished successfully for trout in a bright, miniature lake whose surface was between ten and eleven thousand feet above the level of the sea; cooling ourselves during the hot August noons by sitting on snowbanks ten feet deep, under whose sheltering edges *fine grass and dainty flowers flourished luxuriously;* and at night entertaining ourselves by almost freezing to death. Then we returned to Mono Lake, and finding that the cement excitement was over for the present, packed up and went back to Esmeralda. Mr. Ballou reconnoitered awhile, and not liking the prospect, set out alone for Humboldt.

Between 1860 and 1864, William Brewer travelled extensively in California as the first assistant to James D. Whitney, Director of the Geological Survey of California. A Yale graduate and a member of the first class at the Sheffield Scientific School, Brewer was by training an agricultural chemist. He later taught for many years at Yale.

Brewer published the results of his explorations and geological investigations of the western frontier area in an engaging book, Up and Down California, 1860–1864. *Our sample from Brewer is of a visit to Mono Lake, a saline body of water in eastern California. For another view of the same area and for a contrast in style and interpretation, compare the description of the inimitable Mark Twain.*

MONO LAKE—AURORA—SONORA PASS

William H. Brewer

We descended safely, and camped in the high grass and weeds by a stream a short distance south of Lake Mono. This camp had none of the picturesque beauty of our mountain camps, and a pack of coyotes barked and howled around us all night.

July 8 Hoffmann and I visited a chain of extinct volcanoes which stretches south of Lake Mono. They are remarkable hills, a series of truncated cones, which rise about 9,700 feet above the sea. Rock peeps out in places, but most of the surface is of dry, loose, volcanic ashes, lying as steep as the material will allow. The rocks of these volcanoes are a gray lava, pumice stone so light that it will float on water, obsidian or volcanic glass, and similar volcanic products. It was a laborious climb to get to the summit. We sank to the ankles or deeper at every step, and slid back most of each step. But it was easy enough getting down—one slope that took three hours to ascend we came down leisurely in forty-five minutes. The scene from the top is desolate enough—barren volcanic mountains standing in a desert cannot form a cheering picture. Lake Mono, that American "Dead Sea," lies at the foot. Between these hills and our camp lie about six miles of desert, which is very tedious to ride over—dry sand, with pebbles of pumice, supporting a growth of crabbed, dry sagebrushes, whose yellow-gray foliage does not enliven the scene.

July 9 we came on about ten miles north over the plain and camped at the northwest corner of Lake Mono. This is the most remarkable lake I have ever seen. It lies in a basin at the height of 6,800 feet above the sea. Like the Dead Sea, it is without an outlet. The streams running into it all evaporate from the surface, so of course it is very salt—not common salt. There are hot springs in it, which feed it with peculiar mineral salts. It is said that it contains borax, also boracic acid, in addition to the materials generally found in saline lakes. I have bottled water for analysis and hope to know some time.

July 11 we were up at dawn, a clear, calm morning. Clouds of gulls screamed around us. An early breakfast, then a tour of examination. These islands are entirely volcanic, and in one place the action can hardly be said to have ceased, for there are hundreds of hot springs over a surface of many acres. Steam and hot gases issue from fissures in the rocks, and one can hear the boiling and gurgling far beneath. Some of the springs are very copious, discharging large quantities of hot water with a very peculiar odor. Some boil up in the lake, near the shore, so large that the lake is warmed for many rods—no wonder that the waters hold such strange mineral ingredients. The rock is all lava, pumice, and cinders. At the northeast corner of the island are two old craters with water in them. The smaller, or north island, has no fresh water—it looks scathed and withered by fire. One volcanic cone, three hundred or four hundred feet high, looked more recent than any other I have seen in the state.

Water

Nothing under heaven is softer or more yielding than water, but when it attacks things hard and resistant there is not one of them that can prevail.

—"Tao Te Ching"

That T. E. Lawrence (1888–1935), archeologist, scholar, soldier, adventurer, writer, and lover of the desert should have a poetic sense for landscape is not unexpected. Lawrence first visited the Middle East in 1909 when he went to Syria and Palestine to study the architecture of Crusader castles. He returned as an officer in the British Army during World War I, where he organized Arab groups in the desert fighting with Turkey. He published an account of his wartime experiences, Seven Pillars of Wisdom, *which is his greatest work. The account is laden with descriptions of dunes, desert pavement, oases, wadis, lava fields, watersheds, and escarpments. We include a passage from Lawrence, in which he relates the crossing of a rough lava plateau.*

SEVEN PILLARS OF WISDOM

T. E. Lawrence

Dawn found us crossing a steep short pass out of Wadi Kitan into the main drainage valley of these succeeding hills. We turned aside into Wadi Reimi, a tributary, to get water. There was no proper well, only a seepage hole in the stony bed of the valley; and we found it partly by our noses: though the taste, while as foul, was curiously unlike the smell. We refilled our waterskins. Arslan baked bread, and we rested for two hours. Then we went on through Wadi Amk, an easy green valley which made comfortable marching for the camels.

When the Amk turned westward we crossed it, going up between piles of the warped grey granite (like cold toffee) which was common up-country in the Hejaz. The defile culminated at the foot of a natural ramp and staircase: badly broken, twisting, and difficult for camels, but short. Afterwards we were in an open valley for an hour, with low hills to the right and mountains to the left. There were water pools in the crags, and Merawin tents under the fine trees which studded the flat. The fertility of the slopes was great: on them grazed flocks of sheep and goats. We got milk from the Arabs: the first milk my Ageyl had been given in the two years of drought.

The track out of the valley when we reached its head was execrable, and the descent beyond into Wadi Marrakh almost dangerous; but the view from the crest compensated us. Wadi Marrakh, a broad, peaceful avenue, ran between two regular straight walls of hills to a circus four miles off where valleys from left, right and front seemed to meet. Artificial heaps of uncut stone were piled about the approach. As we entered it, we saw that the grey hill-walls swept back on each side in a half-circle. Before us, to the south, the curve was barred across by a straight wall or step of blue-black lava, standing over a little grove of thorn trees. We made for these and lay down in their thin shade, grateful in such sultry air for any pretence of coolness.

The day, now at its zenith, was very hot; and my weakness had so increased that my head hardly held up against it. The puffs of feverish wind pressed like scorching hands against our faces, burning our eyes. My pain made me breathe in gasps through the mouth; the wind cracked my lips and seared my throat till I was too dry to talk, and drinking became sore; yet I always needed to drink, as my thirst would not let me lie still and get the peace I longed for. The flies were a plague.

The bed of the valley was of fine quartz gravel and white sand. Its glitter thrust itself between our eyelids; and the level of the ground seemed to dance as the wind moved the white tips of stubble grass to and fro. The camels loved this grass, which grew in tufts, about sixteen inches high, on slate-green stalks. They gulped down great quantities of it until the men drove them in and couched them by me. At the moment I hated the beasts, for too much food made their breath stink; and they rumblingly belched up a new mouthful from their stomachs each time they had chewed and swallowed the last, till a green slaver flooded out between their loose lips over the side teeth, and dripped down their sagging chins.

Lying angrily there, I threw a stone at the nearest, which got up and wavered about behind my head; finally it straddled its back legs and staled in wide, bitter jets; and I was so far gone with the heat and weakness and pain that I just lay there and cried about it unhelping. The men had gone to make a fire and cook a gazelle one of them had fortunately shot; and I realized that on another day this halt would have been pleasant to me; for the hills were very strange and their colours vivid. The base had the warm grey of old stored sunlight; while about their crests ran narrow veins of granite-coloured stone, generally in pairs, following the contour of the skyline like the rusted metals of an abandoned scenic railway. Arslan said the hills were combed like cocks, a sharper observation.

After the men had fed we re-mounted, and easily climbed the first wave of the lava flood. It was short, as was the second, on the top of which lay a broad terrace with an alluvial plot of sand and gravel in its midst. The lava was a nearly clean floor of iron-red rock-cinders, over which were scattered fields of loose stone. The third and other steps ascended to the south of us: but we turned east, up Wadi Gara.

Gara had, perhaps, been a granite valley down whose middle the lava had flowed, slowly filling it, and arching itself up in a central heap. On each side were deep troughs, between the lava and the hill-side. Rain water flooded these as often as storms burst in the hills. The lava flow, as it coagulated, had been twisted like a rope, cracked, and bent back irregularly upon itself. The surface was loose with fragments through which many generations of camel parties had worn an inadequate and painful track.

We struggled along for hours, going slowly, our camels wincing at every stride as the sharp edges slipped beneath their tender feet. The paths were only to be seen by the droppings along them, and by the slightly bluer surfaces of the rubbed stones. The Arabs declared them impassable after dark, which was to be believed, for we risked laming our beasts each time our impatience made us urge them on. Just before five in the afternoon, however, the way got easier. We seemed to be near the head of the valley, which grew narrow. Before us on the right, an exact cone-crater, with tidy furrows scoring it from lip to foot, promised good going; for it was made of black ash, clean as though sifted, with here and there a bank of harder soil, and cinders. Beyond it was another lava-field, older perhaps than the valleys, for its stones were smoothed, and between them were straths of flat earth, rank with weeds. In among these open spaces were Beduin tents, whose owners ran to us when they saw us coming; and, taking our head-stalls with hospitable force, led us in.

They proved to be Sheikh Fahad el Hansha and his men: old and garrulous warriors who had marched with us to Wejh, and had been with Garland on that great occasion when his first automatic mine had succeeded under a troop train near Toweira station. Fahad would not hear of my resting quietly outside his tent, but with the reckless equality of the desert men urged me into an unfortunate place inside among his own vermin. There he plied me with bowl after bowl of diuretic camel-milk between questions about Europe, my home tribe, the English camel-pasturages, the war in the Hejaz and the wars elsewhere, Egypt and Damascus, how Feisal was, why did we seek Abdulla, and by what perversity did I remain Christian, when their hearts and hands waited to welcome me to the Faith?

In the morning we woke early and refreshed, with our clothes stinging-full of fiery points feeding on us. After one more bowl of milk proffered us by the eager Fahad, I was able to walk unaided to my camel and mount her actively. We rode up the last piece of Wadi Gara to the crest, among cones of black cinders from a crater to the south. Thence we turned to a branch valley, ending in a steep and rocky chimney, up which we pulled our camels.

Beyond we had an easy descent into Wadi Murrmiya, whose middle bristled with lava like galvanized iron, on each side of which there were smooth sandy beds, good going. After a while we came to a fault in the flow, which served as a track to the other side. By it we crossed over, finding the lava pocketed with soils apparently of extreme richness, for in them were leafy trees and lawns of real grass, starred with flowers, the best grazing of all our ride, looking the more wonderfully green because of the blue-black twisted crusts of rock about. The lava had changed its character. Here were no piles of loose stones, as big as a skull or a man's hand, rubbed and rounded together; but bunched and crystallized fronds of metallic rock, al-together impassable for bare feet.

Another watershed conducted us to an open place where the Jeheina had ploughed some eight acres of the thin soil below a thicket of scrub. They said there were like it in the neighbourhood other fields, silent witnesses to the courage and persistence of the Arabs. It was called Wadi Chetf, and after it was another broken river of lava, the worst yet encountered. A shadowy path zigzagged across it. We lost one camel with a broken fore-leg, the result of a stumble in a pot-hole; and the many bones which lay about showed that we were not the only party to suffer misfortune in the passage. However, this ended our lava, according to the guides, and we went thence forward along easy valleys with finally a long run up a gentle slope till dusk. The going was so good and the cool of the day so freshened me that we did not halt at nightfall, after our habit, but pushed on for an hour across the basin of Murrmiya into the basin of Wadi Ais, and there, by Tleih, we stopped for our last camp in the open.

Time

If I were asked as a geologist what is the single greatest contribution of the science of geology to modern civilized thought, the answer would be the realization of the immense length of time. So vast is the span of time recorded in the history of the earth that it is generally distinguished from the more modest kinds of time by being called "geologic time."

—Adolph Knopf

No anthology of the Earth could be considered complete without an extract from the writings of John Muir, mountain man, environmentalist (before they became known as such), fighter for National Parks, friend of Presidents, founder of the Sierra Club—a man who understood the beauty of solitude. Born in Scotland, Muir (1838–1914) attended the University of Wisconsin but never graduated. With a degree of financial independence, Muir spent most of his life in the mountains of California. In addition to his many journeys afoot and mostly alone in the then little known Sierras of California, John Muir also visited Alaska and in 1880 the Arctic.

Our selection is a delightful excerpt from "The Yosemite" in which the Ritter glacier is poetically described. John Muir not only documents the large and spectacular landforms, glaciers, and rugged mountain crests, but he exhibits the naturalist's eye for smaller features as rocks, plants, clouds, and birds as well. His powers of description and his fundamental knowledge of geology are clearly shown in the passages that are presented here.

TRIP TO THE MIDDLE AND NORTH FORKS OF SAN JOAQUIN RIVER

John Muir

> Camp at head of main
> Joaquin Canyon.
> Altitude 9800 feet.

Here the river divides, one fork coming in from the north and the other from Ritter. The Ritter fork comes down the mountain-side here in a network of cascades, wonderfully woven, as are all slate cascades of great size near summits, when the slate has a cleavage well pronounced. The mountains rise in a circle, showing their grand dark bosses and delicate spires on the starry sky. Down the canyon a company of sturdy, long-limbed mountain pines show nobly. All the rest of the horizon is treeless, because moraineless. How fully are all the forms and languages of waters, winds, trees, flowers, birds, rocks, subordinated to the primary structure of the mountains ere they were ice-sculptured! When all was planned and ready, snow-flowers were dusted over them, forming a film of ice over the mountain plate. And so all this development—the photography of God.

Linnæus says Nature never leaps, which means that God never shouts or spouts or speaks incoherently. The rocks and sublime canyons, and waters and winds, and all life structures—animals and ouzels, meadows and groves, and all the silver stars—are words of God, and they flow smooth and ripe from his lips.

The branches of no pine have so long a downward swooping curve as those of *Pinus monticola*. The ends are tufted and conspicuous, the bark redder the higher you go. And it is nobler in form, the colder and balder the rocks and mountains about it.

Never saw so many large gentians as today (*Gentiana frigida*). I counted thirty-three flowers on a patch not much larger than my hand—some flowers one and a quarter inches in diameter.

Ursa Major nearly horizontal, and has gone to rest. So must I. I am in a small grove of mountain hemlock. Their feathery boughs are extended above my head like hands of gentle spirits. Good night to God and His stars and mountains.

August 21 (?).

Clouds, dense and black, come from the southwest, over the black crests and peaks of Ritter, the lowest torn and raked to shreds. The cold wind is tuned to the opaque, somber sky. Now a little rain, snow, and hail.

I wish to cross to the east side of Ritter today, visiting the main north glacier on my way. I have never crossed these summits, and fear to try on so dark a day. . . . At noon, the clouds breaking, I decide to make the attempt, climbing up the fissured and flowery edge of the rock near Ritter Cascade. At the top of the Cascade I find a grand amphitheater, where glaciers from Ritter and peaks to the north once congathered—now meadow and lake and red and black walls with wild undressed falls and rapids and a few hardy flowers. From the rim of this I have a glorious view of high fountain glaciers and névés, with pointed spears and tapering towers and spires innumerable. Here, too, I find a white stemless thistle, larger than the purple one of Mono.

Yet higher, following the wild streams and climbing around inaccessible gorges, at length I am in sight of the lofty top crest of Ritter and know I am exactly right. The clouds grow dark again and send hail, but I know the rest of my way too well to fear. I can push down to the tree-line if need be, in any kind of storm. A glacier which I had not before seen heaves in sight. It is on the north side of the highest of the many spires that run off from Ritter Peak as a center. It is one of the best I've seen in this wild region, occupying a most delicately curved basin and discharging into a lake.

I come in sight of the main North Ritter Glacier, its snout projecting from a narrow opening gap into a lake about three hundred yards in diameter. Its waters are intensely blue. The basin was excavated by the Ritter Glacier and those of Banner Peak to the north at the time of their greater extension. The glacier enters on the east side. On the west is a splendid frame of pure white névé, abounding in deep caves with arched openings. One of these has a span of forty feet, and a height of thirty. This frame of névé comes down to the water's edge and in places reaches out over the lake. Most of its face is made precipitous by sections breaking off into the lake after the manner of icebergs. These are snowbergs. The strength of the glacier itself is so nearly spent ere it reaches the lake, it does not break off in bergs, but melts with many a rill, giving a network of sweet music.

Wishing to reach the head of this glacier, I follow up the left lateral moraine that extends from the flank of the mountain to the lake until I reach the top. Then I try to cross the glacier itself where it is not so steep. But I find it bare, unwalkable ice. The least slip would send me down a slope of thirty-five degrees, a distance of three hundred yards, into the lake. I begin to cut steps, but the snout of the glacier is so hard, and every step has to be so perfect, that I make slow and wearisome progress with blankets on my back. I soon realize it will be dark ere I succeed in crossing, therefore I resolve to retrace my steps and attempt crossing at the foot of the glacier close to the lake where it is narrower, and if I find that too dangerous from crevasses, to go around the lake, which last alternative, from the roughness of its shore, would be no easy thing. I find the ice on the end of the glacier softer and cross, then push rapidly up the right moraine to where the glacier becomes much more easily accessible on account of the lowness of its slope and because its surface is roughened with ridges and rocks.

In going up between the edge of the glacier and the moraine I discover the main surface stream of the glacier has cut a large cave, which I enter and find is made of clear-veined ice—a very unhuman place, with strange water gurgles and tinkles.

The lower steep portion of the glacier is alive with swift-running rills. A larger stream begins its course at the head of the glacier, and runs in a westerly direction along near the right side. It has cut for itself a strangely curved and scalloped channel in the solid green ice, in which it glides with motions and tones and gestures I never before knew water to make.

When my observations on this most interesting glacier (the main North Ritter Glacier) were finished, it was near sunset, and I had to make haste down to the tree-line. Yet I lingered reveling in the grandeur of the landscape. I was on the summit of the pass, looking upon Ritter Lake with its snowy crags and banks, and many a wide glacial fountain beyond, rimmed with peaks, the wind making stern music among their thousand spires, the sky with grand openings in the huge black clouds—openings jagged, walled, and steep like the passes of the mountains beneath them. Eastward lay Islet Lake with its countless little rocky isles; to the left, the splendid architecture of Mammoth Mountain, and in the distance range on range of mountains yet unnamed, with Mono plains and the magnificent lake and volcanic cones, sunlit and warm, between. To the eastward over the Great Basin swelled a range of alabaster cumuli, presenting a series of precipices deeply cleft with shadowy canyons, the whole fringed about their bases with a grand talus of the same alabaster material. Here and there occurred black masses with clearly defined edges like metamorphic slate in granite. Beneath these noble cloud mountains were horizontal bars and feathery touches of rose and crimson with clear sky between, of that exquisite spiritual kind that is connected in some way with our other life, and never fails, wherever we chance to be, to produce a hush of all cares and a longing, longing, longing. . . .

The Sensation of the Mystical

The most beautiful and most profound emotion we can experience is the sensation of the mystical. It is the sower of all true science. He to whom this emotion is a stranger, who can no longer wonder and stand rapt in awe, is as good as dead.

—Albert Einstein

A definitive scientific study of the Gulf of California is by Steinbeck and Ricketts and it is the John Steinbeck, *contemporary American author and Pulitzer and Nobel Prize winner. In two of his lighter novels,* Cannery Row *and* Sweet Thursday, *the central character is a marine scientist who has a Ph.D. and is known as Doc. There was a real-life Doc, Ed Ricketts, and he and Steinbeck studied the Sea of Cortez and recorded it in a leisurely journal of travel and research. We present a description of a visit to Guardian Angel Island in the Gulf of California.*

SEA OF CORTEZ

John Steinbeck and E. F. Ricketts

The long, snake-like coast of Guardian Angel lay to the east of us; a desolate and fascinating coast. It is forty-two miles long, ten miles wide in some places, waterless and uninhabited. It is said to be crawling with rattlesnakes and iguanas, and a persistent rumor of gold comes from it. Few people have explored it or even gone more than a few steps from the shore, but its fine harbor, Puerto Refugio, indicates by its name that many ships have clung to it in storms and have found safety there. Clavigero calls the island both "Angel de la Guardia" and "Angel Custodio," and we like this latter name better.

The difficulties of exploration of the island might be very great, but there is a drawing power about its very forbidding aspect—a Golden Fleece, and the inevitable dragon, in this case rattlesnakes, to guard it. The mountains which are the backbone of the island rise to more than four thousand feet in some places, sullen and desolate at the tops but with heavy brush on the skirts. Approaching the northern tip we encountered a deep swell and a fresh breeze. The tides are very large here, fourteen feet during our stay, and that not an extreme tide at all. It is probable that a seventeen-foot tide would not be unusual here. Puerto Refugio is really two harbors connected by a narrow channel. It is a safe, deep anchorage, the only danger lying in the strength and speed of the tidal current, which puts a strain on the anchor tackle. It was so strong, indeed, that we were not able to get weighted nets to the bottom; they pulled out sideways in the water and sieved the current of weed and small animals, so that catch was fairly worthwhile anyway.

We took our time getting firm anchorage, and at about three-thirty P.M. rowed ashore toward a sand and rubble beach on the southeastern part of the bay. Here the beach was piled with debris: the huge vertebrae of whales scattered about and piles of broken weed and skeletons of fishes and birds. On top of some low bushes which edged the beach there were great nests three to four feet in diameter, pelican nests perhaps, for there were pieces of fish bone in them, but all the nests were deserted—whether they were old or it was out of season we do not know. We are so used to finding on the beaches evidence of man that it is strange and lonely and frightening to find no single thing that man has touched or used. Tiny and Sparky made a small excursion inland, not over several hundred yards from the shore, and they came back subdued and quiet. They had not seen any rattlesnakes, nor did they want to. The beach was alive with hoppers feeding on the refuse, but the coarse sand was not productive of other animal life. The tide was falling, and we walked around a rocky point to the westward and came into a bouldery flat where the collecting was very rich. The receding water had left many small tide pools. The smoothness of the rocks indicated a fairly strong surf; they were dangerously slippery, and Sally Lightfoots and *Pachygrapsus* both scuttled about. As we moved out toward the entrance of the harbor, the boulders became larger and smoother, and then there was a sudden change to unbroken reef, and the smooth rocks gave way to barnacle- and weed-covered stones. The tide was down about ten feet now, exposing the lower tide pools, rich and beautiful with sponges and corals and small pleasant algae. We tried to cover as much territory as possible, but again and again found ourselves fascinated by some small and perfect pool, like a set stage, peopled with broken-back shrimps and small masked crabs.

The point itself was jagged volcanic rock in which there were high mysterious caves. Entering one, we noticed a familiar smell, and a moment later recognized it. For the sound of our voices alarmed myriads of bats, and their millions of squeaks sounded like rushing water. We threw stones in to try to dislodge some, but they would not brave the daylight, and only squeaked more fiercely.

As evening approached, it grew quite cold. Our hands were torn from the long collecting day and we were glad when it was too dark to work any more. We had taken great numbers of animals. There was an echiuroid worm with a spoon-shaped proboscis, found loose under the rocks; many shrimps; an encrusting coral (*Porites* in a new guise); many chitons, some new; and several octopi. The most obvious animals were the same marine pulmonates we had found at Angeles Bay, and these must have been strong and tough, for they were in the high rocks, fairly dry and exposed to the killing sun. The rocky ledge was covered with barnacles. The change in animal sizes on different levels was interesting. In the high-up pools there were small animals, mussels, snails, hermits, limpets, barnacles, sponges; while in the lower pools, the same species were larger. Among the small rocks and coarse gravel we found a great many stinging worms and a type of ophiuran new to us—actually it turned out to be the familiar *Ophionereis* in its juvenile stage. These high tide pools can be regarded as nurseries for more submerged zones. There were urchins, both club- and sharp-spined, and, in the sand, a few heart-urchins. The caverns under the rocks, exposed by the receding tide, were beautiful with many species of sponge, some pure white, some blue, and some purple, encrusting the rock surface. These under-rock caverns were as beautiful as those near Point Lobos in Central California. It was a long job to lay out and list the animals taken; meanwhile the crab-nets meant for the bottom were straining the current. In them we caught a number of very short fat stinging worms (*Chloeia viridis*), a species we had not seen before, probably a deep-water form torn loose by the strength of the currrent. With a hand-net we took a pelagic nudibranch *Chioraera leonina*, found also in Puget Sound. The water swirled past the boat at about four miles an hour and we kept the dip-nets out until late at night. This was a strange collecting place. The water was quite cold, and many of the members of both the northern and the southern fauna occurred here. In this harbor there were conditions of stress, current, waves, and cold which seemed to encourage animal life. And it is reasonable that this should be so, for active, churning water means not only a strong oxygen content, but the constant movement of food. And in addition, the very difficulties involved in such a position— necessity for secure footing, crowding, and competition—seem to encourage a ferocity and a tenacity in the animals which go past survival and into successful reproduction. Where there is little danger, there seems to be little stimulation. Perhaps the pattern of struggle is so deeply imprinted in the genes of all life conceived in this benevolently hostile planet that the removal of obstacles automatically atrophies a survival drive. With warm water and abundant food, the animals may retire into a sterile sluggish happiness. This has certainly seemed true in man. Force and cleverness and versatility have surely been the children of obstacles. Tacitus, in the *Histories*, places as one of the tactical methods advanced to be used against the German armies their exposure to a warm climate and a soft rich food supply. These, he said, will ruin troops quicker than anything else. If these things are true in a biologic sense, what is to become of the fed, warm, protected citizenry of the ideal future state?

Islands have always been fascinating places. The old storytellers, wishing to recount a prodigy, almost invariably fixed the scene on an island—Faëry and Avalon, Atlantis and Cipango, all golden islands just over the horizon where anything at all might happen. And in old days at least it was rather difficult to check up on them. Perhaps this quality of potential prodigy still lives in our attitude toward islands. We want very much to go back to Guardian Angel with time and supplies. We wish to go over the burned hills and snake-ridden valleys, exposed to heat and insects, venom and thirst, and we are willing to believe almost anything we hear about it. We believe that great gold nuggets are found there, that unearthly animals make their homes there, that the mountain sheep, which is said never to drink water, abounds there. And if we were told of a race of troglodytes in possession, we should think twice before disbelieving. It is one of the golden islands which will one day be toppled by a mining company or a prison camp.

In the decade before World War II, Antoine de St. Exupéry was a best-selling author in France, England and the United States. An aviator who wrote and wrote very well, he also was a keen observer of the geologic features of the areas over which he flew. Born in Lyons in 1900, St. Exupéry studied architecture for 15 months at the Ecole de Beaux Arts but these studies were terminated in 1921 when he was conscripted into the French Air Force. From his experience flying the mail, he wrote in 1929, Southern Mail, *and this was followed by* Night Flight; Wind, Sand and Stars; *and* Flight to Arras. *The latter was written in 1942, two years before his death.*

As a writer, one of his biographers noted his unique blend of graphic narrative, rich imagery, and of poetic meditation on human values. The excerpt presented is from a chapter entitled "Plane and Planet", and illustrates these talents, as well as a superb ability to describe landscape.

WIND, SAND AND STARS

Antoine de St. Exupéry

The pilot flying towards the Straits of Magellan sees below him, a little to the south of the Gallegos River, an ancient lava flow, an erupted waste of a thickness of sixty feet that crushes down the plain on which it has congealed. Farther south he meets a second flow, then a third; and thereafter every hump on the globe, every mound a few hundred feet high, carries a crater in its flank. No Vesuvius rises up to reign in the clouds; merely, flat on the plain, a succession of gaping howitzer mouths.

This day, as I fly, the lava world is calm. There is something surprising in the tranquillity of this deserted landscape where once a thousand volcanoes boomed to each other in their great subterranean organs and spat forth their fire. I fly over a world mute and abandoned, strewn with black glaciers.

South of these glaciers there are yet older volcanoes veiled with the passing of time in a golden sward. Here and there a tree rises out of a crevice like a plant out of a cracked pot. In the soft and yellow light the plain appears as luxuriant as a garden; the short grass seems to civilize it, and round its giant throats there is scarcely a swelling to be seen. A hare scampers off; a bird wheels in the air; life has taken possession of a new planet where the decent loam of our earth has at last spread over the surface of the star.

Finally, crossing the line into Chile, a little north of Punta Arenas, you come to the last of the craters, and here the mouths have been stopped with earth. A silky turf lies snug over the curves of the volcanoes, and all is suavity in the scene. Each fissure in the crust is sutured up by this tender flax. The earth is smooth, the slopes are gentle; one forgets the travail that gave them birth. This turf effaces from the flanks of the hillocks the sombre sign of their origin.

We have reached the most southerly habitation of the world, a town born of the chance presence of a little mud between the timeless lava and the austral ice. So near the black scoria, how thrilling it is to feel the miraculous nature of man! What a strange encounter! *Who knows how, or why, man visits these gardens ready to hand, habitable for so short a time—a geologic age—for a single day blessed among days?*

I landed in the peace of evening. Punta Arenas! I leaned against a fountain and looked at the girls in the square. Standing there within a couple of feet of their grace, I felt more poignantly than ever the human mystery.

But by the grace of the airplane I have known a more extraordinary experience than this, and have been made to ponder with even more bewilderment the fact that this earth that is our home is yet in truth a wandering star.

A minor accident had forced me down in the Rio de Oro region, in Spanish Africa. Landing on one of those table-lands of the Sahara which fall away steeply at the sides, I found myself on the flat top of the frustrum of a cone, an isolated vestige of a plateau that had crumbled round the edges. In this part of the Sahara such truncated cones are visible from the air every hundred miles or so, their smooth surfaces always at about the same altitude above the desert and their geologic substance always identical. The surface sand is composed of minute and distinct shells; but progressively as you dig along a vertical section, the shells become more fragmentary, tend to cohere, and at the base of the cone form a pure calcareous deposit.

Without question, I was the first human being ever to wander over this . . . this iceberg: its sides were remarkably steep, no Arab could have climbed them, and no European had as yet ventured into this wild region.

I was thrilled by the virginity of a soil which no step of man or beast had sullied. I lingered there, startled by this silence that never had been broken. The first star began to shine, and I said to myself that this pure surface had lain here thousands of years in sight only of the stars.

But suddenly my musings on this white sheet and these shining stars were endowed with a singular significance. I had kicked against a hard, black stone, the size of a man's fist, a sort of moulded rock of lava incredibly present on the surface of a bed of shells a thousand feet deep. A sheet spread beneath an apple-tree can receive only apples; a sheet spread beneath the stars can receive only star-dust. Never had a stone fallen from the skies made known its origin so unmistakably.

And very naturally, raising my eyes, I said to myself that from the height of this celestial apple-tree there must have dropped other fruits, and that I should find them exactly where they fell, since never from the beginning of time had anything been present to displace them.

Excited by my adventure, I picked up one and then a second and then a third of these stones, finding them at about the rate of one stone to the acre. And here is where my adventure became magical, for in a striking foreshortening of time that embraced thousands of years, I had become the witness of this miserly rain from the stars. The marvel of marvels was that there on the rounded back of the planet, between this magnetic sheet and those stars, a human consciousness was present in which as in a mirror that rain could be reflected.

Empty Spaces

They cannot scare me with their empty spaces
Between stars—on stars where no human race is.
I have it in me so much nearer home
To scare myself with my own desert places.

—Robert Frost

After reading the Snows of Kilimanjaro, The Short, Happy Life of Francis Macomber, *and* The Old Man and the Sea, *one is convinced that during his schooling Ernest Hemingway (1898–1961) was a student in one or more geology courses and had a singularly fine teacher. The Pulitzer Prize winner of 1953 and the master of tight, polished prose is also an expert in terrain description. In all of his short stories and novels set in the out-of-doors, Hemingway accurately describes landscape and does so utilizing the proper technical terms. Almost invariably some aspect of the terrain is integral to the story as it unfolds. In the segment presented from the* Green Hills of Africa, *Hemingway describes a hunting party in search of kudu and sable. He starts, "In the morning Karl and his outfit started for the salt-lick and Garrick, Abdullah, M'Cola and I crossed the road, angled behind the village up a dry watercourse and started climbing the mountains in a mist. We headed up a pebbly, boulder-filled, dry stream bed. . . ." The hunt continues and we see and hear more of a hilly terrain in Africa in the typically Hemingway sample that follows.*

GREEN HILLS OF AFRICA

Ernest Hemingway

We kept along the face of this hill on a pleasant sort of jutting plateau and then came out to the edge of the hill where there was a valley and a long open meadow with timber on the far side and a circle of hills at its upper end where another valley went off to the left. We stood in the edge of the timber on the face of this hill looking across the meadow valley which extended to the open out in a steep sort of grassy basin at the upper end where it was backed by the hills. To our left there were steep, rounded, wooded hills, with outcroppings of limestone rock that ran, from where we stood, up to the very head of the valley and there formed part of the other range of hills that headed it. Below us, to the right, the country was rough and broken in hills and stretches of meadow and then a steep fall of timber that ran to the blue hills we had seen to the westward beyond the huts where the Roman and his family lived. I judged camp to be straight down below us and about five miles to the northwest through the timber.

The husband was standing, talking to the brother and gesturing and point-ing out that he had seen the sable feeding on the opposite side of the meadow valley and that they must have fed either up or down the valley. We sat in the shelter of the trees and sent the Wanderobo-Masai down into the valley to look for tracks. He came back and reported there were no tracks leading down the valley below us and to the westward, so we knew they had fed on up the meadow valley.

Now the problem was to so use the terrain that we might locate them, and get up and into range of them without being seen. The sun was coming over the hills at the head of the valley and shone on us while everything at the head of the valley was in heavy shadow. I told the outfit to stay where they were in the woods, except for M'Cola and the husband who would go with me, we keeping in the timber and grading up our side of the valley until we could be above and see into the pocket of the curve at the upper end to glass it for the sable.

Elemental Sounds

The three great elemental sounds in nature are the sound of rain, the sound of wind in a primeval wood, and the sound of outer ocean on a beach.

—Henry Beston

A runaway best-seller in the early 1970s, The French Lieutenant's Woman
*is a novel of mid-Victorian England. The life and mores of that era have
rarely been recreated in better fashion than by John Fowles. That the
hero, Charles, is a self-trained paleontologist is not by chance nor are the
references to Darwinism. Gentlemen of that era were attracted to nature,
so much so that geology was a popular avocation, and Darwinism was
a topic for discussion and argumentation by most educated people.*

*Here then is a novel built around a paleontologist and fossil collecting
and including sea urchin (*Micraster*) tests, ammonites, and* The Old Fossil
Shop.

THE FRENCH LIEUTENANT'S WOMAN

John Fowles

Two days passed during which Charles's hammers lay idle in his rucksack.
He banned from his mind thoughts of the tests lying waiting to be discovered:
and thoughts, now associated with them, of women lying asleep on sunlit
ledges. But then, Ernestina having a migraine, he found himself unexpectedly
with another free afternoon. He hesitated a while; but the events that passed
before his eyes as he stood at the bay window of his room were so few, so
dull. The inn sign—a white lion with the face of an unfed Pekinese and a
distinct resemblance, already remarked on by Charles, to Mrs. Poulteney—
stared glumly up at him. There was little wind, little sunlight . . . a high gray
canopy of cloud, too high to threaten rain. He had intended to write letters,
but he found himself not in the mood.

To tell the truth he was not really in the mood for anything; strangely
there had come ragingly upon him the old travel-lust that he had believed
himself to have grown out of those last years. He wished he might be in
Cadiz, Naples, the Morea, in some blazing Mediterranean spring not only
for the Mediterranean spring itself, but to be free, to have endless weeks of
travel ahead of him, sailed-towards islands, mountains, the blue shadows of
the unknown.

Half an hour later he was passing the Dairy and entering the woods of
Ware Commons. He could have walked in some other direction? Yes, indeed
he could. But he had sternly forbidden himself to go anywhere near the cliff-
meadow; if he met Miss Woodruff, he would do, politely but firmly, what
he ought to have done at that last meeting—that is, refuse to enter into
conversation with her. In any case, it was evident that she resorted always
to the same place. He felt sure that he would not meet her if he kept well
clear of it.

Accordingly, long before he came there he turned northward, up the
general slope of the land and through a vast grove of ivyclad ash trees. They
were enormous, these trees, among the largest of the species in England,

with exotic-looking colonies of polypody in their massive forks. It had been their size that had decided the encroaching gentleman to found his arboretum in the Undercliff; and Charles felt dwarfed, pleasantly dwarfed as he made his way among them towards the almost vertical chalk faces he could see higher up the slope. He began to feel in a better humor, especially when the first beds of flint began to erupt from the dog's mercury and arum that carpeted the ground. Almost at once he picked up a test of *Echinocorys scutata*. It was badly worn away . . . a mere trace remained of one of the five sets of converging pinpricked lines that decorate the perfect shell. But it was better than nothing and thus encouraged, Charles began his bending, stopping search.

Gradually he worked his way up to the foot of the bluffs where the fallen flints were thickest, and the tests less likely to be corroded and abraded. He kept at this level, moving westward. In places the ivy was dense—growing up the cliff face and the branches of the nearest trees indiscriminately, hanging in great ragged curtains over Charles's head. In one place he had to push his way through a kind of tunnel of such foliage; at the far end there was a clearing, where there had been a recent fall of flints. Such a place was most likely to yield tests; and Charles set himself to quarter the area, bounded on all sides by dense bramble thickets, methodically. He had been at this task perhaps ten minutes, with no sound but the lowing of a calf from some distant field above and inland, the clapped wings and cooings of the wood pigeons; and the barely perceptible wash of the tranquil sea far through the trees below. He heard then a sound as of a falling stone. He looked, and saw nothing, and presumed that a flint had indeed dropped from the chalk face above. He searched on for another minute or two; and then, by one of those inexplicable intuitions, perhaps the last remnant of some faculty from our paleolithic past, knew he was not alone. He glanced sharply around.

She stood above him, where the tunnel of ivy ended, some forty yards away. He did not know how long she had been there; but he remembered that sound of two minutes before. For a moment he was almost frightened; it seemed uncanny that she should appear so silently. She was not wearing nailed boots, but she must even so have moved with great caution. To surprise him; therefore she had deliberately followed him.

"Miss Woodruff!" He raised his hat. "How come you here?"

"I saw you pass."

He moved a little closer up the scree towards her. Again her bonnet was in her hand. Her hair, he noticed, was loose, as if she had been in wind; but there had been no wind. It gave her a kind of wildness, which the fixity of her stare at him aggravated. He wondered why he had ever thought she was not indeed slightly crazed.

"You have something . . . to communicate to me?"

Again that fixed stare, but not through him, very much down at him. Sarah had one of those peculiar female faces that vary very much in their attractiveness; in accordance with some subtle chemistry of angle, light, mood. She was dramatically helped at this moment by an oblique shaft of wan sunlight that had found its way through a small rift in the clouds, as not infrequently happens in a late English afternoon. It lit her face, her figure standing before the entombing greenery behind her; and her face was suddenly very beautiful, truly beautiful, exquisitely grave and yet full of an inner, as well as outer, light. Charles recalled that it was just so that a peasant near Gavarnie, in the Pyrenees, had claimed to have seen the Virgin Mary standing on a *déboulis* beside his road . . . only a few weeks before Charles once passed that way. He was taken to the place; it had been most insignificant. But if such a figure as this had stood before him!

However, this figure evidently had a more banal mission. She delved into the pockets of her coat and presented to him, one in each hand, two excellent *Micraster* tests. He climbed close enough to distinguish them for what they were. Then he looked up in surprise at her unsmiling face. He remembered— he had talked briefly of paleontology, of the importance of sea urchins, at Mrs. Poulteney's that morning. Now he stared again at the two small objects in her hands.

"Will you not take them?"

She wore no gloves, and their fingers touched. He examined the two tests; but he thought only of the touch of those cold fingers.

"I am most grateful. They are in excellent condition."

"They are what you seek?"

"Yes indeed."

"They were once marine shells?"

He hesitated, then pointed to the features of the better of the two tests: the mouth, the ambulacra, the anus. As he talked, and was listened to with a grave interest, his disapproval evaporated. The girl's appearance was strange; but her mind—as two or three questions she asked showed—was very far from deranged. Finally he put the two tests carefully in his own pocket.

"It is most kind of you to have looked for them."

"I had nothing better to do."

"I was about to return. May I help you back to the path?"

But she did not move. "I wished also, Mr. Smithson, to thank you . . . for your offer of assistance."

"Since you refused it, you leave me the more grateful."

There was a little pause. He moved up past her and parted the wall of ivy with his stick, for her to pass back. But she stood still, and still facing down the clearing.

"I should not have followed you."

He wished he could see her face, but he could not.

"I think it is better if I leave."

Educated in Leipzig and Strassburg, Johann Goethe (1749–1832), the greatest German writer, had a passionate interest in matters scientific, including anatomy, botany, physics and geology. While in his thirties, Goethe sought relaxation, solace and fulfillment on several trips to the southern part of Europe. His observations and feelings are expressed in a book, Letters from Switzerland and Travels in Italy. *One is amazed at the constant references to terrain and landforms, minerals and mineral deposits, rock types, volcanoes and volcanic activity and the many speculations as to the origin of the features he encountered during his travels.*

In our selection you will read of an eruption of Vesuvius in March of 1787 complete with a bombardment of lapilli, flowing lava, sulfurous gas, and pithy observations on volcanic activity. All of this and more, from the ever-surprising Goethe.

LETTERS FROM SWITZERLAND AND TRAVELS IN ITALY

Johann Goethe

BOLOGNA, Oct. 20, 1786
Evening.

The whole of this bright and beautiful day I have spent in the open air. I scarcely ever come near a mountain, but my interest in rocks and stones again revives. I feel as did Antæus of old, who found himself endued with new strength as often as he was brought into fresh contact with his mother earth. I rode towards Palermo, where is found the so-called Bolognese sulphate of barytes, out of which are made the little cakes, which, being calcined, shine in the dark, if previously they have been exposed to the light, and which the people here call, shortly and expressively, "phosphori."

On the road, after leaving behind me a hilly track of argillaceous sandstone, I came upon whole rocks of selenite, quite visible on the surface. Near a brick-kiln a cascade precipitates its waters, into which many smaller ones also empty themselves. At first sight the traveller might suppose he saw before him a loamy hill, which had been worn away by the rain: on closer examination, I discovered its true nature to be as follows: the solid rock of which this part of the line of hills consists is schistous, bituminous clay of very fine strata, and alternating with gypsum. The schistous stone is so intimately blended with pyrites, that, exposed to the air and moisture, it wholly changes its nature. It swells, the strata gradually disappear, and there is formed a kind of potter's clay, crumbling, shelly, and glittering on the surface like stone-coal. It is only by examining large pieces of both (I myself broke several, and observed the forms of both), that it is possible to convince one's self of the transition and change. At the same time we observed the shelly strata studded with white points, and occasionally, also, variegated with yellow particles. In this way, by degrees, the whole surface crumbles away; and the hill looks like a mass of weatherworn pyrites on a large scale. Among the lamina some are harder, of a green and red color. Pyrites I very often found disseminated in the rock.

I now passed along the channels which the last violent gullies of rain had worn in the crumbling rock, and, to my great delight, found many specimens of the desired barytes, mostly of an imperfect egg-shape, peeping out in several places of the friable stone, some tolerably pure, and some slightly mingled with the clay in which they were embedded. That they have not been carried hither by external agency, any one may convince himself at the first glance. Whether they were contemporaneous with the schistous clay, or whether they first arose from the swelling and dissolving of the latter, is matter calling for further inquiry. Of the specimens I found, the larger and smaller approximated to an imperfect egg-shape: the smallest might be said to verge upon irregular crystalline forms. The heaviest of the pieces I brought away weighed seventeen *loth* (eight ounces and a half). Loose in the same clay, I also found perfect crystals of gypsum. Mineralogists will be able to point out further peculiarities in the specimens I bring with me. And I was now again loaded with stones! I have packed up at least half a quarter of a hundred-weight.

Naples, March 6, 1787.

Most reluctantly, yet for the sake of good-fellowship, Tischbein accompanied me to Vesuvius. To him—the artist of form, who concerns himself with none but the most beautiful of human and animal shapes, and one also whose taste and judgment lead to humanize even the formless rock and landscape—such a frightful and shapeless conglomeration of matter, which, moreover, is continually preying on itself, and proclaiming war against every idea of the beautiful, must have appeared utterly abominable.

We started in two calèches, as we did not trust ourselves to drive through the crowd and whirl of the city. The drivers kept up an incessant shouting at the top of their voice whenever donkeys, with their loads of wood or rubbish, or rolling calèches, met us, or else warning the porters with their burdens, or other pedestrians, whether children or old people, to get out of the way. All the while, however, they drove at a sharp trot, without the least stop or check.

As you get into the remoter suburbs and gardens, the road soon begins to show signs of a Plutonic action. For as we had not had rain for a long time, the naturally ever-green leaves were covered with a thick gray and ashy dust; so that the glorious blue sky, and the scorching sun which shone down upon us, were the only signs that we were still among the living.

At the foot of the steep ascent, we were received by two guides, one old, the other young, but both active fellows. The first pulled me up the path, the other, Tischbein,—pulled I say: for these guides are girded round the waist with a leathern belt, which the traveller takes hold of; and when drawn up by his guide, he makes his way the more easily with foot and staff. In this manner we reached the flat from which the cone rises. Towards the north lay the ruins of the Somma.

A glance westwards over the country beneath us, removed, as well as a bath could, all feeling of exhaustion and fatigue; and we now went round the ever-smoking cone, as it threw out its stones and ashes. Wherever the space allowed of our viewing it at a sufficient distance, it appeared a grand and elevating spectacle. In the first place, a violent thundering resounded from its deepest abyss; then stones of larger and smaller sizes were showered into the air by thousands, and enveloped by clouds of ashes. The greatest part fell again into the gorge: the rest of the fragments, receiving a lateral inclination, and falling on the outside of the crater, made a marvellous rumbling noise. First of all, the larger masses plumped against the side, and rebounded with a dull, heavy sound; then the smaller came rattling down; and last of all, a shower of ashes was trickling down. All this took place at regular intervals, which, by slowly counting, we were able to measure pretty accurately.

Between the Somma, however, and the cone, the space is narrow enough: moreover, several stones fell around us, and made the circuit anything but agreeable. Tischbein now felt more disgusted than ever with Vesuvius; as the monster, not content with being hateful, showed inclination to become mischievous also.

As, however, the presence of danger generally exercises on man a kind of attraction, and calls forth a spirit of opposition in the human breast to defy it, I bethought myself, that, in the interval of the eruptions, it would be possible to climb up the cone to the crater, and to get back before it broke out again. I held a council on this point with our guides, under one of the overhanging rocks of the Somma, where, encamped in safety, we refreshed ourselves with the provisions we had brought with us. The younger guide was willing to run the risk with me. We stuffed our hats full of linen and silk handkerchiefs, and, staff in hand, prepared to start, I holding on to his girdle.

The little stones were yet rattling round us, and the ashes still drizzling, as the stalwart youth hurried forth with me across the hot, glowing rubble. We soon stood on the brink of the vast chasm, the smoke of which, although a gentle air was bearing it away from us, unfortunately veiled the interior of the crater, which smoked all round from a thousand crannies. At intervals, however, we caught sight, through the smoke, of the cracked walls of the rock. The view was neither instructive nor delightful: but for the very reason that one saw nothing, one lingered in the hope of catching a glimpse of something more; and so we forgot our slow counting. We were standing on a narrow ridge of the vast abyss: of a sudden the thunder pealed aloud; we ducked our heads involuntarily, as if that would have rescued us from the precipitated masses. The smaller stones soon rattled; and, without considering that we had again an interval of cessation before us, and only too much rejoiced to have outstood the danger, we rushed down, and reached the foot of the hill, together with the drizzling ashes, which pretty thickly covered our heads and shoulders.

Tischbein was heartily glad to see me again. After a little scolding and a little refreshment, I was able to give my especial attention to the old and new lava. And here the elder of the guides was able to instruct me accurately in the signs by which the ages of the several strata were indicated. The older were already covered with ashes, and rendered quite smooth: the newer, especially those which had cooled slowly, presented a singular appearance. As, sliding along, they carried away with them the solid objects which lay on the surface, it necessarily happened, that, from time to time, several would come into contact with each other; and these again being swept still farther by the molten stream, and pushed one over the other, would eventually form a solid mass, with wonderful jags and corners, still more strange even than the somewhat similarly formed piles of the icebergs. Among this fused and waste matter I found many great rocks, which, being struck with a hammer, present on the broken face a perfect resemblance to the primeval rock formation. The guides maintained that these were old lava from the lowest depths of the mountain, which are very often thrown up by the volcano.

NAPLES,
Tuesday, March 20, 1787.

The news that an eruption of lava had just commenced, which, taking the direction of Ottajano, was invisible at Naples, tempted me to visit Vesuvius for the third time. Scarcely had I jumped out of my cabriolet (zweirädrigen einpferdigen Fuhrwerk), at the foot of the mountain, when immediately appeared the two guides who had accompanied us on our previous ascent. I had no wish to do without either, but took one out of gratitude and custom, the other for reliance on his judgment, and the two for the greater convenience. Having ascended the summit, the older guide remained with our cloaks and refreshment, while the younger followed me; and we boldly went straight towards a dense volume of smoke, which broke forth from the bottom of the funnel: then we quickly went downwards by the side of it, till at last, under the clear heaven, we distinctly saw the lava emitted from the rolling clouds of smoke.

We may hear an object spoken of a thousand times, but its peculiar features will never be caught till we see it with our own eyes. The stream of lava was narrow, not broader perhaps than ten feet, but the way in which it flowed down a gentle and tolerably smooth plain was remarkable. As it flowed along, it cooled both on the sides and on the surface, so that it formed a sort of canal, the bed of which was continually raised in consequence of the molten mass congealing even beneath the fiery stream, which, with uniform action, precipitated right and left the scoria which were floating on its surface. In this way a regular dam was at length thrown up, which the glowing stream flowed on as quietly as any mill-stream. We passed along the tolerably high dam, while the scoria rolled regularly off the sides at our feet. Some cracks in the canal afforded opportunity of looking at the living stream from below; and, as it rushed onward, we observed it from above.

A very bright sun made the glowing lava look dull, but a moderate steam rose from it into the pure air. I felt a great desire to go nearer to the point where it broke out from the mountain: there, my guide averred, it at once formed vaults and roofs above itself, on which he had often stood. To see and experience this phenomenon, we again ascended the hill, in order to come from behind to this point. Fortunately at this moment the place was cleared by a pretty strong wind, but not entirely, for all round it the smoke eddied from a thousand crannies; and now we actually stood on the top of the solid roof, which looked like a hardened mass of twisted dough, but projected so far outward, that it was impossible to see the welling lava.

We ventured about twenty steps farther; but the ground on which we stepped became hotter and hotter, while around us rolled an oppressive steam, which obscured and hid the sun. The guide, who was a few steps in advance of me, presently turned back, and, seizing hold of me, hurried out of this Stygian exhalation.

After we had refreshed our eyes with the clear prospect, and washed our gums and throat with wine, we went round again to notice any other peculiarities which might characterize this peak of hell, thus rearing itself in the midst of a paradise. I again observed attentively some chasms, in appearance like so many Vulcanic forges, which emitted no smoke, but continually shot out a steam of hot, glowing air. They were all tapestried, as it were, with a kind of stalactite, which covered the funnel to the top with its knobs and chintz-like variation of colors. In consequence of the irregularity of the forges, I found many specimens of this sublimation hanging within reach, so that, with our staves and a little contrivance, we were able to hack off a few and secure them. I had seen in the shop of the lava-dealer similar specimens, labelled simply "Lava;" and was delighted to have discovered that it was volcanic soot precipitated from the hot vapor, and distinctly exhibiting the sublimated mineral particles it contained.

The most glorious sunset, a heavenly evening, refreshed me on my return: still, I felt how all great contrasts confound the mind and senses. From the terrible to the beautiful—from the beautiful to the terrible: each destroys the other, and produces a feeling of indifference. Assuredly, the Neapolitan would be quite a different creature, did he not feel himself thus hemmed in between Elysium and Tartarus.

CALTANISETTA, Saturday, April 28, 1787.

Geology by way of an appendix! From Girgenti, the muschelkalk rocks. There also appeared a streak of whitish earth, which afterwards we accounted for. The older limestone formation again occurs, with gypsum lying immediately upon it. Broad flat valleys, cultivated almost up to the top of the hillside and often quite over it, the older limestone mixed with crumbled gypsum. After this appears a looser, yellowish, easily crumbling, limestone: in the arable fields you distinctly recognize its color, which often passes into darker, indeed occasionally violet, shades. About half-way the gypsum again recurs. On it you see growing, in many places, sedum, of a beautiful violet, almost rosy red; and on the limestone rocks, moss of a beautiful yellow.

The former crumbling limestone often shows itself; but most prominently in the neighborhood of Caltanisetta, where it lies in strata, containing a few fossils: there its appearance is reddish, almost of a vermilion tint, with little of the violet hue which we formerly observed near San Martino.

Pebbles of quartz I only observed at a spot about half-way on our journey, in a valley which, shut in on three sides, is open towards the east, and consequently also towards the sea.

On the left, the high mountain in the distance, near Camerata, was remarkable, as also was another, looking like a propped up cone. For the greatest half of the way not a tree was to be seen. The crops looked glorious, 'though they were not so high as they were in the neighborhood of Girgenti and near the coast; however, as clean as possible. In the fields of corn, which stretched farther than the eye could reach, not a weed to be seen. At first we saw nothing but green fields; then some ploughed lands; and lastly, in the moister spots, little patches of wheat, close to Girgenti. We saw apples and pears everywhere else; on the heights, and in the vicinity of a few little villages, some fig-trees.

These thirty miles, together with all that I could distinguish either on the right or left of us, was limestone of earlier or later formations, with gypsum here and there. It is to the crumbling and elaboration of these three together by the atmosphere that this district is indebted for its fertility. It must contain but very little sand, for it scarcely grates between the teeth. A conjecture with regard to the river Achates must wait for the morrow to confirm it.

The valleys have a pretty form; and although they are not flat, still one does not observe any trace of rain gullies,—merely a few brooks, scarcely noticeable, ripple along them, for all of them flow direct to the sea. But little of the red clover is to be seen; the dwarf palm also disappears here, as well as all the other flowers and shrubs of the south-western side of the island. The thistles are permitted to take possession of nothing but the waysides:

every other spot is sacred to Ceres. Moreover, this region has a great similarity to the hilly and fertile parts of Germany,—for instance, the tract between Erfurt and Gotha,—especially when you look out for points of resemblance. Very many things must combine in order to make Sicily one of the most fertile regions of the world.

On our whole tour we have seen but few horses: ploughing is carried on with oxen, and a law exists which forbids the killing of cows and calves. Goats, asses, and mules we met in abundance. The horses are mostly dapple-gray, with black feet and manes. The stables are very splendid, with well-paved and vaulted stalls. For beans and flax the land is dressed with dung: the other crops are then grown after this early one has been gathered in. Green barley in the ear, done up in bundles, and red clover in like fashion, are offered for sale to the traveller as he goes along.

On the hill above Caltanisetta I found a hard limestone with fossils: the larger shells lay lowermost, the smaller above them. In the pavement of this little town, we noticed a limestone with pectinites.

Behind Caltanisetta the hill subsided suddenly into many little valleys, all of which pour their streams into the river Salso. The soil here is reddish and very loamy, much of it unworked: what was in cultivation bore tolerably good crops, though inferior to what we had seen elsewhere.

St. Ronan's Well

And some rin up hill and down dale, knapping the chucky stanes to pieces wi' hammers, like sa mony road-makers run daft. They say 'tis to see how the world was made!

—Sir Walter Scott

Born and educated in Edinburgh, Sir Arthur Conan Doyle (1859–1930) practiced medicine until 1891 when he published his first book, a historical novel entitled, The White Company. *His fame is chiefly based on Sherlock Holmes and Dr. Watson, the activities at 221–B Baker Street and the sinister Professor Moriarty. Yet Sir Arthur was almost as well-received as one of the first science fiction writers, or more appropriately, writer of geological romances. Two examples are the* Marcot Deep *and the more familiar,* The Lost World, *published in 1912. In the extract that follows, Doyle's imagination soars as he describes a pterodactyl rookery on an isolated basalt plateau in South America.*

THE LOST WORLD

Sir Arthur Conan Doyle

Creeping to his side, we looked over the rocks. The place into which we gazed was a pit, and may, in the early days, have been one of the smaller volcanic blowholes of the plateau. It was bowl-shaped, and at the bottom, some hundreds of yards from where we lay, were pools of green-scummed, stagnant water, fringed with bulrushes. It was a weird place in itself, but its occupants made it seem like a scene from the Seven Circles of Dante. The place was a rookery of pterodactyls. There were hundreds of them congregated within view. All the bottom area round the water-edge was alive with their young ones, and with hideous mothers brooding upon their leathery, yellowish eggs. From this crawling, flapping mass of obscene reptilian life came the shocking clamour which filled the air and the mephitic, horrible, musty odour which turned us sick. But above, perched each upon its own stone, tall, grey, and withered, more like dead and dried specimens than actual living creatures, sat the horrible males, absolutely motionless save for the rolling of their red eyes or an occasional snap of their rat-trap beaks as a dragonfly went past them. Their huge, membranous wings were closed by folding their forearms, so that they sat like gigantic old women, wrapped in hideous web-coloured shawls, and with their ferocious heads protruding above them. Large and small, not less than a thousand of these filthy creatures lay in the hollow before us.

Our professors would gladly have stayed there all day, so entranced were they by this opportunity of studying the life of a prehistoric age. They pointed out the fish and dead birds lying about among the rocks as proving the nature of the food of these creatures, and I heard them congratulating each other on having cleared up the point why the bones of this flying dragon are found in such great numbers in certain well-defined areas, as in the Cambridge Green-sand, since it was now seen that, like penguins, they lived in gregarious fashion.

Finally, however, Challenger, bent upon proving some point which Summerlee had contested, thrust his head over the rock and nearly brought destruction upon us all. In an instant the nearest male gave a shrill, whistling cry, and flapped its twenty-foot span of leathery wings as it soared up into the air. The females and young ones huddled together beside the water, while the whole circle of sentinels rose one after the other and sailed off into the sky. It was a wonderful sight to see at least a hundred creatures of such enormous size and hideous appearance all swooping like swallows with swift, shearing wing strokes above us; but soon we realized that it was not one on which we could afford to linger. At first the great brutes flew round in a huge ring, as if to make sure what the exact extent of the danger might be. Then, the flight grew lower and the circle narrower, until they were whizzing round and round us, the dry, rustling flap of their huge slate-coloured wings filling the air with a volume of sound that made me think of Hendon aerodrome upon a race day.

"Make for the wood and keep together," cried Lord John, clubbing his rifle. "The brutes mean mischief."

The moment we attempted to retreat the circle closed in upon us, until the tips of the wings of those nearest to us nearly touched our faces. We beat at them with the stocks of our guns, but there was nothing solid or vulnerable to strike. Then suddenly out of the whizzing, slate-coloured circle a long neck shot out, and a fierce beak made a thrust at us. Another and another followed. Summerlee gave a cry and put his hand to his face, from which the blood was streaming. I felt a prod at the back of my neck, and turned, dizzy with the shock. Challenger fell, and as I stooped to pick him up I was again struck from behind and dropped on the top of him. At the same instant I heard the crash of Lord John's elephant gun, and, looking up, saw one of the creatures with a broken wing struggling upon the ground, spitting and gurgling at us with a wide-opened beak and bloodshot, goggled eyes, like some devil in a mediæval picture. Its comrades had flown higher at the sudden sound, and were circling above our heads.

"Now," cried Lord John, "now for our lives!"

We staggered through the brushwood, and even as we reached the trees the harpies were on us again. Summerlee was knocked down, but we tore him up and rushed among the trunks. Once there we were safe, for those huge wings had no space for their sweep beneath the branches. As we limped homewards, sadly mauled and discomfited, we saw them for a long time flying at a great height against the deep blue sky above our heads, soaring round and round, no bigger than wood pigeons, with their eyes no doubt still following our progress. At last, however, as we reached the thicker woods they gave up the chase, and we saw them no more.

"A most interesting and convincing experience," said Challenger, as we halted beside the brook and he bathed a swollen knee. "We are exceptionally well informed, Summerlee, as to the habits of the enraged pterodactyl."

CHAPTER 10

Geology and the Arts

Graham Sutherland, one of the great British and European painters of the last 200 years, died in 1980. In assessing his career, Robert Waterhouse wrote of the various influences that had shaped his style. "A chance trip to Pembrokeshire in 1934 was a revelation: 'I learnt that landscape was not necessarily scenic,' he wrote later, 'and found that its parts have an individual figurative detachment.' Subjects came together in a clearly resolved sequence of images. In spite of his attachment to the Home Counties and to the Kent coast in particular, Sutherland found his real impetus in the ordered magnificence of Pembroke."

Sutherland was not alone in this, for a host of other artists have been influenced by landscapes from which they drew inspiration, and to which they contributed new vision. Monet and Giverney, other impressionists and Argenteuil, Hornfleur and Auvers-sur-Oise, Gauguin and Tahiti, Constable and southern England, Van Goyen and Ruysdael and Holland, all spring readily to mind. Sometimes an area can give inspiration and unity to a whole 'school' of art. The Hudson Valley School of American landscape painters flourished in the mid-nineteenth century and included such artists as Thomas Doughty, Thomas Cole and Frederick Church. The artist always draws experience from his subject, and in turn, contributes to his subject, often expanding his new perception to other areas and subjects. One artist once exclaimed, "I begin by observing, I end by interpreting."

This creative search has ancient roots. Hand tracings on the walls of caves and cave paintings and statues represent the Upper Paleolithic period of some 20,000 years ago. Whether magic, symbolic, religious, aesthetic, or all of these things, man expressed his innermost feelings with sensitivity and skill in which we can still delight. Nor is landscape the only influence on painting. Materials were of prime importance, for with chalk and charcoal, ocher and oxide, early man created works of art.

Sculpture and architecture are even more strongly influenced by materials and location, as the selections by Jack Burnham and Jacquetta Hawkes reveal.

251

Jacquetta Hawkes, archeologist, writer, lecturer and broadcaster was born in 1910 and educated at Newnham College, Cambridge. Among her better known books are Early Britain *and the* White Countess, *co-authored with her husband, J. B. Priestley. In 1951 she received the Kensley Award for an exceptional book,* A Land. *This is a marvelous book, in which Jacquetta Hawkes shows the close relationship that exists among geology, archeology and landscape on the one hand and architecture and art on the other. Chapter 7 of* A Land *is titled, "Digression on Rocks, Soil and Men" from which we include two excerpts. The first deals with rocks, fossils and sculpture and more particularly the work of Rodin and Henry Moore. The second relates both rock type and age with architectural style and mood in the British Isles. The author writes authoritatively but with unusual warmth and feeling. One quotation will prepare the reader for the treat in store: "The dour grey and brown rocks of the Carboniferous Age which are so apt an expression of the stubborn civic pride, the puritanical distrust of elegance and light of our northern industrialists are followed by a return of the warm colours that commemorate the Permian and Triassic Ages, the renewed denudation of mountains in desert and heat."*

A LAND

Jacquetta Hawkes

Life has grown from the rock and still rests upon it; because men have left it far behind, they are able consciously to turn back to it. We do turn back, for it has kept some hold over us. A liberal rationalist, Professor G. M. Trevelyan, can write of 'the brotherly love that we feel . . . for trees, flowers, even for grass, nay even for rocks and water' and of 'our brother the rock'; the stone of Scone is still used in the coronation of our kings.

The Church, itself founded on the rock of Peter, for centuries fought unsuccessfully against the worship of 'sticks and stones'. Such pagan notions have left memories in the circles and monoliths that still jut through the heather on our moorlands or stand naked above the turf of our downs. I believe that they linger, too, however faintly, in our churchyards—for who, even at the height of its popularity, ever willingly used cast-iron for a tombstone?

It is true that these stones were never simply themselves, but stood for dead men, were symbols of fertility, or, as at Stonehenge, were primarily architectural forms. But for worshippers the idea and its physical symbol are ambivalent; peasants worship the Mother of God and the painted doll in front of them; the peasants and herdsmen of prehistoric times honoured the Great Mother or the Sky God, the local divinities or the spirits of their ancestors and also the stones associated with them. The Blue Stones of Stonehenge, for example, were evidently laden with sanctity. It seems that these slender monoliths were brought from Pembrokeshire to Salisbury Plain because in Wales they had already absorbed holiness from their use in some other sacred structure. There is no question here that the veneration must have been in part for the stones themselves.

Up and down the country, whether they have been set up by men, isolated by weathering, or by melting ice, conspicuous stones are commonly identified with human beings. Most of our Bronze Age circles and menhirs have been thought by the country people living round them to be men or women turned to stone. The names often help to express this identification and its implied sense of kinship; Long Meg and her Daughters, the Nine Maidens, the Bridestone and the Merry Maidens. It is right that they should most often be seen as women, for somewhere in the mind of everyone is an awareness of woman as earth, as rock, as matrix. In all these legends human beings have seen themselves melting back into rock, in their imaginations must have pictured the body, limbs and hair melting into smoke and solidifying into these blocks of sandstone, limestone and granite.

Some feeling that represents the converse of this idea arises from sculpture. I have never forgotten my own excitement on seeing in a Greek exhibition an unfinished statue in which the upper part of the body was perfect (though the head still carried a mantle of chaos) while the lower part disappeared into a rough block of stone. I felt that the limbs were already in existence, that the sculptor had merely been uncovering them, for his soundings were there—little tunnels reaching towards the position of the legs, feeling for them in the depths of the stone. The sculptor is in fact doing this, for the act of creation is in his mind, from his mind the form is projected into the heart of the stone, where then the chisel must reach it.

Rodin was one of the sculptors most conscious of these emotions, and most ready to exploit them. He expressed both aspects of the process—man merging back into the rock, and man detaching himself from it by the power of life and mind. He was perhaps inclined to sentimentalize the relationship by dwelling on the softness of the flesh in contrast with the rock's harshness. This was an irrelevance not dreamt of by the greatest exponent of the feeling—Michelangelo. It is fitting that the creator of the mighty figures of Night and Day should himself have spent many days in the marble quarries of Tuscany supervising the removal of his material from the side of the mountain. So conscious was he of the individual quality of the marble and of its influence on sculpture and architecture that he was willing to endure a long struggle with the Pope and at last to suffer heavy financial loss by maintaining the superiority of Carrara over Servezza marble. Michelangelo was an Italian working with Italian marble and Italian light; with us it has been unfortunate that since medieval times so many of our sculptors have sought the prestige of foreign stones rather than following the idiom of their native rock. It is part of the wisdom of our greatest sculptor, Henry Moore, to have returned to English stones and used them with a subtle sensitiveness for their personal qualities. He may have inherited something from his father who, as a miner spending his working life in the Carboniferous horizons of Yorkshire, must have had a direct understanding not only of coal but of the sandstones and shales in which it lies buried and on which the life of the miner depends. Henry Moore has himself made studies of miners at work showing their bodies very intimate with the rock yet charged with a life that separates them from it. (Graham Sutherland in his studies of tin mines became preoccupied with the hollow forms of the tunnels and in them his men appear almost embryonic.)

Henry Moore uses his understanding of the personality of stones in his sculpture, allowing their individual qualities to contribute to his conception. Indeed, he may for a moment be regarded in the passive role of a sympathetic agent giving expression to the stone, to the silting of ocean beds shown in those fine bands that curve with the sculpture's curves, and to the quality of the life that shows itself in the delicate markings made by shells, corals and sea-lilies.

It would certainly be inappropriate to his time if Moore habitually used the Italian marbles so much in favour since the Renaissance. For this fashion shows how man in his greatest pride of conscious isolation wanted stone which was no more than a beautiful material for his mastery. Now when our minds are recalling the past and our own origins deep within it, Moore reflects a greater humility in avoiding the white silence of marble and allowing his stone to speak. That is why he has often chosen a stone like Hornton, a rock from the Lias that is full of fossils all of which make their statement when exposed by his chisel. Sometimes the stone may be so assertive of its own qualities that he has to battle with it, strive against the hardness of its shells and the softness of adjacent pockets to make them, not efface themselves, but conform to his idea, his sense of a force thrusting from within, which must be expressed by taut lines without weakness of surface.

Moore uses Hornton stone also because it has two colours, a very pale brown and a green with deeper tones in it. The first serves him when he is conscious of his subject as a light one, the green when it must have darkness in it. Differences in climate round the shores of the Liassic lakes probably caused the change in colour of Hornton stone, and so past climates are reflected in the feeling of these sculptures. As for the sculptor's sense of light or darkness inherent in his subjects, it is my belief that it derives in large part from the perpetual experience of day and night to which all consciousness has been subject since its beginnings. The sense of light and darkness seems to go to the depths of man's mind, and whether it is applied to morality, to aesthetics or to that more general conception—the light of intellectual processes in contrast with the darkness of the subconscious—its symbolism surely draws from our constant swing below the cone of night.

It is hardly possible to express in prose the extraordinary awareness of the unity of past and present, of mind and matter, of man and man's origin which these thoughts bring to me. Once when I was in Moore's studio and saw one of his reclining figures with the shaft of a belemnite exposed in the thigh, my vision of this unity was overwhelming. I felt that the squid in which life had created that shape, even while it still swam in distant seas was involved in this encounter with the sculptor; that it lay hardening in the mud until the time when consciousness was ready to find it out and imagination to incorporate it in a new form. So a poet will sometimes take fragments and echoes from other earlier poets to sink them in his own poems where they will enrich the new work as these fossil outlines of former lives enrich the sculptor's work.

Rodin pursued the idea of conscious, spiritual man emerging from the rock; Moore sees him rather as always a part of it. Through his visual similes he identifies women with caverns, caverns with eye-sockets; shells, bones, cell plasm drift into human form. Surely Mary Anning might have found one of his forms in the Blue Lias of Lyme Regis? That indeed would be fitting, for I have said that the Blue Lias is like the smoke of memory, of the subconscious, and Moore's creations float in those depths, where images melt into one another, the direct source of poetry, and the distant source of nourishment for the conscious intellect with its clear and fixed forms. I can see his rounded shapes like whales, his angular shapes like ichthyosaurs, surfacing for a moment into that world of intellectual clarity, but plunging down again to the sea bottom, the sea bottom where the rocks are silently forming.

Men know their affinity with rock and with soil, but they also use them, at first as simply as coral organisms use calcium, or as caddis-worms use shell and pebbles, but soon also consciously to express imagined ideas.

Building is one of the activities relating men most directly to their land. Everyone who travels inside Britain knows those sudden changes between region and region, from areas where houses are built of brick or of timber and daub and fields are hedged, to those where houses are of stone and fields enclosed by drystone walling. Everywhere in the ancient mountainous country of the west and north stone is taken for granted; where the sudden appearance of walls instead of hedges catches the eye is along the belt of Jurassic limestones, often sharply delimited. The change is most dramatic in Lincolnshire where the limestone of Lincoln Edge is not more than a few miles wide and the transformation from hedges to the geometrical austerity of drywalling, from the black and white, red and buff of timber and brick to the melting greys of limestone buildings, is extraordinarily abrupt.

The distinctive active qualities of the stones of each geological age and of each region powerfully affect the architecture raised up from them; if those qualities precisely meet particular needs then, of course, the stones are carried out of their own region. Since the eighteenth century the value of special qualities in building material has greatly outweighed the labour of transport, and stones of many kinds have not only been carried about Britain to places far from those where they were originally formed, but have been sent overseas to all parts of the world. Men, in fact, have proved immensely more energetic than rivers or glaciers in transporting and mixing the surface deposits of the planet.

Now the process has gone too far; what was admirable when it concerned only the transport of the finest materials to build the greatest buildings has become damnable when dictated by commercial expediency. The cheapness of modern haulage has blurred the clear outlines of locality in this as in all other ways; slate roofs appear among Norfolk reed beds, red brick and tile in the heart of stone country, while cities weigh down the land with huge masses of stone, brick, iron, steel, and artificial marble dragged indiscriminately from far and near.

Nevertheless, there are still regional differences that will hardly disappear. Britain would sink below the sea before a Yorkshireman would buy Scottish granite to build his town hall, or an Aberdonian outrage his granite city with a bank of Millstone Grit. The danger is that Britain will not sink below the sea, but simply into a new form of undifferentiated chaos, when both Yorkshireman and Scot adopt artificial stone and chromium hung on boxes of steel and concrete.

While, on the one hand, it is admitted that even in the twentieth century regional differences still persist, it would, on the other, be false to suggest that even when all transport was by wind or muscle stone was not sometimes moved about the country. If the Blue Stones of Stonehenge are the most startling prehistoric instance, for the early Middle Ages it is the importation of Caen stone. Very many cargoes of this oolitic limestone were shipped from Normandy to build our abbeys and cathedrals. Often it was ordered by the great Norman clerics who, in a hostile land, found reassurance in building with their native rock. The genes of the Norman conquerors are now mingled with those of most of our royal and noble families, and through them also Caen stone has been incorporated in our most sacred national buildings—old St. Paul's, Canterbury Cathedral, Westminster Abbey.

It was of course most usually for ecclesiastical buildings and for castles that stone was shipped and carted about Britain, particularly to those youthful parts of lowland England south-east of the Jurassic belt. The material for Ely Cathedral and other great East Anglian churches came from Barnack in Northamptonshire, as did that for Barnwell, Romsey and Thorney abbeys and many of the early college buildings at Cambridge. The lower courses of King's Chapel at Cambridge, the foundation stone laid by Henry VI, came from the Permian Limestone of Yorkshire, while, after the long interruption in building, the upper courses were constructed of Jurassic stone from Northamptonshire, the personal gift to the college of Henry VII. The fan vaulting of the roof, however, those exquisite artificial stalactites, is again carved from a Permian deposit—the noted Roche Abbey quarries in Yorkshire. So in one building the Permian and Jurassic ages, the north and the Midlands have been made tributary to royal and scholastic pride, the service of God and the imagination of man. I have brought in these facts far from their proper place, to suggest a truth which is perhaps too obvious to need such attention. That the centre of gravity of a people in any age may be expected to be found in the objects for which they will transport great quantities of building material. Neolithic communities hauled megalithic blocks to their communal tombs, Bronze Age men did the same for their temples, the Iron Age Celts amassed materials for their tribal strongholds, the Romans for their military works and public buildings; medieval society sweated for its churches, colleges and castles.

What are the qualities which have attracted men to the different stones formed in such varied conditions—on ocean floors, in salt lakes and lagoons, by the eruption of volcanoes?

Granite—it seems inevitable to begin with granite, even though so many people have ended with it, lying under those glossy pinkish slabs labelled in gold or black and sometimes crowned with a stony wreath. The royal mausoleum at Frogmore is built of Dartmoor granite, while inside it, the thirty-ton sarcophagus in which the anxieties of the Prince Consort were laid to rest is of granite too, but Scottish granite, a genuine blue Peterhead from Aberdeen. The Queen loved granite because she hated change; at her express wish nineteen varieties representing the principal Aberdeen granites were used to ornament the pulpit at Balmoral. The mausoleum and sarcophagus at Frogmore represent the two main sources of granite: Scotland and the south-western peninsula. Detached blocks of granite have been used in their own localities since the beginning of our architecture—for megalithic tombs, for standing stones and sacred avenues, for early Christian crosses—but it was not until the eighteenth century that it was quarried for anything more than rough local purposes, and even then with great difficulty. Some of the first from Cornwall was used for the outside of Smeaton's ill-fated Eddystone lighthouse, then both Devon and Aberdeen granite went into Rennie's Waterloo Bridge (I seem to remember that on its demolition the balusters were sold as mementoes, and must now be scattered up and down the country). With the Victorians, and how appropriately, granite came into its own; the substance of wild moorlands was transformed into kerbstones, railway bridges, into post offices, public fountains and public houses, family fish-shops, and above all, into banks.

Ironically, this rock, the pillar of Victorianism, a symbol for endurance, can also remind us of insecurity and impermanence, coming as it does directly from the restless quag beneath us, the molten sea on which our wafer floats. For the basalts and other igneous rocks that are the products of actual volcanic eruption little use has been found except for road making. Visitors to the Lake District may know that Keswick is largely built of the volcanic ash of Borrowdale, the particles now held in the walls of its cottages and flower gardens once having fallen in fiery cascades onto an Ordovician sea.

For the mason there is an important distinction between granite and other igneous rocks and the sedimentary deposits of whatever age. Normally the layers of silt forming the sedimentary rocks have given them a grain which the masons must study almost as carefully as a carpenter studies his wood. These layers, having been laid down horizontally, must be kept horizontal in their human setting, for in this position they can better throw off rainwater. Rocks of so fine and close a grain that no layering is visible are called freestones, for the mason is free to cut and set them as he will. There are perfectionists, however, who maintain that even with freestones every block should be marked in the quarry so that it may be kept in its natural plane.

There is another characteristic of the sedimentary rocks of very great significance for mason and builder. All newly cut stone is permeated with 'quarry water' which holds various minerals either dissolved or in suspension. On exposure the quarry water is drawn gradually to the surface where it evaporates, depositing the minerals near the surface and so forming a tough outer skin. It is therefore most desirable that every block should first be cut into its final shape and then be seasoned to allow it to go into the building with the skin unbroken. In this way it is assured that mouldings, leaves, noses and other excrescences have the best possible chance to survive weathering. Christopher Wren, an artist properly sensitive to his materials, would use no block in St. Paul's unless it had been exposed for at least three years.

The dour grey and brown rocks of the Carboniferous Age which are so apt an expression of the stubborn civic pride, the puritanical distrust of elegance and light of our northern industrialists are followed by a return of the warm colours that commemorate the Permian and Triassic Ages, the renewed denudation of mountains in desert and heat. The New Red Sandstone glows pleasantly in many local churches and had its fling in the astonishing pile of the St. Pancras Hotel. For this period it is the Magnesian Limestones that have had a national currency. Although they have given the material for many successful buildings up and down the country, they have also been responsible for one conspicuous, notorious and expensive failure. This was in the very shrine of our democratic institutions, the Houses of Parliament. While Barry and Pugin were not very happily seeking to agree over the principles of Gothic design and the details of Gothic ornament, a Royal Commission was responsible for the choice of the stone in which their ideas were to be embodied. After earnest weighing of evidence, during which

Portland stone most unfortunately was rejected, the Commission reported that 'for crystalline character, combined with a close approach to the equivalent proportions of carbonate of lime and carbonate of magnesia; for uniformity of structure, facility and economy in conversion, and for advantage of colour, the Magnesian Limestone or Dolomite of Bolsover Moor, and its neighbourhood, is, in our opinion, the most fit and proper material to be employed in the proposed new Houses of Parliament'. But alas for semi-scientific pomposity when it is not anchored to either real knowledge or an intimate understanding of particular facts. Before the buildings had risen above the level of their basements the Bolsover quarries were exhausted, and had in truth never yielded blocks large enough to serve the ambitions of Barry and Pugin. After further solemn and ill-formed discussion it was decided to transfer to the Anston Stone Quarries, a few miles from Bolsover, but across the Derbyshire boundary into Yorkshire. As a result of the profound irresponsibility, the lack of contact with reality, characteristic of the 'sound' civil servant, the 'stone was quarried and delivered indiscriminately, without regard to the nature of the bed, the lie of the rock in the quarry, or the necessary seasoning of the stone'. A few decades of exposure to the climate of London and particularly to its acid-charged rain, and the whole of that vast display of Gothic revivalism began to crumble and dissolve. All the heraldic and architectural detail, the coats of arms, crowns, gargoyles, canopies and finials, turned leprous, flaked and peeled, and, as dust or in solution, found their way into the Thames and so back to the sea. The nation has had to meet a heavy bill and is now confronted with an unsightly, mottled façade with its pale patches of Jurassic Clipsham among the darker Anston stone. For more than ten years the Victoria tower has stood above Westminster as a scaffold ruin. There can hardly be a more revealing example of the disaster that threatens remote control, when men deal with practical matters on a diet of words. These solemn Commissioners, and these civil servants, who doubtless haggled over sixpences with the most conscientious futility, had never touched, seen, studied or understood the land they were attempting to use. That the catastrophe could have been avoided is proved by the old Geological Museum. The stone for this building came at much the same time from the identical source, but its quarrying was supervised by Mary Anning's old friend, de la Beche, a man who knew his rocks both scientifically and humanly. Hardly a block has decayed.

In forming the oolitic limestones, the wide sea of the Jurassic Age made the greatest and fairest contribution to the buildings which in time were to be raised in Britain. For the last two thousand years, since masoned stone was introduced by the Roman conquerors, the rocks of the Jurassic belt have been used for buildings of every kind, culminating in many of our finest mansions, colleges, churches and cathedrals. It was the most easterly deposit of any great extent old enough to make a hard building stone. For this reason it was not only quarried by its own population but was also sought by the increasing millions who lived on the younger formations to the east of it, people who had no native building stone equal to expressing their imagination, wealth and ambition. Through centuries, carts, barges, ships, railway trucks and lorries have gone to the Jurassic belt and carried away heavy cargoes to embody the architectural aspirations of the lowland English. The architect who held shaped in his mind a pier cluster, a crocketed finial or a west front, a pediment, acanthus leaf or colonnade, whatever was appropriate to his moment of time, would seek to give it substance in these limestones with their great range of colours and textures.

The oldest of the Jurassic deposits, the soft and crumbly Lias, had little value until our synthetic age when it has come into its own for the making of lime and cement. But from Somerset to Yorkshire the overlying oolites have been so much quarried that many of the best varieties are now exhausted. In the extreme south-west the Doulting quarries gave the material for Wells Cathedral and for Glastonbury, but Gloucestershire is the region where these limestones have done most to create an entire countryside. Men and sheep and the limestone hills have together made the Cotswold realm, with its small unchanging towns and church-proud villages, its hamlets and country houses, surely one of the most lovely stretches of rural urbanity in the world. All buildings from the low gabled cottages to the huge Perpendicular churches are walled, and should be roofed, with the stone on which they rest. All reflect its faint golden light, though the dry-stone walls seem to assume a greyer tone in contrast with the ruddy browns and russet of the Cotswold soil.

I have said 'should be roofed' because some buildings have now lost the stone tiles that are their proper covering. There are several places in the Cotswolds where the special limestones necessary for these tiles can be quarried, but perhaps none is so well known as the Stonesfield pits near Oxford. There the tile beds were very thin and had to be pursued by means of shafts and horizontal tunnels. This work was done between Michaelmas and Christmas, but once the pendle, as it was called, had been quarried, it had to be put in clamps until the first sharp frost. Extraordinary as it may seem, it could not be artificially split and the whole industry depended entirely on the help of frost. When it did come, and it was hoped for in January, every man in the village rallied to spread out the slabs of pendle; if it fell suddenly during the night the church bells were rung to summon the villagers. They must often have hurried up the street while the bell was still ringing through the frosty air; then, dark figures in the moonlight, they attacked the clamps and strewed the big slabs on the stiffening grass. If the frost had done its work, the men gave the summer to shaping and piercing the thin sheets, each sitting in a little shelter of hurdles or waste stone. If the frost failed then the industry was at a standstill and the pendle had to be buried deeply in cool soil, for if once the 'quarry water' was allowed to escape the slabs became 'bound' and could never be split.

The Stonesfield beds are full of fossils from a warm shallow Jurassic sea: corals mingled with the spines and shells of sea-urchins, molluscs, sea reptiles and turtles. They also yield a few land creatures—even, though very rarely, the teeth and jaws of early mammals. Like Mr. Anning of Lyme, the Stonesfield workers knew the value of these fossils and displayed them for sale in their cottage windows where they might be seen by learned men from Oxford.

The demand for Stonesfield tiles is still so great that the roofs of cottages and barns have been stripped to sell them for cash; the present villagers, too, recall how much the old men loved their work, knowing the characters of their pits as intimately as those of their wives. Yet now the pits and spoil heaps are overgrown and it is many years since the tapping of pick and hammer was heard during the summer, or the village was roused on a frosty night by the sound of bells. The industry has died partly because earnings were poor, and partly, so it is said, because in this century the winters have too often been mild. The last tile worker, Thomas Griffin, died recently as a very old man.

Portland alone surpasses Bath stone as a medium in which Renaissance architecture could achieve perfection. It is the paler of the two with more grey and less yellow in its tone. The Isle of Portland is ancient Crown property and when Inigo Jones was chief architect and Surveyor-General to James I he was charged as a part of his routine duties to make a survey of the island. So he was led to an intimate knowledge of the oolite of Portland, and intimacy resulted (as it occasionally does) in deep admiration. He himself used it for the Great Banqueting Hall in Whitehall and its reputation was established. Some doggerel verses were composed at the time by one Farley who claimed a great familiarity with Portland stones and to know 'as much of their mindes as any man'.

> *Ere since the Architect of Heaven's fair frame*
> *Did make the World and man to use the same;*
> *In Earth's wide womb as in our nat'ral bed,*
> *We have been hid, conceal'd and covered . . .*
> *We were discovered and to London sent*
> *And by good Artistes tried incontinent;*
> *Who (finding us in all things firm and sound*
> *Fairer and greater than elsewhere are found;*
> *Fitter for carriage and more sure for weather*
> *Than Oxford, Ancaster or Beer-stone eyther)*
> *Did well approve our worth above them all*
> *Unto the King for service at Whitehall.*

It was inevitable for Jones's successor in office, Sir Christopher Wren, to succeed him also as patron of the Portland quarries.

The Fire of London opened the way for Portland stone and transformed the Isle into a vast stone-mason's yard, with its own cottages and wharves. Boatload after boatload of huge blocks were brought along the south coast and up the Thames to rebuild the gutted capital. After its long passivity, after one hundred and fifty million years untouched by consciousness, this stone was now to spring up in the rich variety of Wren's towers and steeples, so urbane and yet so fired with the idiosyncrasy of his genius, gleaming like lilies among the rose-red brick of Canaletto's paintings, and now tottering but still gracious in our philistine and ruined city. As its greatest glory, the stone was to grow, to blossom, into St. Paul's, that incomparable building which has endured all our latter-day barbarities.

As fortune in all things favours a woman in love until it seems that she can do nothing wrong, a nation and country in a certain state of vitality and enthusiasm will be consistently fortunate. Seventeenth-century England still held some of this vitality. Although Puritanism had already sown its seeds of materialism and joylessness, the plants had not yet grown large. The intellectual fire and clear light of the Renaissance was burning among a people who still had some of the poetic insights of theMiddle Ages, and some of their earthiness. It was one of the happy chances of a fortunate age that the nation, as personified in its king, should have at its command both an architect of genius and a material fitted to give that genius its finest expression. As a final stroke of good fortune the Great Fire came to give it room.

There is of course the cruel reverse of this state. Again like a woman, when a country is out of love with itself the whole of life conspires against it. So at the present time if we have architects of genius we also have means for preventing them from being used; we are addicted to concrete and artificial stone, and in the office of the Minister of Works, instead of a Wren succeeding an Inigo Jones, an individual who can build only mud pies is likely to be succeeded by one who has no accomplishments.

The seventeenth century did not waste the chances offered to it. At the king's command the quarries in the Isle of Portland were put under Wren's control to be exclusively used for the rebuilding of St. Paul's. With an artist's understanding of his material, Wren scrupulously supervised the selection, cutting and seasoning of the stone. All went well: the quarries were not exhausted, Wren did not die and money was not withheld. A considerable part of the Isle of Portland, milliards of oolites which had once rolled softly on the sea floor, were raised by king, people and architect into our one great Renaissance cathedral. That building owes much of its quality to the subtle shading of the Portland stone as it passes from the rain and wind-bleached points of exposure to the sooty darkness of its most sheltered coigns and hollows.

From the British Museum to the Cambridge Senate House, the substance of many of our best known or finest classical buildings has come from the distant Dorset quarries that have contributed more than any others to the personality of our architecture.

In 1934 the English sculptor, Henry Moore, prepared a program for himself, which—almost without exception—he followed throughout his career. Two portions of this program are reproduced for your reading, Truth to Material *and* Observation of Natural Objects. *Both show Moore's concern with pebbles, rocks, shells, bones, and rounding and polishing by erosive processes. Also included in this excerpt from Jack Burnham's excellent book is a paragraph relating some of the most powerful works of Henry Moore to the rock formations of Yorkshire and the marine stacks and chimneys off the rugged coast of Brittany.*

The author, Jack Burnham, is a sculptor and is a member of the faculty at Northwestern University. In his writing he expresses concern for the loss of spiritual qualities in art and the influence of science and technology on the artist.

BEYOND MODERN SCULPTURE

Jack Burnham

No sculptor has brought out with such bluntness, or described with greater precision, the vitalist position than Henry Moore. Where previous sculptors felt content to imbue themselves with a tacit aesthetic based on the organic influence, Moore has always been quite candid about the origins and techniques of his synthesis of natural elements. He has never regarded himself as a magician but simply as a sensitive and observant artist obeying the characteristics of his materials. If, strictly speaking, vitalist doctrine is concerned with the infusion of "life" into the materials of sculpture, then it should be emphasized that Moore has been quite specific in recommending that the word *vital* be used to describe this infusion, rather than *organic*. On the surface the difference may seem small, but *organic* implies a functionalism, an application of materials to various duties, which runs well beyond the range of vitalism.

Vitalism, based as it is on nonphysical substances and states of life, is a metaphysical doctrine concerned with the irreducible effects and manifestations of living things. It was the great discovery of twentieth-century sculpture that these did not have to be appreciated through strict representationalism. Visual biological metaphors exist on many levels besides the obvious total configuration of an animal or human. The aesthetic of true organicism, on the other hand, is not grounded in the appearances of natural forms and their carryover into sculptural materials, but is concerned with the organization of processes and interacting systems.

Moore's vitalism, in no way scientifically analytical, is the vitalism of the naturalist and sensitive craftsman. Besides the sculpture itself, the most obvious example of this lies in Moore's writings. A statement in the anthology

Unit I (1934) describes the program which Moore set for himself, and which, with a few interruptions, he has followed for the past thirty-five years. Some essential parts read as follows:

Truth to material. Every material has its own individual qualities. It is only when the sculptor works direct, when there is an active relationship with his material, that the material can take its part in the shaping of an idea. Stone, for example, is hard and concentrated and should not be falsified to look like soft flesh—it should not be forced beyond its constructive build to a point of weakness. It should keep its hard tense stoniness.

Observation of Natural Objects. The observation of nature is part of an artist's life, it enlarges his form-knowledge, keeps him fresh and from working only by formula, and feeds inspiration.

The human figure is what interests me most deeply, but I have found principles of form and rhythm from the study of natural objects such as pebbles, rocks, bones, trees, plants, etc.

Pebbles and rocks show Nature's way of working stone. Smooth, sea-worn pebbles show the wearing away, rubbed treatment of stone and principles of asymmetry.

Rocks show the hacked, hewn treatment of stone, and have a jagged nervous block rhythm.

Bones have marvelous structural strength and hard tenseness of form, subtle transition of one shape into the next and great variety in section.

Trees (tree trunks) show principles of growth and strength of joints, with easy twisting movement.

Shells show Nature's hard but hollow form (metal sculpture) and have a wonderful completeness of single shape.

There is in Nature a limitless variety of shapes and rhythms (and the telescope and microscope have enlarged the field) from which the sculptor can enlarge his form-knowledge experience.

Vitality and Power of Expression. For me a work must first have a vitality of its own. I do not mean a reflection of the vitality of life, of movement, physical action, frisking, dancing figures and so on, but that work can have in it a pent-up energy, an intense life of its own, independent of the object it may represent.

In several essays Herbert Read has already elaborated on this and other statements by Moore. But, in truth, what Moore has to say is quite clear to anyone who has taken the trouble to observe his work. Still, at least two things could be said about Moore's doctrine which are only realizable since the early 1960's.

"Truth to material," first of all, is an ambiguous premise. However, during the 1930's, the 1940's and a good deal of the 1950's, it approached the stature of a universal dictum for a number of sculptors other than Moore. "Truth to material" followed to its ultimate conclusions is a *reductio ad absurdum.* Any forming or shaping must take advantage of the plasticity of each material, and, more importantly, no material will do what it is not meant to do. Moore, as a reaction to the falsity of much Neoclassical and Romantic carving, was only setting up an antidote, not an iron law, for the use of materials. The attraction for a great many sculptors to "truth to material" is its ring of moral equilibrium and natural propriety. But, like Humpty-Dumpty's assertion to Alice over the meaning of words, the post-vitalist in reaction to "truth to material" has declared, I can make materials mean anything I want them to mean; after all, who is to be the master?

Another observation has to do with "vitality and power of expression," Moore's limiting criterion of what is *vital* in sculpture. Vitalism, by its very purpose of seeking to imbue physical mass with psychic energy, is an idealistic mechanism for controlling attitudes about inert, nonmoving matter. It is a surrogate for actual physical vitality. The trend toward Kineticism in the 1960's, inelegant as it may be aesthetically, cannot help but undermine and change our attitudes and sensory apprehension of nonmoving art. Increasingly we will regard the traits of vitalistic sculpture with the same amused tolerance that we now reserve for obvious antiquing in nineteenth-century Neoclassical sculpture.

"Natural" is a diabolical term because it can be used with almost unlimited intentions. The biologist C. H. Waddington, in a remarkably penetrating essay, presents a statistically oriented view of what is "natural" and its applications to Moore's vitalistic sculpture. He makes the point that abstract forms which seem to be the most accurate expression of life are lifelike because they display a multifunction visual complexity: inorganic forms, on the other hand, are single purposed and show a lack of multiple purpose. Moore, working with natural forms, both inorganic and organic, shifts back and forth to all magnitude of sizes for his ideas—then merges the results. Waddington explains this in terms of one of Moore's favorite subjects, beach pebbles. Tidal and rhythmic forces of the ocean, the various consistencies of stone, their texture and grain, etc., all subject stone fragments along a beach to many formative abrasive pressures. In Waddington's words:

> It represents not the equilibrium or balancing of many conflicting tendencies, but the chance outcome of a series of random and unrelated events, in fact Whitehead's mere confusion of detail. . . . But if the number of detailed events (forms of abrasion) are large enough, and if they are approximately the same magnitude, a certain statistical regularity will emerge, and there will be a tendency for the production of some reasonably definite shape, which may simulate an organic shape produced by internal equilibrium.

Not surprisingly, Moore's sculpture has only become devitalized and weakened when he has strayed too far from the tenets of his 1934 statement. This is true of some of his work just prior to the Second World War and particularly during the 1950's. Various half-hearted attempts at formalism, totemism, and realism are in no way consistent with the strengths of Moore's sensibilities.

During the present decade Moore has produced some of the most powerful work of his career—much of it in very large bronze castings. No longer are the plaster models finished in smooth perfection, but instead the plaster rasp and the pick hammer simulate the kind of graininess and sedimentary stratification which Moore relates to the rock formations in the Yorkshire countryside. At the same time, the tendency of bones to appear smooth-hard or porous, depending upon the function of a bone at a given area, is of more importance to Moore. Some of the figurative conceptions relate to earlier two-piece and three-piece compositions, and, with the aid of polychrome patinas, completely detached sets of shoulders and knees jut up from a plinth's flat surface like the upright rock promontories off the coast of Brittany.

Moore's sculpture at the present time involves some of the most literal adaptations to be found in his work. In these new locking pieces the sculptor looks more closely at the various hinge mechanisms and contact points between bones in a skeletal position. Included between the larger segments are thinner members, shock cushions so to speak, resembling the cartilage wrapped around the contacting surfaces of bones. Earlier in the interview with David Sylvester, Moore mentions the truncated surfaces produced by sawed cross-sections of the discovered bones. These truncated surfaces are a new aspect of an idea already adapted by Arp, Brancusi, and others. In sum, the locking forms are Moore's lexicon of touching surfaces, some close together in fine, barely perceivable seams, while others open, tapering toward rounded raised edges. It is interesting that the vitalism in these later pieces has come full circle through abstraction to a kind of transcendent realism: a realism not too far removed from the realm of the naturalist and classical biologist.

Sherlock Holmes

You know my method. It is founded upon the observance of trifles.

—Sir Arthur Conan Doyle

Thomas Moran (1837–1926), one of the leading landscape artists of the nineteenth century, was a member of two of the great geological and geographical surveys of the West. He accompanied Dr. F. V. Hayden to the Yellowstone and, following an extended visit to Yosemite, was a member of Major John Wesley Powell's expedition to southern Utah and Arizona. Among his best known landscapes are the Lower Falls of the Yellowstone, Glory of the Canyon, Pictured Rocks of Lake Superior, and Lower Geyser Basin. Partly because of the public exposure his paintings provided and because of the authenticity and excitement of his scenes of the American West, a number of his sketching sites have since been made National Parks.

In writing of the Hayden Survey, R. A. Bartlett provides us in the following citation an interesting glimpse of Moran.

THOMAS MORAN: AMERICAN LANDSCAPE PAINTER

R. A. Bartlett

With Hayden that year as a guest was one of the outstanding landscape painters of the nineteenth century, Thomas Moran. Today, the modernists have tried to forget him. His paintings are in storage in art museum basements. It is said that his attempts to paint the realities of the Grand Canyon of the Yellowstone, the Grand Canyon of the Colorado, and the Mount of the Holy Cross were all failures, and anyway, painting has now passed far beyond attempts merely to convey an accurate scene. Yet in the 1870's, Moran's work was judged among the finest, and Congress twice appropriated $10,000 for the purchase of his western scenes. One of them, of Yellowstone Falls and the Grand Canyon, hung for years in the Capitol. "It captured more than any other painting I know, the color and atmosphere of spectacular nature," said Jackson, himself an artist of some ability. But the paintings deteriorated and, though restored, no longer hang in the Capitol.

If painters ever return to reality, to a respect for the man who can make authentic paintings of authentic subjects, then a reappraisal of Moran may bring him back from the limbo of discarded artists. He boldly tried to convey on canvas the most majestic scenes in the entire American West, and some of his work—his painting of Yellowstone Falls and the Grand Canyon of the Yellowstone, for example—is breath-taking in concept, majesty, and color. Writing to Hayden about the Yellowstone painting early in 1872, he expressed his objectives. Most artists, he said, had considered it next to impossible to make good pictures of strange and wonderful scenes in nature, and the most that could be done with such material was to give it topographical or geological characteristics. "But I have always held that the grandest, most beautiful, or wonderful in nature, would, in capable hands, make the grandest, most beautiful, or wonderful pictures," he wrote, "and that the business of a great painter should be the representation of great scenes in nature. All the above characteristics attach to the Yellowstone region, and if I fail to prove this, I fail to prove myself worthy the name of painter. I cast all my claims to being an artist, into this one picture of the Great Canyon and am willing to abide by the judgement upon it."

Moran was thirty-four years old in 1871, tall, gaunt, and cadaverous, but compassionate, helpful, and level headed. He had never mounted a horse before, but he rode the creatures all day long from the very beginning and accepted his sore posterior as part of his sacrifices for the sake of art. He was one of that rare breed of men that have some knowledge about almost everything. He taught the old camp men how to wrap trout in wet paper, bury them beneath the fire, and uncover them baked and delicious half an hour later. He was interested in photography, and he gave unstintingly of his artistic knowledge, especially with regard to the problems of composition.

Digging Deep

To take what there *is*, and use it, without waiting forever in vain for the preconceived—to dig deep into the actual and get something out of *that*—this doubtless is the right way to live.

—**Henry James**

Geology can also sometimes assist the professional studies of art histo-
rians. There were four major sites in ancient Greece from which marble
was quarried. By using modern geochemical techniques, it is possible to
assign a piece of Grecian marble to its place of origin. Archeologists and
historians find this geologic approach of great value, not only in detecting
forgeries and copies, but, since it is known when the various ancient
quarries were in operation, once the source of a marble fragment is known,
a date can be assigned. There is also a lesson for acquisitive tourists. As
the Craigs note, "Athenians scatter fragments of Pentetic marble around
the Parthenon each winter, in order to provide material for the insatiable
pillage by tourists. This marble is from modern quarries and is isotrop-
ically distinct from that of the classical (ancient) quarries."

In the article that follows, the husband-wife team at the Scripps Insti-
tution of Oceanography report on the method, validity, and several of the
results of this cooperative venture between science and the arts.

GREEK MARBLES: DETERMINATION OF
PROVENANCE BY ISOTOPIC ANALYSIS

Harmon and Valerie Craig

The problem of identifying the locality from which a given specimen of
Greek marble was quarried has been studied for almost 100 years. In addition
to the obvious applications such as detection of copies and forgeries there
are more general problems of great importance which could be solved by a
method capable of uniquely assigning a piece of marble to its place of origin.
The major quarrying localities in use from preclassical to Hellenistic and
Roman periods shifted successively from the Cycladic islands of Naxos, and
later Paros, to Mount Pentelikon on the Attic mainland, and finally during
Roman times, to nearby Mount Hymettus. Thus in many instances the spec-
ification of the provenance of a piece of statuary or a building can be used
to date the object from the known periods of quarrying operations in each
area. Finally, there are geological problems at hand such as the question of
whether the Pentelic and Hymettian marbles represent two formations or a
single formation repeated by folding or faulting.

Methods applied to Greek marbles for provenance determination in the
past include petrographic techniques, petrofabric studies, and Na and Mn
analysis by neutron activation. Although Herz and Pritchett were able to
solve a problem of a mismatch of inscription fragments by a foliation study,
and Ryback and Nissen were able to characterize Marmara and Parian marble
as low in Mn, none of these methods has been generally very successful.
Trace element studies suffer inherently from the high variability in concentration (factors of over 100 for Na and Mn) due to highly localized interaction

with inclusions and surrounding rocks. On the other hand, petrofabric studies are difficult and time consuming, require relatively large amounts of material, and depend on statistical treatments of what are, in these marbles, rather poorly developed fabrics.

Our approach to this problem was the idea that the $^{13}C/^{12}C$ and $^{18}O/^{16}O$ variations in Greek marbles probably provided the best chance for unique characterization by locality. The isotopes ^{13}C and ^{18}O vary by at least 10 and 20 per mil (1 to 2 percent), respectively, in marbles from different areas, so that with two isotopes there is a reasonable probability of obtaining some distinctive differences. Moreover, we expected that the very process which originally formed the marbles might well smooth out the variations within a single quarrying area by local isotopic equilibration with fluid phases. In general both these expectations were fulfilled.

We made detailed collections in the four major quarrying areas which provided marble for statuary and building purposes in ancient Greece. In each area we concentrated our sampling at the sites of ancient quarries, as indicated by old chisel marks, worked faces, grooves, and other evidence. Samples were collected over the maximum extent of each locality and from representative phases, in order to determine the variability in each area. A total of 170 quarry samples and some 20 archeological samples were taken.

It appears that the isotopic method for determining the origin of Greek marbles will probably be the most useful of the tests so far devised, especially if used in conjunction with other techniques.

To a Lady on Her Passion for Old China (23-6)

Fossils give joy to *Galen's* soul,
He digs for knowledge, like a mole;
In shells so learn'd, that all agree
No fish that swims knows more than he!

—John Gay

CHAPTER 11

A Geologic Detective Story or Two

Hercule Poirot, Philip Marlowe, Sherlock Holmes, Nero Wolfe, Lew Archer: all detectives of fictional renown. They observe, collect data, hypothesize, deduce, test hypotheses and invariably name the culprit in the last chapter. Many geologic studies require the same mental procedures and the results often compare favorably with the best of the mystery-detective genre. Two samples to illustrate high level geologic sleuthing and sharp deductive reasoning are offered in this section.

The first selection has all the elements of a first-class detective tale; a setting in flamboyant San Francisco in the immediate post-Civil War era, two mysterious prospectors, a bagful of uncut diamonds, involvement of a prominent mining consultant, a huge public stock offering, and a detective-geologist—Clarence King —who was to become the first Director of the United States Geological Survey.

THE GREAT DIAMOND HOAX

W. H. Goetzman

In the spring and summer of 1872, a series of events were taking place that would come to a climax as the capstone of King's entire career. Early in 1872, two grizzled prospectors, Philip Arnold and John Slack, had appeared at a San Francisco bank and attempted to deposit a bagful of uncut diamonds. Having done so, they disappeared. But the diamonds came into instant prominence. They were shown to William Ralston, a director of the Bank of California, and then to a number of other capitalists in the city, who were all intrigued as to their origin. A search was instituted and Arnold and Slack were quickly located. When quizzed, they reported that they had found the diamonds at an undisclosed site. They consented to take a representative of the capitalists to the site, but insisted on blindfolding him first. After disembarking from the Union Pacific Railroad, they led General David Colton on a four-day trip through mountains and canyons to the site of their discovery, where he, too, found diamonds, and rubies, and a number of other precious stones.

Extremely cautious, the California businessmen then took Arnold and his diamonds to New York, where Charles Tiffany, upon examination, pronounced the gems "a rajah's ransom." They had, as one writer put it, discovered "the American Golconda"—a vast field of gems virtually untouched by the hand of man, just waiting to be exploited by sound business enterprise. If they succeeded, Amsterdam itself might well move West. Immediately they formed a nationwide syndicate, the New York and San Francisco Mining and Commercial Company, which bought the rights to the diamond fields from Arnold and Slack for about $600,000, and they prepared to float a public stock issue of $12,000,000. To insure that everything would be absolutely safe and sure, the syndicate hired the most cautious mining engineer in the West, Henry Janin, and he too examined and was dazzled by the diamond field, reporting in the public print that he had personally invested in it and that he considered "any investment at the rate of forty dollars per share or at the rate of four million dollars for the whole property, a safe and attractive one."

During this time, the syndicate had managed to keep the location of the diamond fields a strict secret, and as the news of the discovery leaked out, men with imaginations and the strong acquisitive instincts of the day went wild. Most people located Golconda in Arizona, others in Nevada, and almost every day prospectors wandered into Denver announcing they had found gems. Not since Coronado had followed Cibola halfway across North America had so many men so assiduously pursued windfall wealth. All along the transcontinental railroad, spies watched the trains for suspicious parties of prospectors. Even the telegraphers were not to be trusted.

King and othe other members of the Fortieth Parallel Survey were of course fascinated by the rumors of the diamond discovery, particularly inasmuch as it very probably lay in the heart of country they had crossed many times. They were suspicious. Never in all the weary years of surveying and geologizing had they found gems or even a formation that looked as if it might contain such precious stones. When King visited Emmons and Wilson at camp near Fort Bridger in midsummer, they begged to be allowed to institute a search of their own for the stones, but King, with one eye on the War Department, put it off until the end of the season, when they had completed their routine work. Meanwhile he did some speculating on his own. The location of the diamond field could not be in Arizona, he knew, because the flooded condition of the Snake [Little Snake], Bear [Yampa], and Green rivers precluded any foray in that direction during the past summer. Piecing this fact together with Janin's description of the location of the field "upon a mesa near pine timber," King realized that "there was only one place in that country which answers to the description, and . . . that place lay within the limits of the Fortieth Parallel Survey."

When King, Emmons, Gardner, and A. D. Wilson all met at the office in Montgomery Block on the eighteenth of October, they compared the results of their thinking and, King remembered: "Curiously enough Mr. Gardner and I had without formerly mentioning the subject reached an identical conclusion as to where the spot was."

Actually, some ingenious detective work had been done by Emmons and Wilson, as well as Gardner and King. On the fifth of October, while on the westbound train just out of Battle Mountain, Nevada, Emmons and Gardner had noticed a party of Eastern surveyors who were joined at Alta Station in the Sierras by Henry Janin. They were obviously returning from the diamond fields, so Gardner and Emmons struck up a conversation and gleaned what bits of information they could. They didn't get much out of Janin and his friends, but they did learn that Janin's earlier trip to the diamond fields had taken only three weeks, not enough time to go to Arizona. A. D. Wilson, who had been in the Green River country that summer, reported that the party had disembarked from the train somewhere between Green River and Rawlins. Putting this together with the data about the flooded Green, Yampa, and Snake rivers, they narrowed the field of possibilities considerably. Emmons had further learned that Janin had camped at the foot of a pine-covered mountain which had snow on its slopes as late as June, and Gardner pried from his surveyor the observation that their camp had been on the northeast side of a mountain from which they could see no other high peaks either to the north or to the east. Emmons, Gardner, Wilson, and Hague, who joined the game, concluded that there was only one place the diamonds could be—a peak just north of Brown's Hole, some forty miles east of Green River, right in the country where Utah, Wyoming, and Colorado came together. King agreed completely, and they made plans for an immediate departure for the site.

Everything had to be kept secret, of course. King did not even inform General Humphreys. Instead, just before they boarded the train—in separate parties, with sieves and shovels—he sent a dispatch from Fort Bridger to the General, on October 28, telling him he was off on a reconnaissance of Brown's Hole.

They departed from Fort Bridger on a bitter-cold day. It was October 29. For the next few days they worked their way north through the most unpleasant, freezing weather imaginable. Emmons reported that the horses "became encased in balls of ice." Four days out, they crossed the icy Green River and struck a deep gulley cutting into the side of the mountain they were aiming for. There, out of the biting wind, they made camp and the search began. A few hundred yards out of camp, they found a mark on a tree, then Henry Janin's official water claim, then more mining notices, and finally a small sandstone mesa stained with iron deposits. This must be the spot. Quickly they got down on hands and knees. They probed until dark, finding one diamond apiece on the first try. "That night," remembered Emmons, "we were full-believers in the verity of Janin's reports, and dreamed of untold wealth that might be gathered."

The next day, ignoring the sweeping wind and the shriveling cold, they surveyed the entire area, picking up quantities of stones. They learned, however, that the diamonds did not occur anywhere else but in the mesa where they first found them. By the afternoon they began to get suspicious. King discovered that the gems were always found in a ratio of twelve rubies to one diamond. Then they noticed that most of the stones were in anthills and places where the earth had been ever so slightly disturbed. Their suspicions heightened when one man found a diamond delicately perched atop an anthill where, had it been there long, it would surely have been washed away. Later they found evidence that the stones had been pushed into the anthills with a stick, and some disturbed places even had human footprints around them, though this was not conclusive since Janin and others had worked the site. To be sure, on the third day they checked the stream beds and other places where diamonds must surely have settled, and they dug a deep pit far into the sandstone formation. All this yielded nothing. Golconda was a fraud.

Just as they concluded this, a stranger appeared. He was an unscrupulous New York City diamond dealer who had followed their labors for days through field glasses. When he learned that the diamonds were a hoax, he declared: "What a chance to sell short." This reminded King of the grave responsibility they bore, and that night they made plans to slip out and beat the diamond dealer back to civilization. The next day King and Wilson left before dawn and rode all day, forty-five miles through the badlands, directly to Black Butte Station. They arrived just in time to catch the train for San Francisco.

In San Francisco, King first sought out his friend Janin and confronted him with the evidence. Then they went to the company directors, and King presented a letter with his findings. Ralston and Company stalled for time and begged King to keep quiet until they could sell out, but he refused. They did gain a few days, however, by persuading King to take Janin and the others to the diamond fields for a final check. When they returned, crowds had gathered outside the bank waiting for a dramatic announcement from Ralston. The San Francisco tycoons had been duped by Arnold and Slack and caught by King. There was nothing to do but reveal the whole story. Ralston and his partners published King's letter and the syndicate assumed full responsibility and repaid the stockholders. There were many who still continued to believe, however, that the scheme had originated with the syndicate in the first place. Slack disappeared, and Arnold, back in Kentucky, bought a safe and coolly stored his loot. In 1879 he was shot in the back and the loot disappeared.

Meanwhile King was the hero of the hour. The San Francisco *Bulletin* called him "a cool headed man of scientific education who esteemed it a duty to investigate the matter in the only right way, and who proceeded about his task with a degree of spirit and strong common sense as striking as his success." The Fortieth Parallel Survey as a scientific endeavor, according to the editor, had proven its practical value. Back East, Whitney, who was still fighting his own battle against the oil swindles, rejoiced at King's success. "Who's the King of Diamonds now?" he asked. "And isn't he trumps?"

Unbelievably, however, General Humphreys was displeased. Upon receiving King's official report, he replied soberly: "As these Fields are situated within the limits of the Survey you have charge of, it was eminently proper that they should be included in your operations and an exhaustive examination of them be made." But, he added: "The manner of publicly announcing the results of the examination should I think have been different." In short, King should have gone through channels. But perhaps General Humphreys's stiffness can be excused, for by 1873 the Army was beginning to feel the competition from rival civilian surveys under the Interior Department. In view of increased Congressional opposition, it had to be jealous of whatever laurels fortune cast its way.

In 1911 near Piltdown, England, on the Sussex Downs, a portion of a human skull was recovered from a gravel pit. Additional skull fragments, several humanlike teeth and other mammalian remains were found during the following three or four years. These fragments comprise the corpus delecti *of the second geological detective story. A controversy raged for four decades over the Piltdown discoveries and their significance, but late in 1953, the remains of the "earliest Englishman" were shown to be a hoax. Just how this scientific forgery was exposed is deftly treated by William L. Straus, Jr., of the Johns Hopkins University.*

THE GREAT PILTDOWN HOAX

William L. Straus, Jr.

When Drs. J. S. Weiner, K. P. Oakley, and W. E. Le Gros Clark recently announced that careful study had proven the famous Piltdown skull to be compounded of both recent and fossil bones, so that it is in part a deliberate fraud, one of the greatest of all anthropological controversies came to an end. Ever since its discovery, the skull of "Piltdown man"—termed by its enthusiastic supporters the "dawn man" and the "earliest Englishman"—has been a veritable bone of contention. To place this astounding and inexplicable hoax in its proper setting, some account of the facts surrounding the discovery of the skull and of the ensuing controversy seems in order.

Charles Dawson was a lawyer and an amateur antiquarian who lived in Lewes, Sussex. One day, in 1908, while walking along a farm road close to nearby Piltdown Common, he noticed that the road had been repaired with peculiar brown flints unusual to that region. These flints he subsequently learned had come from a gravel pit (that turned out to be of Pleistocene age) in a neighboring farm. Inquiring there for fossils, he enlisted the interest of the workmen, one of whom, some time later, handed Dawson a piece of an unusually thick human parietal bone. Continuing his search of the gravel pit, Dawson found, in the autumn of 1911, another and larger piece of the same skull, belonging to the frontal region. His discoveries aroused the interest of Sir Arthur Smith Woodward, the eminent paleontologist of the British Museum. Together, during the following spring (1912), the two men made a systematic search of the undisturbed gravel pit and the surrounding spoil heaps; their labors resulted in the discovery of additional pieces of bone, comprising—together with the fragments earlier recovered by Dawson—the larger part of a remarkably thick human cranium or brain-case and the right half of an apelike mandible or lower jaw with two molar teeth *in situ*. Continued search of the gravel pit yielded, during the summer of 1913, two human nasal bones and fragments of a turbinate bone (found by Dawson), and an apelike canine tooth (found by the distinguished archeologist, Father

Teilhard de Chardin). All these remains constitute the find that is known as Piltdown I.

Dawson died in 1916. Early in 1917, Smith Woodward announced the discovery of two pieces of a second human skull and a molar tooth. These form the so-called Piltdown II skull. The cranial fragments are a piece of thick frontal bone representing an area absent in the first specimen and a part of a somewhat thinner occipital bone that duplicates an area recovered in the first find. According to Smith Woodward's account, these fragments were discovered by Dawson early in 1915 in a field about two miles from the site of the original discovery.

The first description of the Piltdown remains, by Smith Woodward at a meeting of the Geological Society of London on December 18, 1912, evoked a controversy that is probably without equal in the history of paleontological science and which raged, without promise of a satisfactory solution, until the studies of Weiner, Oakley, and Clark abruptly ended it. With the announcement of the discovery, scientists rapidly divided themselves into two main camps representing two distinctly different points of view (with variations that need not be discussed here).

Smith Woodward regarded the cranium and jaw as belonging to one and the same individual, for which he created a new genus, *Eoanthropus*. In this monistic view toward the fragments he found ready and strong support. In addition to the close association within the same gravel pit of cranial fragments and jaw, there was advanced in support of this interpretation the evidence of the molar teeth in the jaw (which were flatly worn down in a manner said to be quite peculiar to man and quite unlike the type of wear ever found in apes) and, later, above all, the evidence of a second, similar individual in the second set of skull fragments and molar tooth (the latter similar to those imbedded in the jaw and worn away in the same unapelike manner). A few individuals (Dixon, Kleinschmidt, Weinert), moreover, have even thought that proper reconstruction of the jaw would reveal it to be essentially human, rather than simian. Reconstructions of the skull by adherents to the monistic view produced a brain-case of relatively small cranial capacity, and certain workers even fancied that they had found evidences of primitive features in the brain from examination of the reconstructed endocranial cast—a notoriously unreliable procedure; but subsequent alterations of reconstruction raised the capacity upward to about 1400 cc—close to the approximate average for living men.

A number of scientists, however, refused to accept the cranium and jaw as belonging to one and the same kind of individual. Instead, they regarded the brain-case as that of a fossil but modern type of man and the jaw (and canine tooth) as that of a fossil anthropoid ape which had come by chance to be associated in the same deposit. The supporters of the monistic view, however, stressed the improbability of the presence of a hitherto unknown ape in England during the Pleistocene epoch, particularly since no remains of fossil apes had been found in Europe later than the Lower Pliocene. An anatomist, David Waterston, seems to have been the first to have recognized the extreme morphological incongruity between the cranium and the jaw. From the announcement of the discovery he voiced his disbelief in their anatomical association. The following year (1913) he demonstrated that superimposed tracings taken from radiograms of the Piltdown mandible and the mandible of a chimpanzee were "practically identical"; at the same time he noted that the Piltdown molar teeth not only "approach the ape form, but in several respects are identical with them." He concluded that since "the cranial fragments of the Piltdown skull, on the other hand, are in practically all their details essentially human . . . it seems to me to be as inconsequent to refer the mandible and the cranium to the same individual as it would be to articulate a chimpanzee foot with the bones of an essentially human thigh and leg".

In 1915, Gerrit Miller, curator of mammals at the United States National Museum, published the results of a more extensive and detailed study of casts of the Piltdown specimens in which he concluded that the jaw is actually that of a fossil chimpanzee. This view gradually gained strong support, e.g., from Boule and Ramström. Miller, furthermore, denied that the manner of wear of the molar teeth was necessarily a peculiarly human one; he stated that it could be duplicated among chimpanzees. That some other workers (Friederichs; Weidenreich) have ascribed the jaw to a fossil ape resembling the orangutan, rather than to a chimpanzee, is unimportant. What is important, in the light of recent events, is that the proponents of the dualistic theory agreed in pronouncing the jaw as that of an anthropoid ape, and as unrelated to the cranial fragments. Piltdown II remained a problem; but there was some ambiguity about this discovery, which was announced after the death of Dawson "unaccompanied by any direct word from him". Indeed, Hrdlička, who studied the original specimens, felt convinced that the isolated molar tooth of Piltdown II must have come from the original jaw and that there was probably some mistake in its published history.

A third and in a sense neutral point of view held that the whole business was so ambiguous that the Piltdown discovery had best be put on the shelf, so to speak, until further evidence, through new discoveries, might become available. I have not attempted anything resembling a thorough poll of the literature, but I have the distinct impression that this point of view has become increasingly common in recent years, as will be further discussed. Certainly, those best qualified to have an opinion, especially those possessing a sound knowledge of human and primate anatomy, have held largely—with a few notable exceptions—either to a dualistic or to a neutral interpretation of the remains, and hence have rejected the monistic interpretation that led to the reconstruction of a "dawn man." Most assuredly, and contrary to the impression that has been generally spread by the popular press when reporting the hoax, "Eoanthropus" has remained far short of being universally accepted into polite anthropological society.

An important part of the Piltdown controversy related to the geological age of the "Eoanthropus" fossils. As we shall see, it was this aspect of the controversy that eventually proved to be the undoing of the synthetic Sussex "dawn man." Associated with the primate remains were those of various other mammals, including mastodon, elephant, horse, rhinoceros, hippopotamus, deer, and beaver. The Piltdown gravel, being stream-deposited material, could well contain fossils of different ages. The general opinion, however, seems to have been that it was of the Lower Pleistocene (some earlier opinions even allocated it to the Upper Pliocene), based on those of its fossils that could be definitely assigned such a date. The age of the remains of "Piltdown man" thus was generally regarded as Lower Pleistocene, variously estimated to be from 200,000 to 1,000,000 years. To the proponents of the monistic, "dawn-man" theory, this early dating sufficed to explain the apparent morphological incongruity between cranium and lower jaw.

In 1892, Carnot, a French mineralogist, reported that the amount of fluorine in fossil bones increases with their geological age—a report that seems to have received scant attention from paleontologists. Recently, K. P. Oakley, happening to come across Carnot's paper, recognized the possibilities of the fluorine test for establishing the relative ages of bones found within a single deposit. He realized, furthermore, that herein might lie the solution of the vexed Piltdown problem. Consequently, together with C. R. Hoskins, he applied the fluorine test to the "Eoanthropus" and other mammalian remains found at Piltdown. The results led to the conclusion that "all the remains of *Eoanthropus* . . . are contemporaneous"; and that they are, "at the earliest, middle Pleistocene." However, they were strongly indicated as being of late or Upper Pleistocene age, although "probably at least 50,000 years" old. Their fluorine content was the same as that of the beaver remains but significantly less than that of the geologically older, early Pleistocene mammals of the Piltdown fauna. This seemed to increase the probability that cranium and jaw belonged to one individual. But at the same time, it raised the enigma of the existence in the late Pleistocene of a human-skulled, large-brained individual possessed of apelike jaws and teeth—which would leave "Eoanthropus" an anomaly among Upper Pleistocene men. To complete the dilemma, if cranium and jaw were attributed to two different animals—one a man, the other an ape—the presence of an anthropoid ape in England near the end of the Pleistocene appeared equally incredible. Thus the abolition of a Lower Pleistocene dating did not solve the Piltdown problem. It merely produced a new problem that was even more disturbing.

As the solution of this dilemma, Dr. J. S. Weiner advanced the proposition to Drs. Oakley and Clark that the lower jaw and canine tooth are actually those of a modern anthropoid ape, deliberately altered so as to resemble fossil specimens. He demonstrated experimentally, moreover, that the teeth of a chimpanzee could be so altered by a combination of artificial abrasion and appropriate staining as to appear astonishingly similar to the molars and canine tooth ascribed to "Piltdown man." This led to a new study of all the "Eoanthropus" material that "demonstrated quite clearly that the mandible and canine are indeed deliberate fakes". It was discovered that the "wear" of the teeth, both molar and canine, had been produced by an artificial planing down, resulting in occlusal surfaces unlike those developed by normal wear. Examination under a microscope revealed fine scratches such as would be caused by an abrasive. X-ray examination of the canine showed that there was no deposit of secondary dentine, as would be expected if the abrasion had been due to natural attrition before the death of the individual.

An improved method of fluorine analysis, of greater accuracy when applied to small samples, had been developed since Oakley and Hoskins made their report in 1950. This was applied to the Piltdown specimens. The results clearly indicate that whereas the Piltdown I cranium is probably Upper Pleistocene in age, as claimed by Oakley and Hoskins, the attributed mandible and canine tooth are "quite modern." As for Piltdown II, the frontal fragment appears to be Upper Pleistocene (it probably belonged originally to Piltdown I cranium), but the occipital fragment and the isolated molar tooth are of recent or modern age.

Weiner, Oakley, and Clark also discovered that the mandible and canine tooth of Piltdown I and the occipital bone and molar tooth of Piltdown II had been artificially stained to match the naturally colored Piltdown I cranium and Piltdown II frontal. Whereas these latter cranial bones are all deeply stained, the dark color of the faked pieces is quite superficial. The artificial color is due to chromate and iron. This aspect of the hoax is complicated by the fact that, as recorded by Smith Woodward, "the colour of the pieces which were first discovered was altered a little by Mr. Dawson when he dipped them in a solution of bichromate of potash in the mistaken idea that this would harden them." The details of the staining, which confirm the conclusions arrived at by microscopy, fluorine analysis, and nitrogen estimation, need not be entered into here.

In conclusion, therefore, the *disjecta membra* of the Piltdown "dawn man" may now be allocated as follows: (1) the Piltdown I cranial fragments (to which should probably be added Piltdown II frontal) represent a modern type of human brain-case that is in no way remarkable save for its unusual thickness and which is, at most, late Pleistocene in age; (2) Piltdown I mandible and canine tooth and Piltdown II molar tooth are those of a modern anthropoid ape (either a chimpanzee or an orangutan) that have been artificially altered in structure and artificially colored so as to resemble the naturally colored cranial pieces—moreover, it is almost certain that the isolated molar of

Piltdown II comes from the original mandible, thus confirming Hrdlička's earlier suspicion; and (3) Piltdown II occipital is of recent human origin, with similar counterfeit coloration.

Weiner, Oakley, and Clark conclude that "the distinguished palaeontologists and archaeologists who took part in the excavations at Piltdown were the victims of a most elaborate and carefully prepared hoax" that was "so extraordinarily skilful" and which "appears to have been so entirely unscrupulous and inexplicable, as to find no parallel in the history of palaeontological discovery."

The Forge of Science

How many learned men are working at the forge of science—laborious, ardent, tireless Cyclopes, but one-eyed!

—**Joseph Joubert**

CHAPTER 12

Earth Scientists Also Known As...

War, it is said, is too important to be left to generals. So also with geology: it is much too important to be left only to professionals. It is refreshing to discover that individuals who achieved fame in other fields were also avid students of the Earth. The list is long, but as an appetizer seven examples are given of men renowned for their endeavors in other fields who were also meritorious earth scientists. Consider the following eminent personages: two presidents of the United States, a 19th century English artist, author, critic, and social reformer, a statesman of the American Revolution and Founding Father of his country, two of a handful of the acknowledged geniuses of human history, and one of America's greatest painters. All shared a common interest, geology.

The third president of the United States also was a reputable scientist, with interests that ranged from agronomy to botany and chemistry to paleontology. As a geologist, Thomas Jefferson's major contributions were in the field of vertebrate paleontology, but in his writings he speculated on the origin of the earth, the formation of mountain ranges, the origin of the Natural Bridge of Virginia, the occurrence of shells far above sea level and he also involved himself in the Plutonist-Neptunist controversy. While president of the American Philosophical Society of Philadelphia, the foremost scientific organization in America, he presented several learned geologic papers, one of which, "A Memoir on the Discovery of Certain Bones of a Quadruped of the Clawed Kind in the Western Part of Virginia," caused a mild furor in geologic circles. Later he commisioned General Clark, of Lewis and Clark fame, to excavate a mammoth in Kentucky. Our extract describes the Great-claw Megalonyx of Jefferson, mammoth bones in the presidential mansion, and details concerning the pioneer American vertebrate paleontologist.

MEGALONYX, MAMMOTH, AND MOTHER EARTH

E. T. Martin

In 1796 Jefferson came into possession of some fossilized bones of a large animal heretofore unknown to American scientists. The bones had been found in a saltpeter cave in Greenbriar County, Virginia, now West Virginia. Upon receipt of them Jefferson sent a brief description to David Rittenhouse, President of the American Philosophical Society. Rittenhouse was actually dead at the date of this letter, July 3, 1796, though the news had not yet reached Monticello. Jefferson ascribed the bones to "the family of the lion, tyger, panther etc. but as preeminent over the lion in size as the Mammoth is over the elephant." In fact, the size of the claw and general bulk of the animal led him to name this unknown creature "the Great-claw, or, Megalonyx." He expressed his ultimate intention to deposit the bones with the Society.

Jefferson's discovery naturally created great interest among his scientific friends in Philadelphia. Benjamin Smith Barton wrote him on August 1 that his account of these bones would be "very acceptable" to the American Philosophical Society for publication in the forthcoming volume of the *Transactions*, if it were received in time.

Jefferson's excitement over his new discovery and his eagerness to have his account published in the *Transactions* glows through his letter to Dr. Benjamin Rush of January, 1797: "What are we to think of a creature whose claws were eight inches long, when those of the lion are not 1½ inches; whose thighbone was 6¼ diameter; when that of the lion is not 1½ inches?

Were not the things within the jurisdiction of the rule and compass, and of ocular inspection, credit to them could not be obtained. . . . I wish the usual delays of the publications of the Society may admit the addition to our new volume, of this interesting article, which it would be best to have first announced under the sanction of their authority."

When Jefferson rode to Philadelphia to assume his duties as Vice-President of the United States, he carried with him his collection of bones and the article he had written about them. In Philadelphia, he discovered something that challenged his identification of this unknown animal, his *Megalonyx,* as a carnivore belonging to the lion or tiger family. This was "an account published in Spain of the skeleton of an enormous animal from Paraguay, of the clawed kind, but not of the lion class at all; indeed, it is classed with the sloth, ant-eater, etc., which are not of the carnivorous kinds; it was dug up 100 feet below the surface, near the river La Plata. The skeleton is now mounted at Madrid, is 12 feet long and 6 feet high."

The discovery Jefferson refers to had been made in South America in 1789. It was, in actuality, an extinct ground sloth of herbivorous habits whose habitat had been both North and South America. Its remains have been found in Pleistocene deposits.

Though his own identification was now in doubt, Jefferson went ahead with his original plans. His paper was read before the American Philosophical Society on March 10, 1797, and was published in the fourth volume of its *Transactions,* in 1799. The bones which he had—a radius, an ulna, the second, third, and fifth metacarpals, a second phalanx of the index finger, and a third phalanx of a thumb—he deposited with the Society. Jefferson added a postscript to his paper, commenting on the similarities between his animal and the giant sloth *(Megatherium),* and concluded that positive identification must await the discovery of missing parts of the skeleton. In the meantime he was unwilling to change his own identification.

This decision was consistent with Jefferson's own demand that scientific conclusions be reached with greatest caution after the most careful investigation of all facts. Perhaps he considered his postscript sufficient notice that the identification was tentative. Interestingly enough, an essentially accurate identification of Jefferson's animal as a species of ground sloth was made by Dr. Caspar Wistar, Philadelphia physician, later professor of anatomy at the University of Pennsylvania and a recognized authority in American vertebrate paleontology during this period, and was also published in the fourth volume of the *Transactions* of the American Philosophical Society.

Jefferson's greatest popular fame in the field of paleontology came from his work with the so-called "mammoth," a huge prehistoric animal whose remains attracted international attention during Jefferson's day. Jefferson's interest in the mammoth appears to have begun when he was collecting material for his *Notes on Virginia,* that is, about 1780.

In his *Notes,* Jefferson points out that the mammoth must have been the largest of American quadrupeds. Remains "of unparalleled magnitude," he says, have been found rather recently on the Ohio River and in many parts of America further north. Indian tradition, Jefferson noted with satisfaction, has it that such a huge, carnivorous creature still roams the northern parts of America. Jefferson then argues that, contrary to certain European opinion, ascribing these remains to the hippopotamus or elephant, or to both, they cannot belong to either.

Though they definitely belong to one animal only, the tusks, frame, and teeth of these remains prove that this animal was not a hippopotamus. Nor could it have been an elephant since (1) its skeleton bespeaks an animal "five or six times the cubic volume of the elephant," (2) the "grinders are five times as large, are square, and the grinding surface studded with four or five rows of blunt points" and hence quite different from those of the elephant, (3) no elephant grinder has been found in America, (4) the elephant could not have lived in those regions where these remains are now discovered—unless one is willing to suppose some great difference in the earth in past ages, which Jefferson feels no one given to "cautious philosophy" could do. The habitat of the elephant, he points out, is from 30° south of the equator to 30° north of the equator; that of the mammoth must have begun at 36½° north latitude and probably extended to the pole itself. This "largest of all terrestrial beings" was built upon a scale so vast that it should never have been confused with the elephant by European naturalists.

The mammoth's great superiority over the elephant in size was insisted upon by Jefferson time and again during his ambassadorship to France. In arguing that the remains of this huge creature could not have belonged to the hippopotamus or to the elephant, he challenged the opinion of Daubenton, Buffon, and other leading naturalists of this period.

Further evidence of Jefferson's interest in this great creature and other prehistoric animals appeared during the early days of his presidency of the American Philosophical Society. At its first meeting over which he presided (May 19, 1797), "A Plan for Collecting Information Respecting the Antiquities of N. A." came under consideration. On April 6, 1798, a second report upon this plan was gone into in great detail and a committee consisting of Jefferson, George Turner, Caspar Wistar, Dr. Adam Seybert, C. W. Peale, James Wilkinson, and John Williams was appointed. By the end of the year this committee reported that a circular letter had been "extensively distributed" throughout the country soliciting help in accomplishing, among other things, the following: the procuring of "one or more entire skeletons of the Mammoth, so called, and of such other unknown animals as either have been, or hereafter may be discovered in America."

Fossils were among the treasures brought back by the Lewis and Clark Expedition of 1804-1806. At about this time Jefferson and Dr. Caspar Wistar also tried, but without success, to procure for the American Philosophical

Society a large collection of bones dug out by a Dr. William Goforth at Big Bone Lick, famous source of prehistoric animal remains located in what is now the State of Kentucky. Jefferson expected Goforth's findings to include the skull of a mammoth—so desperately needed by the American Philosophical Society to complete its collection of the bones of this huge vertebrate. He was also hopeful (as he wrote to Dr. Samuel Brown, in July, 1806) that there would be bones of the Megalonyx. But Goforth sold his collection to an adventurer who took them to England. Therefore President Jefferson had to undertake a private paleontological venture. He arranged with the owner of the Big Bone Lick for Clark to go there in 1807 and do some excavating at Jefferson's expense. Clark "employed ten laborers several weeks" and unearthed some three hundred bones, which he shipped down the Mississippi to New Orleans, from whence they were forwarded to Jefferson in Washington.

On receiving Clark's letter with the news of the collection he had made, Jefferson immediately invited Dr. Caspar Wistar to Washington to see the bones as soon as they arrived. Jefferson confides to him his plans for their disposal. There is to be "a tusk and femur which General Clarke [*sic*] procured particularly at my request, for a special kind of cabinet I have at Monticello." But his primary purpose has been "to procure for the society as complete a supplement to what is already possessed as that Lick can furnish at this day." Hence Wistar is to select what bones the American Philosophical Society may desire, and the rest are to go to the National Institute of France.

The odd cargo arrived at the Presidential mansion in Washington early in 1803. "The bones are spread in a large room [the unfinished East Room]," Jefferson immediately informs Wistar, "where you can work at your leisure, undisturbed by any mortal, from morning till night, taking your breakfast and dinner with us. It is a precious collection, consisting of upwards of three hundred bones, few of them of the large kinds which are already possessed. There are four pieces of the head, one very clear, and distinctly presenting the whole face of the animal. The height of his forehead is most remarkable. In this figure, the indenture of the eye gives a prominence of six inches to the forehead. There are four jaw bones tolerably entire, with several teeth in them, and some fragments; three tusks like elephants; one ditto totally different, the largest probably ever seen, being now from nine to ten feet long, though broken off at both ends; some ribs; an abundance of teeth studded, and also those of the striated or ribbed kind; a fore-leg complete; and then about two hundred small bones, chiefly of the foot. This is probably the most valuable part of the collection, for General Clarke, aware that we had specimens of the larger bones, has gathered up everything of the small kind. There is one horn of a colossal animal. . . . Having sent my books to Monticello, I have nothing here to assist you but the 'Encyclopédie Methodique.'" All this during the desperate days of the Embargo!

In addition to this interest in prehistoric forms of life (though to him they were not strictly prehistoric, since he thought it likely that the mammoth and *Megalonyx* still existed), Jefferson also had some interest in the history of the earth itself—this despite two objections he had to geology—its lack of utility and its uncertainty. His objections have already been touched upon. He thought geology useful, however, as an aid in prospecting for minerals.

Furthermore the range of Jefferson's scientific interest was too great to leave out geology. We find him speculating on the formation of mountain ranges and whether they are generally parallel with the earth's axis; on the origin of the Natural Bridge; on the possibility of American rivers having once been inland lakes which broke through their mountain dams and made their ways to the ocean (as in the case of the Potomac, Shenandoah, and Delaware rivers); on the possibility of the Gulf of Mexico having been formed by an ocean current; on the general history of the topography of America; on the Flood recorded in the Bible; on the occurrence of sea shells in places far distant from the ocean; and on the origin and history of the earth itself. In 1797 he urged the American Philosophical Society to promote "researches into the Natural History of the Earth, the changes it has undergone as to Mountains, Lakes, Rivers, Prairies, etc."

Specialization

How did the jolly dinosaur
Improve each shining era?
By getting large and specialized
Extinction drawing nearer.

—Radcliffe "Blue Book"

James McNeill Whistler's official biographer, Joseph Pennell, contended that the world has known but two supreme etchers, Rembrandt and Whistler, and it is Whistler who is considered the greater. Few are aware that James Whistler learned the art of engraving and etching while employed by the United States Coast and Geodetic Survey. His first plate, made in 1854 of the coastline of Boston Bay, was used to demonstrate his aptitude for the craft, and in the same year he completed U. S. Coast and Geodetic Survey Plate 414A of Anacapa Island, one of the Channel Islands of California. William A. Stanley of the Geodetic Survey in a revealing article, Three Short, Happy Months *tells how Whistler became a member of the Survey, the circumstances under which he left the organization, and the influence these experiences had on the career of one of America's greatest artists.*

THREE SHORT, HAPPY MONTHS

William A. Stanley

How would the Coast and Geodetic Survey deal with a draftsman who was habitually late, frequently absent, given to graffiti, and inclined to doodle on its official charts?

You guessed it. But today, after more than a century, such a man remanins one of that august agency's most highly regarded alumni.

He was James McNeill Whistler, famed for his magnificent portrait of his mother and for many other works of art. But to his superiors in the survey, where the first evidence of his artisic potential was expressed in its charts, he was trouble with a capital T.

A native of Lowell, Massachusetts, born July 10, 1834, he belonged to a family of soldiers, Scotch-Irish by descent. At the age of seventeen, after spending some time in St. Petersburg, where his father was carrying on an engineering assignment for the Czar, he entered West Point. He spent three years there, departing in his senior year due mainly to his disinterest in obeying rules, chief of which had been his lack of promptness.

His father's plan was to apprentice him to a friend who had connections with a locomotive works in Baltimore, Maryland, and where his stepbrother George was then employed. Whistler soon saw enough of the locomotive works, however, to know that he did not want to be an apprentice, and it was not long before he left Baltimore and headed for Washington. D.C.

In November 1854, after he tried to enter the Navy, his ability as a draftsman induced another friend of his father's to recommend him for a position in the engraving department of the U.S. Coast Survey. He was appointed on November 7, 1854, and at the age of 20 began a short but colorful employment with the Federal Government.

It has been reported that Whistler remarked, "I shall always remember the courtesy shown me by that fine Southern gentleman Jefferson Davis, through whom I got my appointment with the Coast Survey." Whistler continued, "It was after my little difference with the Professor of Chemistry at West Point that it was suggested—all in the most courteous and correct West Point manner—that perhaps I had better leave the Academy." With an introduction by Jefferson Davis he reported to Captain Benham, his new Coast Survey boss, who assigned him to the position of draftsman at the salary of $1.50 a day.

In the Coast Survey, then located on the northwest corner of New Jersey and C Street, SE., Whistler, the draftsman, was considered a playboy of sorts as he performed the duties of an assistant in the cartographic section of the Survey; even in those permissive days when offices closed at 3:00 p.m. and it was quite the custom to send out the messenger for a bucket of ale. Whistler would engage his time by drawing caricatures and sketches in unconventional places somewhere in the building. It was said that he would bring an extra hat to the office and when a superior came looking for him he would find his hat on the peg, assuming then that James was in the building. However, it is said he was wearing the other hat on his way to a nearby tavern.

During his stay he was regarded as a person who could not adjust himself to any form of regular routine. He would say "It was not that I was late; the office opened too early." He worked only intermittently on his assignments and took great delight in occupying his time with odd sketches on any available fabric that presented a suitable surface—an envelope, or a copperplate upon which he might have been assigned to engrave a chart.

Whistler sometimes drew uncomplimentary images of the Bureau officials on the bare white walls leading to the Superintendent's office. He would not overcome the temptation to stop frequently on his way up or down the stairs to correct, change, or add to his caricatures.

This was a time of uncertainty for him, when visions of his real career were beginning to form. In the brief period of less than four months Coast Survey employment, from November 1854 to February 1855, one fact was certain; he had no ambition to become a permanent government employee.

John Ross Key, fellow draftsman in the Coast Survey, and the grandson of Francis Scott Key, recalled in later years that he and Whistler roomed in a house of Thirteenth Street, near Pennsylvania Avenue and that Whistler usually dined in a restaurant closer by on Pennsylvania Avenue. He also lived for a while in a house at the north-east corner of Twelfth and E Streets, NW., a two-story brick building which is no longer standing. His rent was ten dollars a month at the rooming house, a considerable sum in the light of his small government pay.

He produced two works which have been referred to by historians as

"Coast Survey No. 1" and "Coast Survey No. 2, Anacapa Island" (engraved by J. Whistler, J. Young, and C. A. Knight). The latter, which includes an etching by Whistler of the headland of the eastern extremity of Anacapa Island, California, has been reproduced and issued as Coast and Geodetic Survey plate number 414A, but there is no record of the former.

These two plates are not, however, Whistler's first efforts as a copperplate engraver in the Coast Survey. Upon his entrance to duty he received technical instructions in the art of etching and copper engraving, which are reflected in a series of practice sketches. First place should be accorded to those which are the initial efforts on copper of his fancy and invention, and the first genuine Whistler etchings. The sketches are in the shapes of little heads that intrude on the blank spaces of the copper above and around two neatly engraved views of portions of the coast of Boston Bay. At intervals, while doing the topographical view, he paused to etch on the upper part of the copperplate the vignette of a soldier's head, a suggestion of a portrait of himself as a Spanish hidalgo and an etching of a motherly figure which has attracted considerable interest in recent years. The original copper engraving is regarded as priceless. It was donated to the Freer Gallery of Art, where it is now displayed.

Legend has it that on occasion he would also give play to his genius by inserting drawings of dignified and scientific characters that might not have been in the original plan of a seacoast; however, no one could prove it. It has also been reported, but this cannot be verified either, that late in 1854 he engraved a sketch of a portion of the Atlantic coast. This drawing in copper is said not only to include some graceful sea serpents and beautiful mermaids but also several large and smiling whales. His Coast Survey supervisor reportedly told him that if he ever again desecrated on the Survey's charts with animal life, he would be discharged.

His next assignment was the sketch of Anacapa Island, previously described, showing the natural bridge which the waters of the Pacific Ocean had carved. Whistler did the work with extra care. He finished his pictorial view in approved style, although to him it looked humdrum. With his extreme feeling of personal frustration when not allowed to express himself freely, and with due consideration given to his job, he had to fix it—and fix it he did. He added two flocks of gulls sailing gracefully over the rocky headland as if heading south for winter quarters. These gulls may be seen today on plate 414A. When informed of wrongdoing and confronted with possible discharge Whistler replied, "Surely the birds don't detract from the sketch. Anacapa Island couldn't look as blank as that map did before I added the birds."

Due in large part to tardiness, James McNeill Whistler in mid-February 1855 terminated his employment with the U.S. Coast Survey and went off to Paris and London to achieve undying fame as one of the world's great artists. He is now known the world over, and by some authorities has been acclaimed the greatest of American artists. During his early training in the Coast and Geodetic Survey, he came in contact with some of the most talented engravers in the country.

As his legend is handed down from one generation to another of Coast Survey personnel, there is a feeling of pride that the Agency helped, even briefly, to launch one of the great careers in American art.

Antiquity

How cunningly nature hides every wrinkle of her inconceivable antiquity under roses and violets and morning dew!

—Ralph Waldo Emerson

Over a hundred years after Thomas Jefferson, another scientist-engineer, Herbert Clark Hoover, occupied the White House. Hoover was a superb mining engineer and geologist, who roamed the world finding and developing mineral deposits, a path which led eventually to the presidency. From his autobiography, Hoover's formative years in the first class at Stanford University are described. He worked as a clerk-typist for the chairman of the Geology Department and was guided by such leading geologists of the time as John Branner, Waldemar Lindgren, and Joseph LeConte.

He and his wife, Lou Henry Hoover, also a student of geology, made a significant contribution by translating and publishing De Re Metallica. *Their singular talents in Latin, geology, mining, and metallurgy made possible a translation of Agricola's Latin compendium on mining methods first published in 1556. The passages included here are of Hoover's student days and the reader will find the White House engineer-geologist to be warmer and with more humor than is generally supposed.*

STANFORD UNIVERSITY

1891–1895
Herbert C. Hoover

The University opened formally on October 1, 1891. It was a great occasion. Senator and Mrs. Stanford were present. The speeches of Senator Stanford and Dr. David Starr Jordan, the first President, make dry reading today but they were mightily impressive to a youngster. Dr. John Branner, who was to preside over the Department of Geology and Mining, had not yet arrived, so with Professor Swain's guidance I undertook the preparatory subjects that would lead into that department later on. Upon Dr. Branner's arrival I came under the spell of a great scientist and a great teacher, whose friendship lasted over his lifetime.

My first need was to provide for myself a way of living. I had the $210 less Miss Fletcher's services, together with a backlog of some $600 which had grown from the treasured insurance of my father. The original sum had been safeguarded and modestly increased over the years by the devoted hands of Laurie Tatum. Professor Swain interested himself and secured me a job in the University office at $5 a week—which was enough of a supplement for the time. Soon after, Dr. Branner tendered me a job in his department because I could operate a typewriter. This increased my income to $30 a month. While dwelling on earnings I may mention that with two partners I had established a laundry agency and a newspaper route upon the campus, both of which, being sub-let, brought in constant but very small income.

The first summer vacation Dr. Branner obtained for me a job as an assistant on the Geological Survey of Arkansas where he had been State Geologist. The $60 a month and expenses for three months seemed like a fortune. During my sophomore and junior summer vacations I worked upon the United States Geological Survey in California and Nevada, where I saved all my salary. These various activities and the back-log carried me over the four years at the University. I came out with $40 in my pocket and no debts.

The work in Arkansas consisted of mapping the geologic outcrops on the north side of the Ozarks. I did my job on foot mostly alone, stopping at nights at the nearest cabin. The mountain people were hospitable but suspicious of all "government agents." Some were moonshiners and to them even a gawky boy might be a spy. There were no terms that could adequately explain my presence among them. To talk about the rocks only excited more suspicion. To say that I was making a survey was worse, for they wanted no check-up on their land-holdings. To say I was tracing the zinc- or coal-bearing formations made them fearful of some wicked corporate invasion. I finally gave up trying to explain. However, I never failed to find someone who would take the stranger in at nightfall and often would refuse any payment in the morning.

The living conditions of many of these people were just as horrible as they are today. Generations of sowbelly, sorghum molasses and cornmeal, of sleeping and living half a dozen in a room, had fatally lowered their vitality and ambitions. The remedy—then as today—is to regenerate racial vitality in the next generation through education and decent feeding of the children.

My work on the United States Geological Survey in the glorious High Sierra, the deserts of Nevada, and among the mining camps where vitality and character ran strong, was a far happier job. Dr. Waldemar Lindgren headed these survey parties and I was a cub assistant. When in the high mountains we camped out with teamsters, horses and pack mules, and, of equal importance, a good camp cook.

Most of the work was done on horseback. During those two summers I did my full lifetime mileage with that mode of transportation. In these long mountain rides over trails and through the brush, I arrived finally at the conclusion that a horse was one of the original mistakes of creation. I felt he was too high off the ground for convenience and safety on mountain trails. He would have been better if he had been given a dozen legs so that he had the smooth and sure pace of a centipede. Furthermore he should have had scales as protection against flies, and a larger water-tank like a camel. All these gadgets were known to creation prior to the geologic period when the horse was evolved. Why were they not used?

We had a foreign geologist visiting our party who had never seen a rattlesnake. On one hot day a rattlesnake alarm went off near the trail. The horse notified me by shying violently. I decided to take the snake's corpse to our visitor. I dismounted and carefully hit the snake on the head with a stick, then wrapped him in a bandana handkerchief and hung him on the pommel. Some minutes later while the horse and I were toiling along to camp, half asleep in the sun, the rattlesnake woke up and sounded another alarm. That was too much for the horse. After I got up out of the brush I had to walk five miles to camp. It added to my prejudices against horses in general.

There was some uncertainty one summer as to whether I could get a Geological Survey job. Therefore when vacation came other students and I canvassed San Francisco for work at putting up or painting advertising signs along the roads. Our very modest rates secured a few hundred dollars of contracts, with which we bought a team and camp-outfit. We made for the Yosemite Valley, putting up eyesores, advertising coffee, tea and newspapers along the roads. We pitched our camp in the Valley intending to spend a few days looking the place over. Professor Joseph Le Conte was camped nearby, and I listened spellbound to his campfire talk on the geology of the Valley. A few days later I received a telegram advising me I could join the Survey party. There was not enough money left in the pockets of my sign-painting partners to pay my stage-fare to the railroad. I walked 80 miles in three days and arrived on time.

As the youngest member of the Geological party, I was made disbursing officer. It required a little time for me to realize this was not a distinction but a liability. I had to buy supplies and keep the accounts according to an elaborate book of regulations which provided wondrous safeguards for the public treasury. One morning high in the Sierras we discovered one of the pack-mules dead. I at once read the regulations covering such castastrophes and found that the disbursing officer and two witnesses must make a full statement of the circumstances and swear to it before a notary public. Otherwise the disbursing officer was personally responsible for the value of the animal. I was thus importantly concerned to the extent of $60. The teamsters and I held an autopsy on the mule. We discovered his neck was broken and that the caulk of one loose hind-shoe was caught in the neck-rope with which he was tied to a tree. We concluded that he had been scratching his head with his hind foot, had wedged his halter-rope in the caulk, had jerked back and broken his neck. When we reached civilization we made out an elaborate affidavit to that effect. About two months afterward I was duly advised from Washington that $60 had been deducted from my pay, since this story was too highly improbable. Apparently mules did not, according to the book, scratch their heads with their hind feet. Dr. Lindgren relieved my $60 of misery by taking over the liability, saying he would collect it from some d— bureaucrat when he got back to Washington in the winter.

For years I watched every mule I met for confirmation of my story. I can affirm that they do it. I even bought a statuette of a mule doing it. Some twelve years later I was privileged to engage Dr. Lindgren for an important job in economic geology in Australia. I met him at the steamer in Melbourne. His first words were, "Do you know that that d— bureaucrat never would pay me that $60? And do you know I have since seen a hundred mules scratch their heads with their hind feet?" He would not take the $60 from me.

Stanford is a coeducational institution, but I had little time to devote to coeds. However, a major event in my life came in my senior year. Miss Lou Henry entered Stanford and the geology laboratories, determined to pursue and teach that subject as a livelihood. As I was Dr. Branner's handy boy in the department, I felt it my duty to aid the young lady in her studies both in the laboratory and in the field. And this call to duty was stimulated by her whimsical mind, her blue eyes, and a broad grinnish smile that came from an Irish ancestor. I was not long in learning that she also was born in Iowa, the same year as myself, and that she was the daughter of a hunting-fishing country banker at Monterey who had no sons and therefore had raised his daughter in the out-of-door life of a boy. After I left college she still had three years to complete her college work. I saw her once or twice during this period. We carried on a correspondence.

All of these extra-curricular matters so crowded my life that I neglected to discharge those conditions on entrance "Credits" under which I had entered as a freshman. Had it not been for the active intervention of my friends, Dr. Branner and Professor J. Perrin Smith, who insisted among other things that I could write English, those implacable persons in the University office would have prevented my getting a diploma with my own class. Nevertheless it duly arrived. It has been my lot in life to be the recipient of honorary diplomas (often in exchange for Commencement addresses) but none ever had the sanctity or, in my opinion, the importance of this one.

A giant of the Victorian era in England was John Ruskin, writer, critic, social reformer, and artist. He had such profound influence on the public taste of the people of Britain that he came to be known as "The Great Victorian." Yet his early interests were geological and Ruskin was fascinated by minerals and crystallography, the architecture of the Alps, landscape, and the origin of agate, gneiss, breccias, and puddingstones. He studied with Buckland at Oxford, as a student conversed with Charles Darwin and later in his career published a series of seven articles in the Geological Magazine. In 1869 he was elected Slade Professor of Art at the University of Oxford. His lectures there reached an enormous public. He was later made an Honorary Fellow of Corpus Christi and still later endowed museums at Oxford and at Sheffield. The early training, travels, and discoveries of an "almost-geologist" are described by W. G. Coll-ingwood in the following selections from his Life of Ruskin.

MOUNTAIN-WORSHIP

W. G. Collingwood

More interesting to him than school was the British Museum collection of minerals, where he worked occasionally with his Jamieson's Dictionary. By this time he had a fair student's collection of his own, and he increased it by picking up specimens at Matlock, or Clifton, or in the Alps, wherever he went, for he was not short of pocket-money. He took the greatest pains over his catalogues, and wrote elaborate accounts of the various minerals in a shorthand he invented out of Greek letters and crystal forms.

Grafted on this mineralogy, and stimulated by the Swiss tour, was a new interest in physical geology, which his father so far approved as to give him Saussure's 'Voyages dans les Alpes' for his birthday in 1834. In this book he found the complement of Turner's vignettes, something like a key to the 'reason why' of all the wonderful forms and marvellous mountain-architecture of the Alps.

He soon wrote a short essay on the subject, and had the pleasure of seeing it in print, in Loudon's *Magazine of Natural History* for March, 1834, along with another bit of his writing, asking for information on the cause of the colour of the Rhine-water.

In his second term he had the honour of being elected to the Christ Church Club, a very small and very exclusive society of the best men in the college: 'Simeon, Acland, and Mr. Denison proposed him; Lord Carew and Broadhurst supported.' And he had the opportunity of meeting men of mark, as the following letter recounts. He writes on April 22, 1937:

'MY DEAREST FATHER,

'When I returned from hall yesterday—where a servitor read, or pretended to read, and Decanus growled at him, "Speak out!"—I found a note on my table from Dr. Buckland, requesting the pleasure of my company to dinner, at six, to meet two celebrated geologists, Lord Cole and Sir Philip Egerton. I immediately sent a note of thanks and acceptance, dressed, and was there a minute after the last stroke of Tom. Alone for five minutes in Dr. B.'s drawing-room, who soon afterwards came in with Lord Cole, introduced me, and said that as we were both geologists he did not hesitate to leave us together while he did what he certainly very much required—brushed up a little. Lord Cole and I were talking about some fossils newly arrived from India. He remarked in the course of conversation that his friend Dr. B.'s room was cleaner and in better order than he remembered ever to have seen it. There was not a chair fit to sit upon, all covered with dust, broken alabaster candlesticks, withered flower-leaves, frogs cut out of serpentine, broken models of fallen temples, torn papers, old manuscripts, stuffed reptiles, deal boxes, brown paper, wool, tow and cotton, and a considerable variety of other articles. In came Mrs. Buckland, then Sir Philip Egerton and his brother, whom I had seen at Dr. B.'s lecture, though he is not an undergraduate. I was talking to him till dinner-time. While we were sitting over our wine after dinner, in came Dr. Daubeny, one of the most celebrated geologists of the day—a curious little animal, looking through its spectacles with an air very *distinguée*—and Mr. Darwin, whom I had heard read a paper at the Geological Society. He and I got together, and talked all the evening.'

Of less interest to the general reader, though too important a part of Mr. Ruskin's life and work to be passed over without mention, are his studies in Mineralogy. We have heard of his early interest in spars and ores; of his juvenile dictionary in forgotten hieroglyphics; and of his studies in the field and at the British Museum. He had made a splendid collection, and knew the various museums of Europe as familiarly as he knew the picture-galleries. In the 'Ethics of the Dust' he had chosen Crystallography as the subject in which to exemplify his method of education; and in 1867, after finishing the letters to Thomas Dixon, he took refuge, as before, among the stones, from the stress of more agitating problems.

In the lecture on the Savoy Alps in 1863 he had referred to a hint of Saussure's, that the contorted beds of the limestones might possibly be due to some sort of internal action, resembling on a large scale that separation into concentric or curved bands which is seen in calcareous deposits. The contortions of gneiss were similarly analogous, it was suggested, to those of the various forms of silica. Mr. Ruskin did not adopt the theory, but put it by for examination in contrast with the usual explanation of these phenomena, as the simple mechanical thrust of the contracting surface of the earth.

In 1863 and 1866 he had been among the Nagelflüh of Northern Switzerland, studying the pudding-stones and breccias. He saw that the difference between these formations, in their structural aspect, and the hand-specimens in his collection of pisolitic and brecciated minerals was chiefly a matter of size; and that the resemblances in form were very close. And so he concluded that if the structure of the minerals could be fully understood, a clue might be found to the very puzzling question of the origin of mountain-structure.

Hence his attempt to analyze the structure of agates and similar banded and brecciated minerals, in the series of papers in the *Geological Magazine;* an attempt which, though it was never properly completed, and fails to come to any general conclusion, is extremely interesting as an account of beautiful and curious natural forms too little noticed by mineralogists.

The Lecture

When I think of this lecture, I do not wonder that I determined never to attend to Geology.

—**Charles Darwin**

Truly a man of all seasons, Johann Goethe's interests embraced drama, poetry, painting, philosophy, politics, and various aspects of science. The creator of Faust *and the* Sorrows of Werther *was a member of nearly 30 scientific societies, an active participant in the Plutonist-Neptunist battle over the origin of rocks, foreshadowed portions of Darwin's theory of organic evolution, early recognized the concept of an "Ice Age," established a collection of rocks, minerals, and fossils of over 18,000 items, and it was in his honor that the familiar iron mineral, goethite, was named. The first geologic map of Germany was published in 1821 and the color system adopted was based on Goethe's proposals. Interestingly, modern geologic maps are prepared using essentially the same color code.*

MINERALOGY, GEOLOGY, METEOROLOGY

R. Magnus

"I fear not the reproach that it must have been a spirit of contrariety that has led me from the contemplation and description of the human heart—the youngest, ficklest, most versatile, mobile and changeable aspect of creation—to the observation of nature's oldest, firmest, deepest and most steadfast scion."

Thus wrote Goethe in the year 1784 during his studies of granite. For more than fifty years we find the poet intensely preoccupied with the study of the earth's crust, its structure and origin. To trace down every last scientific detail in this field of research would transcend the framework of this book. All we can do is to give a general picture of his investigations and his views.

As we already know, Goethe's interest in mineralogy sprang from practical causes. There was the problem of reviving the long-dormant mining industry at Ilmenau, a problem which Goethe, as Chief Minister, began to tackle in 1777. He found that in Thuringia the way for geological study had already been paved. Interest in this field was stirring, especially because of the nearness of the Mining Academy at Freiberg and the work of the most noted geologist of the time, Abraham Gottlob Werner.

The mines at Ilmenau exploited seams. This posed the problem of identifying the geological strata carrying the seams. Goethe's attention was drawn to the great regularity in stratification that marked the earth's crust in Thuringia, evidence that to him bore out the teachings of Werner, an adherent of the so-called "Neptunian Theory," which attributed the formation of the earth's crust to the effects of water.

Even during this early period Goethe had occasion to display his practical bent in another direction. At his behest the Duke of Weimar in 1779 acquired the large Walch mineral collection. This was installed in the palace at Jena, under the direction of Lenz, later to become the first professor of mineralogy at Jena. The collection was organized on Werner's system and formed the core of the famous Museum of Mineralogy to come.

In order to secure a trained expert for the mining industry, Goethe persuaded the Duke to send J. C. W. Voigt—later likewise to become famous—to study at the Mining Academy. In 1780 Goethe drafted an elaborate set of instructions under which Voigt was to conduct a geological tour of inspection of the Duchy of Weimar.

Goethe's views on geology, rooted in the soil of Thuringia, greatly expanded on his numerous journeys. As early as his first and second Harz journey he had gathered geological observations and visited the mines at Goslar and Klausthal. His third journey to the Harz mountains, undertaken in the company of the artist Kraus, was wholly devoted to such studies. Goethe's geological diary covering these weeks has been preserved and gives evidence of the seriousness with which these inquiries were conducted. The Goethe House to this day contains the fine drawings by Kraus of the most noteworthy granite formations in the Harz outcroppings—for it was the study of granite that chiefly preoccupied the travelers.

His numerous sojourns in the watering resorts of Bohemia gave Goethe occasion to collect mineralogical data there too. In the course of time he grew intimately familiar with this section of Bohemia. He studied the nature of hot mineral springs, confirmed his "Neptunian" views, traced the distribution of the great coal deposits. In Switzerland he studied the form and distribution of glaciers and the effects of earlier glaciation. In Italy he climbed Aetna and Vesuvius and visited the Phlegraean Plain, thus gaining a firsthand view of volcanic activity. Even during the campaign in France he pursued his mineralogical researches. This, then, was the raw material on which Goethe based his studies in geology. Its significance lies in the fact that it was bound to strengthen the "Neptunian" views which Werner had instilled. Goethe himself pointed out that he might not have become such a strong adherent of this theory, had he conducted his initial studies in the volcanic Auvergne, for example.

He brought home from his journeys numerous specimens which he arranged systematically. Voigt had introduced him to mineralogical nomenclature. Thus there came into being the large collection, numbering more than 18,000 items, that still delights the visitor to the Goethe House by its wealth and the beauty of its rarities. The collection is largely devoted to specimens from Thuringia, the Harz mountains and Bohemia, but other sections of Germany, Italy and many foreign lands are likewise represented. Among other things Goethe kept a careful record of all fossils unearthed in Thuringia.

The collection assembled by one Joseph Müller, a lapidary of Karlsbad, came to have special importance for Goethe. In the course of his trade Müller had collected many mineral specimens from the vicinity of Karlsbad. His collection grew more and more and in 1806 he showed it to Goethe, who began to arrange the minerals in order. Proceeding from granite, Goethe worked out a continuous series of the various deposits and prepared a careful scientific catalogue. He then persuaded Müller to put his collecting activities on a broader basis and to market his specimens by Goethe's system. As early as 1806 Goethe advertised the collection in a literary periodical at Jena, and two years later he published the catalogue in Von Leonhard's "Pocketbook of Mineralogy."

In this way a model collection of minerals was made available to scientists everywhere. Noeggerath, the Bonn mineralogist, described it as ideal for instructional purposes. Using this collection, Goethe published a careful mineralogical account of the region of Karlsbad. In 1821 he followed with a similar account of the region of Marienbad, based on a corresponding collection. He never lost his interest in the Karlsbad collection, and as late as 1832 he advertised a series of thermal-spring sinters marketed by Knoll, Müller's successor.

A collector and scientist himself, Goethe was in close touch with many experts in the field. On the subject of mineralogy he corresponded with his friends Merck and Von Knebel, with Von Trebra, Von Leonhard, August von Herder, Cramer, Count Sternberg, Grüner, and especially with Lenz. There was spirited barter in specimens among these men and even with others more remote. Goethe often figured in these exchanges, thus expanding his own collection and that at Jena. When Jena was unable to acquire the important Cramer collection for lack of funds, Goethe saw to it that it went to Heidelberg.

In 1796–98 Lenz established the Mineralogical Society at Jena, and it elected Goethe its first honorary member. Lenz was extremely active in enrolling new members for the Society and in expanding the Jena collection with their help. It grew to be one of the most important ones of its time, and savants from abroad came in large numbers to admire its treasures.

Mineralogy was at the time enlisting two important auxiliary sciences—analytical chemistry and crystallography. The initiative of the great Scandinavian chemist Berzelius had led to the chemical analysis of minerals. Goethe himself never carried out such analyses, though he followed the progress of the new science with great interest. He was kept abreast of development in the new chemistry first through Göttling and later through Döbereiner. The large collection of handbooks and textbooks of chemistry in his library gives eloquent testimony of his lively interest. His interest in crystallography, aroused especially by Soret, was less intense, though he did make occasional observations on the genesis, growth and size of crystals.

Goethe's researches in mineralogy and geology form no milestone in the history of these sciences, yet "he was up to the best of his contemporaries." In later years at least his observations were held in high esteem by scholars in the field. He collaborated in Von Leonhard's "Pocketbook of Mineralogy," and Noeggerath wrote a most favorable review of his writings in the Jena press. On account of his researches in northern Bohemia he was elected an honorary member of the Bohemian scientific society founded by Count Stern- berg. A scientific society in Edinburgh did likewise.

Goethe devoted special attention to drift boulders and erratic blocks found in the Alps and the North German plains. He ascribed their origin to various causes. The granite blocks of the Rhone valley he thought had been trans- ported to their resting places by glaciers. Many blocks in northern Germany, however, he regarded as remnants of an ancient mountain range that had somehow evaded weathering. As the most important example he cited the so-called Heiligendamm. In addition, he postulated that some of these boul- ders might have drifted across the sea from Scandinavia on ice floes and icebergs. This follows one of Voigt's hypotheses, and it should be noted that Preen actually observed Scandinavian rocks cast up on the Baltic coast by ice floes.

As a result of these studies, Goethe developed the idea of an Ice Age, and it appears that he was indeed the first to assume such an epoch. "It is my conjecture that all of Europe, at least, passed through an epoch of great cold." At the time, he thought, the ocean still extended over the continent to an altitude of a thousand feet, Lake Geneva was directly connected with the sea and the Alpine glaciers flowed down to Lake Geneva. These views reappear in his "Wilhelm Meister."

In his geological studies Goethe seems to have developed a whole series of concepts that did not achieve general recognition until much later. Apart from his views on the historical significance of fossils and his postulation of an Ice Age, there is especially his conviction that the forces that once shaped the earth and the mountains are the very same which we see at work every day, only in modified form. Hence his endeavor to find analogies to geological processes in ordinary observation—such as the formation of glacier ice. "As far as I am concerned, I am quite willing to believe that nature even today is capable of forming precious stones as yet unknown to us."

Indeed, Goethe extended this approach to the very rock that to him was the basis of everything—granite. "It is quite possible that granite may have been formed repeatedly." Since he thought that forces still at work were entirely adequate to explain the formation of the earth and that changes in the earth's surface in our own time take place only very slowly, he was bound to arrive at the conclusion, now generally accepted, that the periods of the earth's history were of extraordinary length. He put this conviction into the mouth of Thales in the second part of "Faust":

> Never did nature with her living might
> Depend on hours, on mere day and night.
> Each form in gentle temperance is wrought.
> Nor even the greatest change with violence fraught.

We see, therefore, that Goethe's geological studies led him to rather modern views on the formation of the earth.

The period that embraced Goethe's preoccupation with geology was dominated by the controversy between the "Neptunians" and the "Vulcanists"—between those who ascribed the main share in shaping the earth to water and those who favored volcanic forces. The controversy has long since been settled. Water and fire have each claimed their appropriate share in geological history. But at the time the battle was joined with the greatest vehemence. By and large Goethe avoided the exaggerations in both doctrines. The cast of his mind inclined him more toward Werner's "Neptunian" view, and in the "Xeniae Tamed" he complained:

> Scarce noble Werner turns about,
> Poseidon's realm falls prey to loot.
> Hephaestus puts them all to rout—
> But not myself, to boot.
> With easy credit I grow pettish;
> Full many a credo have I spurned;
> Full well to hate them have I learned—
> New idol and new fetish.

But he reserved his grandest picture of this scientific controversy for the second part of his "Faust." In the "Classical Walpurgis Night" scene he has Thales and Anaxagoras roam the mountains and seas, hotly debating the issue—the former a "Neptunian," the latter a "Vulcanist." Here too he implied his partisanship on behalf of Thales, rejecting the doctrines of the headlong Vulcanists who would not shrink from having stones rain from the moon. Yet even in "Faust" the issue is not settled. Instead, the origin of the mountains is dramatically recited by Seismos (earthquake) along Vulcanist lines:

> In this, then, you must credit me—
> No other power chalked the score.
> So fair the world would never be,
> Had I not rocked it to the core.

But Thales clings to his Neptunian view, which he expresses in these magnificent lines:

> All sprang from out the briny deep!
> All does the water safe embrace!
> Ocean, grant us thy endless grace!
> If thou thy clouds us didst not send,
> If thou rich torrents didst not spend,
> The curving rivers didst not wend,
> To them thy bosom didst not lend,
> What were the hills, the plains, the world?
> 'Tis thou keeps life all fresh unfurled!

Goethe's own attitude toward the effect and significance of volcanic forces shifted in the course of his researches. He was never able to close his mind to their importance, yet he shrank from the excesses of the Vulcanist school. With Werner he sought to ascribe many alleged products of volcanic eruption to subterranean conflagrations which were supposed to have occurred especially in the great coal beds. He thought there was much evidence of this kind in the Bohemian deposits. The firing of the rocks was supposed to have been facilitated by the interspersed plant remains. Near Grünlass, for example, he found a bituminous shale that could be ignited by a flame. He studied the effect of combustion and annealing on a whole series of rocks— we have a list of thirty-eight different minerals which he had fired in a potter's kiln at Zwätzen in 1820 to study the results. He made repeated tests along these lines and imputed to them great importance in evaluating natural deposits.

Thus in 1824, near Atalbenreuth in Bohemia, he found "newly discovered ancient traces of natural fire and heat." In the vicinity of Karlsbad he studied the effect of such subterranean fires on clay schists and quartz, to see how they ultimately gave rise to earthy slag. He thought such processes very important but never closed his mind to the realization that volcanic forces must also have been at work.

He collected and carefully described the volcanic rock of the Kammerbühl mountain near Eger, and in 1808 he described this peak as an ancient submarine volcano. In 1822 he suggested that this view might be confirmed if a shaft were driven into the mountain from one side so that its internal structure could be studied. Count Sternberg actually carried out such a project after Goethe's death. Yet by 1824 Goethe thought that here too he could recognize pseudovolcanic processes, postulating that the Kammerbühl basalt might have been subsequently modified by the combustion of a superimposed mixture of shale and coal.

In Goethe's view volcanoes have no common origin in a core of liquid fire within the earth, but arise purely locally, when water penetrates into depths where subterranean conflagrations rage. Hence most volcanoes are located close to the sea. In the erupting volcanoes of the High Andes in South America he held that the water was supplied by the melting snow.

One of the problems that preoccupied Goethe and his contemporaries was the genesis of basalt, which was regarded as a very young formation and formed one of the main bones of contention between Neptunians and Vulcanists. Goethe delved into this controversy too, making proposals to reconcile the conflicting opinions, though never reaching a final conclusion. Hence his groan of envy:

> Better, America, thy lot
> Than our old continent's,

> For ruined castles trouble thee not,
> Nor basalt's battlements.

On one point, however, Goethe's anti-Vulcanist position was clear and unequivocal. He utterly rejected the assumption that the earth's surface, having once been formed, was subsequently modified by rising or falling, by folding, cracking or faulting. Today such processes are assumed to have taken very long periods of time, but the Vulcanists of Goethe's time thought they were of cataclysmic character, with whole mountain ranges rising to their full height all at once. To this view Goethe was always firmly opposed. "Be the matter as it may, let it be recorded that I have only contempt for this accursed devil's tattoo theory of creation."

We have already seen that in Goethe's view the mountains in their main outlines were formed coincident with the formation of granite and that he rejected the idea of later mountain formations. It was in this spirit that Faust said:

> Ah, mountain peaks, be noble and be still!
> I ask not whence and why did come this hill.
> When nature to herself her orbit founded,
> The earthly sphere she pure and simple rounded,
> In crest and chasm sought her seemly pleasure,
> Put rock to rock to take the mountain's measure;
> Then, dropping easily from crag and wall,
> Sloped down the hills to where the valleys sprawl,
> Where grows and burgeons green the grass. She asks
> No madcap whirligigs to do her tasks.

The views of his opponents Goethe parodied in Mephistopheles' tale of how the demons, locked inside the earth, wheeze and sneeze, their exhalations altering the earth's surface. Goethe's main reason for rejecting such subsequent changes was the observation he had made early in his geological studies, in central Germany, where strata and seams are arranged with great regularity. He simply applied the evidence of his eyes to other fields. Even where geological strata were more or less inclined rather than horizontal he thought he could postulate that they were originally formed that way.

A main exhibit to show that there had been changes in altitude on earth even in historical times was the Temple of Serapis in Pozzuoli. The columns still stand and halfway up show evidence of having once been attacked by marine paddocks, though now they are inland. Von Heff, in agreement with the view generally held today, took this to prove that this coastal region had been flooded by the sea in the middle ages, later to rise again. Goethe, by means of drawings, tried to show that when the Temple was buried a water-filled depression probably formed in the middle, where the mollusks could have lived, and that there was no need to postulate any subsequent rise.

Just as he had done in the case of volcanoes, Goethe attributed thermal springs to localized causes, deriving them from surface water penetrating deeply. He was convinced that the Karlsbad springs would cease to bubble if the Tepel river were deflected from its bed. In his view the surface water, penetrating into the depths, heated and activated the solid rock by moistening it as in a Galvanic pile. He clung stubbornly to this view, in the face of the generally held theory that the springs issued "from the boiling abyss within the earth's crust."

On one occasion Goethe dealt in a practical sense with "balneology," the science of the therapeutic use of water. This was in 1812, when he wrote a detailed report on whether the sulphur springs near Berka might be exploited by organizing a health resort. The document, based on analytical work by Döbereiner, is again a model of its kind. It contains precise data on the probable yield from the springs, on arrangements to heat the water, to construct steam and mud bath facilities, to ship the water and to plan a practical health resort.

Few geologists, on taking up their fine colored charts and maps, are likely to realize that this color system goes back to Goethe. When Keferstein's geological map of Germany was first published in 1821, the system of coloration—essentially the one still in use today—was adopted according to Goethe's proposals. In making them he had been guided by two considerations: first, to color the different geological strata in such a way that they were as distinct as possible from one another; and second, to keep the color scheme as harmonious as possible. Goethe's studies in the field of color thus continued to play a part in practical geology down to recent times.

Even this brief survey of Goethe's work in mineralogy and geology shows that here too he was a thorough worker, forever deepening and broadening his knowledge. But in contrast to his studies in the field of optics, which he always put forward with great self-assertion and outspokenness, he was much more reserved and modest in his geological writings. He was well aware that the factual material available to him was insufficient to explain the origin of the earth, and he thus often acknowledged the need for resorting to hypotheses in geology. Yet here too it remains characteristic of Goethe that he always made a careful distinction between fact and hypothesis, gathering, sifting and recording the facts as precisely as possible, and reserving the right to change his mind on hypothetical matters.

Printer, author, philanthropist, inventor, statesman, diplomat, Founding Father, and scientist all describe the career of Benjamin Franklin. His interest in science lead him as early as 1737 to write in the Pennsylvania Gazette *of earthquakes and he had equal curiosity about the nature of waves at sea, the cause of springs on mountains, fossil shells, variations in climate, the effect of oil on storm waves, and techniques of swamp drainage. While Postmaster General, Franklin recognized the existence of the Gulf Stream and had a chart engraved and printed by the General Post Office. He made a series of surface temperature measurements while crossing the Atlantic and devised a method to attempt measurement of subsurface water temperatures.*

BENJAMIN FRANKLIN AND THE GULF STREAM

H. Stommel

In 1770 the Board of Customs at Boston complained to the Lords of the Treasury at London that the mail packets usually required two weeks longer to make the trip from England to New England than did the merchant ships. Benjamin Franklin was Postmaster General at the time and happened to discuss the matter with a Nantucket sea captain, Timothy Folger. The captain said he believed that charge to be true.

> 'We are well acquainted with the stream because in our pursuit of whales, which keep to the sides of it but are not met within it, we run along the side and frequently cross it to change our side, and in crossing it have sometimes met and spoke with those packets who were in the middle of it and stemming it. We have informed them that they were stemming a current that was against them to the value of three miles an hour and advised them to cross it, but they were too wise to be councelled [*sic*] by simple American fisherman.'

Franklin had Folger plot the course of the Gulf Stream for him and then had a chart engraved and printed by the General Post Office.
Franklin believed that the Gulf Stream was caused

> by the accumulation of water on the eastern coast of America between the tropics by the trade winds. It is known that a large piece of water 10 miles broad and generally only 3 feet deep, has, by a strong wind, had its water driven to one side and sustained so as to become 6 feet deep while the windward side was laid dry. This may give some idea of the quantity heaped up by the American coast, and the reason of its running down in a strong current through the islands into the Gulf of Mexico and from thence proceeding along the coasts and banks of Newfoundland where it turns off towards and runs down through the Western Islands.

By the time of Maury, in the middle of the nineteenth century, Franklin's estimates of the velocities of the Stream were regarded as excessive, but more recent studies tend to confirm them. Franklin did not give any details concerning the edge of the Stream. Starting in 1775, both Franklin and Charles Blagden independently, conceived the idea of using the thermometer as an instrument of navigation, and each made a series of surface temperature measurements while crossing the Atlantic. On Franklin's last voyage in 1785 he even attempted to measure subsurface temperatures to a depth of about 100 ft., first with a bottle and later with a cask fitted with a valve at each end.

Nature's Way

Nature is often hidden, sometimes
overcome, seldom extinguished...
[and] to be commanded, must
be obeyed.

—Francis Bacon

The great art masterpieces The Last Supper, Mona Lisa, *and* Madonna of the Rocks *and the true nature of fossils, isostatic adjustment, and a modern concept of the immensity of geologic time might appear to be immiscible, but all are products of the genius of Leonardo Da Vinci. A scientist who rigorously applied the scientific method and placed faith only in his own research, Leonardo was centuries ahead of his time as a geologist. His contribution to geology has been less widely recognized than his accomplishments as a painter, canal builder, sculptor, inventor, and military engineer. In a hitherto unpublished article by Dr. Thomas Clements, Emeritus Professor of Geology at the University of Southern California, Da Vinci receives due recognition for his geological endeavors.*

LEONARDO DA VINCI AS A GEOLOGIST

Thomas Clements

Leonardo da Vinci is known to most people as a painter. He was a painter, a great painter; and yet, in his oft-quoted evaluation of his own capabilities, written to Lodovico Sforza, Duke of Milan, in about his thirtieth year, he lists himself first as a military engineer, then as an architect, a canal builder, a sculptor, and last of all as a painter. He says ". . . in painting I can do as well as any other, whoever he may be." The present writer is not attempting to belittle Leonardo as a painter; he simply wishes to show that this many-facetted man was also a great engineer and a great scientist.

It has been argued that Leonardo cannot be rated as a great scientist because he did not make known his works to the world. He did not publish them himself, and his notes are largely written in his peculiar mirror-writing, with many run-together words and abbreviations: notes and sketches on several subjects frequently occurring on a single page. Some believe that this was done purposely, in order to shroud his works in mystery, or perhaps as a safety measure to protect himself from persecution by the church for his unorthodox views. It has also been suggested that he did this as a precaution against plagiarism until such time as he could publish his observations over his own signature.

It is said by those who have made a study of Leonardo's writings and drawings that he was naturally lefthanded, although forced to write with his right hand in youth, and thus apparently becoming ambidextrous. In his maturity, as he became busier with the many works of various natures that he undertook, it seems natural that for his notes he should revert to what probably was for him the faster method of writing: his normal, left-handed method from right to left, and that for the same reason he should abbreviate and run words together. Furthermore, with as active a mind as his, it is not surprising that in the midst of writing notes on one subject something else should suggest itself, and should be jotted down.

There seems to be little question that Leonardo did hope to publish some of his material, for there is record of his beginning to rewrite some of his notes in 1508, eleven years before his death. There are outlines among his notes for several books: one on hydraulics, one on geology, one on anatomy, and many others. The "Treatise on Painting," published many years after his death, was put together from his own notes and notes of his students. That he deliberately endowed the reading of his notebooks with difficulties is certainly open to question, as must be any interpretation of the motives that govern a man's acts. The reason that his works were not published in his lifetime is perhaps because he never ceased working, and therefore never attained the time and leisure necessary to put his notes in order for publication.

For a full appreciation of the scientific work of Leonardo da Vinci, it is necessary to know something of the times in which he lived. He was born in 1452, the year before the fall of Constantinople to the Turks, and a few years after the invention of printing in Europe. In other words, he was born at the beginning of the Renaissance. He lived until the next to the last year of the second decade of the 16th Century.

During his lifetime lived della Robbia, Botticelli, Michaelangelo, Titian, Raphael, Holbein, to name a few of the artists; Columbus, Vespucci, Vasco de Gama, Pizarro, Balboa, Magellan, Cortés, among the explorers; Erasmus, Machiavelli, Rabelais among the philosophers and writers; Savonarola, Luther, Loyola, Calvin among the reformers; and Copernicus among the scientists.

Printing developed tremendously in Italy after its invention, with the establishment by the end of the 15th Century of 532 presses in that country, more than existed in all the rest of Europe. Universities, likewise, were relatively abundant in Italy, and there was ample opportunity to exchange ideas and obtain information throughout most of Leonardo's life. The science of mathematics was well established in Italy, and engineering works of a high order were being undertaken, but science as a whole had advanced little since the time of Aristotle.

Leonardo has sometimes been thought of as dilletante or a dabbler as far as the sciences are concerned, because of his widely scattered interests. In the writer's opinion, this is not a just evaluation of his scientific work. It is true that he delved into many sciences, but Leonardo was an engineer; it was necessary for him to know certain fundamental facts in science, and when the knowledge was not already available, he attempted to work it out for himself. Furthermore, he had a tremendous amount of intellectual curiosity. While most of his research was of a practical nature, or begun with a practical application in view, this intellectual curiosity led him on to fields far from his original starting point. One idea suggested another, as indicated by the notes on varied subjects found on a single page of his manuscripts.

Although an original thinker, Leonardo did not disdain the learning of the past. That he read widely is suggested by lists of books mentioned in his manuscripts: on one page of the Codice Atlantico (Folio 210) there are the names of forty books, and it has been stated that he was familiar to a greater or lesser extent with the works of more than seventy authors of both the Classical and the Middle Ages. Nevertheless, he did not necessarily accept as final the writings of these men on any subject.

It is here that Leonardo is seen as the true scientist. Freely admitting that others had explored all of the various fields in which he himself was interested he insisted on testing their conclusions; he accepted no authority but the results of experimentation. He truly developed the inductive method, that approach to problems today commonly called the scientific method. He performed experiments; he drew conclusions; and he reasoned logically to arrive at general laws.

As a civil engineer, Leonardo found perhaps the greatest outlet for his engineering [geological] talents. He was particularly a hydraulic engineer, and much of his professional engineering had to do with canals and water movement. He was interested in the use of water as a motive force, and also for irrigation, and was particularly active in flood control and the design of canals and locks. It was while he was Ingegnere Camerale of Milan (1498–1503) that he improved the canal system of that city, making an orderly and integrated whole out of what had hitherto been a haphazard system of unrelated canals. In his redesign of the San Marco lock, where the Martesana Canal flowed into the internal canal, he apparently first introduced the mitered lock gate.

He first considered a general plan for control and improvement of the Arno River while at Milan, at which time he made a rough sketch map of the whole watershed. Presumably as a result of this preliminary study, he was officially invited in 1503 by the City of Florence to develop the details, and spent three years in working out a detailed plan. It is said that in the development of this new design "Leonardo gave the most advanced exhibition of creative engineering imagination that so far had been conceived, and it is doubtful when his complete lack of precedent and of accurate knowledge are compared with the wealth of both that has accumulated since, whether his work on the Arno has ever been surpassed as a feat of advance."

He prepared new and careful maps of the whole watershed, proposed dams for flood control, and canals, including locks, for navigation. He made careful cost computations, thus indicating an understanding of engineering economics, and in order to be able to speed up the work he designed machines for raising the earth from the excavations. The cost of putting into effect so bold a design apparently was too great for the city fathers of Florence, and after vainly trying to persuade groups that would profit by the improvements to underwrite it, he returned to Milan.

It is particularly in the field of geology that Leonardo's great powers of observation, his controlled imagination, his sound reasoning, and his immense fund of common sense led him far beyond the beliefs of his time. He apparently early formed the habit of wandering in the fields and hills, seeing all things not only with the eye for beauty of the artist, but also with the inquiring eye of the scientist. His later work with rivers and canals gave him further opportunities to study geologic phenomena. His writings abound with references to geology, and his pictures, such as the *Madonna of the Rocks*, show a sound knowledge of the fundamental geology of landscapes.

Although originally accepting the common belief of the time that shells of marine organisms found far from the sea were the remains of animals left there by the deluge of Noah, his later observations and reasoning showed him the fallacy of such a belief. He argued that cockle shells occurring many miles from the sea could not have travelled that distance in the forty days and nights of the deluge. Nor could they have been washed there by the rising waters of the flood, since they occurred unmixed with other debris that a flood would have carried. That they were not an unnatural creation in place he knew from the presence of paired shells, and shells of varying stages of growth, as shown by their growth lines. He rightly reasoned that the distribution of land and sea had not always been as it was then, but that at one time the sea had covered those parts of the dry land where fossils now occur and ". . . above the plains of Italy, where now birds fly in flocks, fish were wont to wander in large shoals." Parsons has called him the Father of Paleontology.

He made a study of wave action, and he acquired a wide knowledge of the work of running water. He realized that valleys were cut by the rivers occupying them, and varied as the streams varied. He observed that rocks transported by rivers gradually became rounded and were worn smaller and smaller. He knew that the material deposited in the sea by the rivers became cemented together to form the various types of sedimentary rocks, and that these later were uplifted again to form mountains.

His knowledge of the work of rivers led him naturally to a concept of geologic time that like so many other of his conclusions is remarkably modern. The teaching of the church in his day was that the earth was formed some five thousand years before the birth of Christ. Leonardo, however, had learned from observation how slowly river deposits form, and he declared that the deposits of the River Po had required two hundred thousand years to accumulate, and this, to him, by no means meant the whole of geologic time.

His study and experimentation in the field of mechanics led him to another strikingly modern geologic conclusion. He noted that rain and rivers were constantly gnawing at the mountains and carrying material to the sea, and yet the mountains were high, and the basin of the sea low. He says: "That part of the surface (of the earth) of whatever weight, is farthest from the center of gravity (of the earth) which is lightest. Therefore, the place from which weight is removed (by the rivers) is made lighter, and in consequence moves farther from the center of gravity (i.e. is uplifted)." This is a statement of isostatic adjustment, as defined by Dutton in 1889.

In the foregoing it has been shown that Leonardo da Vinci's scientific inquiry led him over a wide range of interests. His early chroniclers were inclined to be apologetic for his scientific work: to them he was a painter and sculptor who let this foolish absorption with other matters interfere with the execution of his art commissions. The wonder is not that he did not finish so many of the paintings and sculpturings that he was commissioned to do, but rather that he completed as many as he did. All honor to Leonardo da Vinci as an artist; but more honor to him as an engineer and scientist!

Science and Religion

Science without religion is lame, religion without science is blind.

—**Albert Einstein**

Part III

GEOLOGY AND SOCIETY

Man is a creature of the earth. All men, like Shakespeare's Caliban, are in a literal sense, "of the earth, earthy." Because we share this small planet with 3.5 billion other members of the human race, dependent alike on its dwindling resources and subject alike to climatic constraints and natural hazards, it is scarcely surprising that geology influences the life of human society on both a local and global scale. Base metals for industry, streams and rivers for transportation and power, coal and petroleum for energy sources, mountain ranges as physical barriers, soil development by means of weathering processes, topographic contol of the location of cities, and the devastation caused by earthquakes, floods, and certain volcanic eruptions are a few obvious relationships that demonstrate the influence of geology on individuals and on society. We have selected three broad areas to illustrate this influence: minerals and world history, geology and political action, and the material basis of our civilization. There are many more, but these three alone establish with clarity the interrelationship of geologic processes and earth materials to human society.

CHAPTER 13

Geology and World History

Topography and mineral resources are as significant in the research of some historians as they are in that of geologists. Topographic features often have dictated where battles were fought, how defenses were planned, what strategy could be applied and why victories were won or lost. Accumulations of iron, silver, gold, copper, coal and petroleum have precipitated wars, disrupted monetary systems, influenced the rise and fall of empires, stimulated exploration, and decided the fate of millions of men and women. Two selections are presented to illustrate the relationship between topography and military activity, and topography and the location of cities. Another depicts the influence of mineral deposits on the course of world history, and the concluding selections show the impact of climatic change on nations and on the individual.

Written in 1917 at the height of World War I, the passions of the time are
evident in the introductory portion of this excerpt. Douglas W. Johnson
(1878–1944), distinguished American geomorphologist, discussed the pro-
found influence of topography on World War I. In this section of his book,
he is concerned with the Paris Basin and the options enforced on defenders
and attackers alike. Terrain in western Europe is such that the flat lands
of the Netherlands and Belgium invite an attacker from the east to approach
Paris from the north. This was true in both of the European conflicts of
the 20th Century. Dr. Johnson explains why.

TOPOGRAPHY AND STRATEGY IN THE WAR

Douglas W. Johnson

The violation of Belgian neutrality was predetermined by events which took
place in western Europe several million years ago. Long ages before man
appeared on the world stage Nature was fashioning the scenery which was
not merely to serve as a setting for the European drama, but was, in fact,
to guide the current of the play into blackest tragedy. Had the land of Belgium
been raised a few hundred feet higher above the sea, or had the rock layers
of northeastern France not been given their uniform downward slope toward
the west, Germany would not have been tempted to commit one of the most
revolting crimes of history and Belgium would not have been crucified by
her barbarous enemy.

For it was, in the last analysis, the geological features of western Europe
which determined the general plan of campaign against France and the detailed movements of the invading armies. Military operations are controlled
by a variety of factors, some of them economic, some strategic, others
political in character. But many of these in turn have their ultimate basis in
the physical features of the region involved, while the direct control of
topography upon troop movements is profoundly important. Geological history had favored Belgium and northern France with valuable deposits of coal
and iron which the ambitious Teuton coveted. At the same time it had so
fashioned the topography of these two areas as to insure the invasion of
France through Belgium by a power which placed "military necessity" above
every consideration of morality and humanity. The surface configuration of
western Europe is the key to events in this theater of war; and he who would
understand the epoch-making happenings of the last few years cannot ignore
the geography of the region in which those events transpired.

What is now the country of northern France was in time long past a part
of the sea. When the sea bottom deposits were upraised to form land, the
horizontal layers were unequally elevated. Around the margins the uplift was
greatest, thus giving to the region the form of a gigantic saucer or basin.

Because Paris today occupies the center of this basin-like structure, it is known to geologists and geographers as "the Paris Basin."

Since the basin was formed it has suffered extensive erosion from rain and rivers. In the central area where the rocks are flat, winding river trenches, like those of the Aisne, Marne, and Seine, are cut from three to five hundred feet below the flat upland surface. To the east and northeast, the gently upturned margin of the basin exposes alternate layers of hard and soft rocks. As one would naturally expect, soft layers like shales have readily been eroded to form broad flat-floored lowlands, like the Woevre district east of Verdun. The harder limestone and chalk beds are not worn so low, and from parallel belts of plateaus, the "côtes" of the French.

The fact that the rock layers dip toward the center of the basin has one striking result of profound military importance. Every plateau belt is bordered on one side by a steep, irregular escarpment, representing the eroded edge of a hard rock layer; while the other side is a gentle slope having about the same inclination as the dip of the beds. The steep face is uniformly toward Germany, the gentle back-slope toward Paris; and the crest of the steep scarp always overlooks one of the broad, flat lowlands to the eastward. The military consequences arising from this peculiar topography will readily appear. It is not difficult to understand why the plateau belts have long been called "the natural defenses of Paris."

Descending the face of the Argonne scarp and crossing the valley of the Aire River, you continue eastward across a minor plateau strip and reach the winding trench of the Meuse. Past immortal Verdun and its outlying forts, you press on to the crest of the next great scarp. What a view here meets the eye! To the north and south stretches the long belt of plateau, cut into parallel ridges by east- and west-flowing streams,—ridges like the Côte du Poivre, whose history is written in the blood of brave men. Below, to the east, lies the flat plain of the Woevre, whose impervious clay soil holds the water on the surface to form marshes and bogs without number. Here the hosts of Prussian militarism fairly tested the strength of the natural defenses of Paris, and suffered disastrous defeat. Moving westward under the hurricane of steel hurled upon them from above, their manoeuvering in the marshes of the plain easily visible to the observant enemy on the crest, the invading armies assaulted the escarpment again and again in fruitless endeavors to capture the plateau. Only at the south where the plateau belt is narrower and the scarp broken down by erosion did the Germans secure a precarious foothold, thereby forming the St. Mihiel salient; while at the north entering by the oblique gateway cut by the Meuse River, they pushed south on either side of the valley only to meet an equally disastrous check at the hands of the French intrenched on the east-and-west cross ridges. Viewing the battle-fields from your vantage point on the plateau crest, you read a new meaning in the Battle of Verdun. You comprehend the full significance of the well-known fact that it was not the artificial fortifications which saved the city. It was the defenses erected by Nature against an enemy from the east, skilfully utilized by the heroic armies of France in making good their battle cry, "They shall not pass." The fortified cities of Verdun and Toul merely defend the two main gateways through this most important escarpment, the river gateway at Verdun being carved by the oblique course of the Meuse, while the famous "Gap of Toul" was cut by a former tributary of the upper Meuse, long ago deflected to join the Moselle. Other fortifications along the crest of the scarp add their measure of strength to the natural barrier.

Once more you resume your eastward progress, traverse the marshy and blood-soaked plain of the Woevre, ascend the gentle back slope of still another plateau belt, and stand at last on the crest of the fifth escarpment. Topographically this is the outermost line of the natural defenses of Paris, and as such might be claimed on geological grounds as the property of France. But since the war of 1870 the northern part of this barrier has been in the hands of Germany, who purposed in 1914 to widen the breach already made in her neighbor's lines of defense. Metz guards a gateway cut obliquely into the scarp, and connects with the Woevre through the Rupt de Mad and other valleys.

Farther south is Nancy, marking the entrance to a double gateway through the same scarp. Here in the first week of September, 1914, under the eyes of the Kaiser, the German armies, moving southward from Metz, where they were already in possession of the natural barrier, attempted to capture the Nancy gateways and the plateau crest to the north and south. Once again the natural strength of the position was better than the Kaiser's best. From the Grand Couronné, as the wooded crest of the escarpment is called, the missiles of death rained down upon the exposed positions of the assaulting legions. The Nancy gateway was saved, and more than three years from that date is still secure in the hands of the French. The test of bitter experience has fully demonstrated to the invading Germans that it was no idle fancy which named the east-facing scarps of northern France "the natural defenses of Paris."

It is but reasonable to expect that many of the rivers of northern France should flow down the dip of the rock layers and converge toward the center of the Paris Basin, where the beautiful city itself is located. Most of the river gateways through the concentric lines of escarpments have been carved by these converging streams, or by streams which did so converge before they were deflected to other courses by drainage rearrangements resulting from the excavation of the parallel belts of broad lowlands. Of course these natural openings through the plateau barriers have great strategic value, and must figure prominently in any military operations in the Paris Basin. Some of them constitute the only feasible routes along which armies and their impedimenta may cross the barriers, as elsewhere steep grades and poor roads are the rule, At each of them a town of greater or less importance has sprung up, and both town and gateway are protected either by permanent forts or hastily constructed field fortifications. So great is the strategic value of the principal gateways, such as those near Toul and Verdun, that we find them marked by some of the most strongly fortified cities in the world. The fortifications dominate the roads, canals, and railway lines which pass through the openings, and must be reduced before the cities can be occupied and the transportation lines freely used. This explains the significance of the frequent mention, especially in the war despatches of the first year, of such towns as La Fère, Laon, Rheims, Epernay, and Sézanne guarding gateways in the first line of cliffs east of Paris; of Rethel and Vitry in the second line; of Bar-le-Duc at the south end of the third; of Verdun and Toul in the fourth; of Metz and Nancy in the fifth; to say nothing of other points in the same and lesser scarps not considered in this volume.

The importance of the strategic gateways will readily appear if we consider their relation to principal railway routes. No railway of eastern France can traverse the country from the German frontier to Paris without seeking out several of these fortified gateways in succession. Take, for example, the main through line from Strassburg to Paris. After crossing the German border it follows the valley of the Meurthe a short distance to reach one of the two Nancy gateways. Turning west through this opening in the Metz-Nancy escarpment, it makes straight for the famous Gap of Toul in the Verdun-Toul escarpment. Once through the Gap, the line bends north to find an opening in the low continuation of the Forêt d'Argonne scarp near the town of Bar-le-Duc. Westward from here the route crosses the Wet Champagne to Vitry, where there is a gateway through the Rethel-Vitry escarpment. At Vitry the line divides, one branch turning northwest to cut through the innermost line of cliffs at the Epernay gateway, the other continuing west to make use of the notch at Sézanne. Evidently there are along this one line at least five strategically important defiles through a corresponding number of military obstacles, all of which defiles must be controlled by the armies which would make free use of the Strassburg-Paris route.

Toward the northwest the several lines of plateau escarpments gradually descend, and ultimately merge with the undulating plain of northern France. The low land between these fading escarpments on the one hand and the rough country of the Ardennes Mountains on the other, gives a roundabout but easy pathway along which one might reach Paris by swinging west beyond the ends of the scarps. Longwy, Montmédy, Sedan, and Mézières are the important points along this route, which is followed by a railway of much strategic value and which was quickly seized by the Germans after they had reduced the antiquated fortifications affording it a poor protection. As a route for an advance on Paris it is too circuitous to be of prime importance.

Historians are captivated by those areas where people gather in large numbers,—our cities. For it is in population centers where many events which changed the course of humankind occurred and where there is a definitive record of the ebb and flow of civilizations. The site of cities is most often determined by the nature of the topography and also to some extent by the rock type present. These and allied geologic factors also often control the direction and ultimate size of population centers. In this extract, Donald F. Eschman and Melvin G. Marcus of the University of Michigan describe the direct relationship of geology to cities. Although the examples cited are for the most part in North America, London, Tokyo, Cairo, Tehran and Rio de Janeiro would have served equally well.

THE GEOLOGIC AND TOPOGRAPHIC SETTING OF CITIES

Donald F. Eschman and Melvin G. Marcus

The geologic and topographic setting of cities plays a major role in their location and growth. First, the physical landscape is a major factor in the initial selection of sites for settlement. Second, topography and landforms strongly influence the early growth and development of settlements, particularly the evolution of their spatial pattern. Last, even though in these days man's technology allows him to move mountains, the economic costs of overcoming geologic and geomorphologic factors continue to impose directional and aerial constraints on urbanization. Thus, although the greatest impact of physical environment on human activities such as urbanization may be found in the historical past or in less developed cultures today, the basic landscape on which cities are situated continues to play an important role in modern urbanization.

This chapter considers the geologic landscape on which cities are built. Because the criteria by which sites for settlement are selected are extremely complex, attention is first directed to the multitude of locational factors in order to place the physical factors in their proper perspective. Basic topographic and geologic questions are then considered. They involve a variety of interactions between cities and their topography, geomorphic processes, and geologic composition. The feedback between human activities and the parent landscape is critical, and therefore examples are given throughout the chapter.

Cities are not simply where you find them. They are located in response to a complex set of interacting processes and forces that encompass a range of factors extending well beyond those presented by the physical landscape. The larger the city, the more complicated are the economic, political, and social factors that influence its location and growth. It is important to recognize, however, that man has only recently acquired the ability to bring a sophisticated technology to bear on the selection and development of his settlements. Although in the long run the survival and growth of urban places must depend upon economic and social factors, success is seldom achieved if rational decisions regarding physical locations are ignored.

The parent material and topography on which cities are built are major site factors, although other elements such as water resources, land-water boundaries, and climate commonly are important. In the case of New York City, for example, the original European settlement was on Manhattan Island, then an area of wooded, bedrock hills, interspersed with low-lying marshes and tidal flats. The island's drainage pattern and shoreline were qualities of site. As the population grew and more space was required for human activities, the site was altered by draining and filling the marshes. The dramatic expansion that New York City experienced, however, was not a simple response to its initial site characteristics. Rather, its broader physical situation—expressed in such features as the Hudson River, good harbors, and access to the interior of North America (via the Hudson River and the Mohawk Lowlands)—provided a physical setting favorable to the growth of industry and commerce. The human situation was even more significant in that it provided the social, economic, and political needs and potentials that allowed a great city to evolve upon and spread from the initial site. In short, the city required a physical base, but it was the broader relationships of that site to human activities and distant landscapes that proved most significant in the city's evolution.

The growth of every city can, in large part, be explained in terms of human determinants. It has been suggested by some urban geographers and sociologists that there is little point in attempts to classify the geological and topographical attributes of urban places, because each city responds to a unique set of environmental and human conditions. According to them, generalized explanations of urbanization that are based on a physical typology must inevitably be fruitless exercises that do not address the major social and economic processes at work. Though these arguments are in some part true, the fact remains that cities are built on earth materials that have topographic expression—and the rational use of this physical base is profitable to man. Conversely, topography and geology may place constraints on human activities that require an expenditure of time, money, and effort to overcome.

Historically, man has sought city locations that provide: (1) access to good water transportation or overland routes; (2) protection from natural hazards such as floods, storms, and landslides; (3) security from enemies; (4) water supply; (5) building materials, fuel, and other usable resources; and (6) a stable base for construction. Other factors, some of which may provide commercial and industrial advantages, such as water power, natural breaks in transportation routes, local food sources, and of course historical accident, are also important. Nearly all of these settlement criteria are dependent on geologic and topographic conditions.

The settlement history of North America clearly reflects the need to satisfy some or all of these environmental conditions. New York, Providence, and Boston are good examples of settlements built at the site of a well-protected harbor, while Baltimore and Charleston, South Carolina, are examples of settlements bordering a large estuary (resulting from the drowning of a river valley). In each of these locations, protection from storm waves is provided by deep coastal indentations and offshore islands—landforms created by the recent post-glacial rise of sea level relative to the land. It is interesting that 22 of the world's 32 largest cities are located on estuaries.

Once a settlement has been established, its continued growth commonly is influenced by the geologic and topographic attributes of its site. In an urbanizing area environmental factors such as steep slopes, poorly drained ground, flowing or standing bodies of water, and aesthetic characteristics constrain or bar development of certain areas. Patterns of urban expansion may develop a bias in a particular direction or favor certain landforms; transportation can be similarly affected. Barriers are most apparent during a city's expansion phases, and considerable technological and economic effort is required to overcome them. As long as the least cost and effort can be achieved by building the city on favorable physical locations, problem terrain is avoided. Eventually, if population pressure and economic and political circumstances demand it, urbanization will spread to less desirable landscapes. It has been economically sensible for cities to expand, especially in their early stages of growth, along the lines of least topographic and geologic resistance. In areas of relatively flat or rolling, well-drained land, cities tend to spread rather evenly in all directions. Examples abound in the Great Plains and Central Lowlands of North America, where only rivers and lakes commonly present major barriers to growth. Houston, Texas, and Saskatoon, Saskatchewan, are good examples of cities where topographic resistance is low and the cities have expanded without interruption and about equally in all directions from the CBD's (central business districts). Chicago and Detroit have sites where only water has been a major obstacle to urban growth; flat, open land has otherwise allowed quite uniform expansion into the rural countryside. In fact, once the lakefront effect has been accounted for in Chicago, the flat terrain so nearly approaches an isotropic surface that that city has become a major focus of studies which attempt to explain patterns of urbanization by various theoretical models. For example, the classical urban studies of Ernest Burgess (1925) and Homer Hoyt (1939) focused on Chicago.

In some regions geologic structure presents formidable obstacles to urban growth. The dramatic relief between ridges and valleys in the Appalachian Mountains has clearly influenced patterns of human settlement. Cities located along the valley floors, on water courses and overland routes, tend to spread longitudinally. Rosary-like strings of towns accompany many rivers, leaving sparsely populated spaces on intervening slopes and ridges. The urban area of Bluefield-South Bluefield, West Virginia, in the Appalachians illustrates the elongation that accompanies urban growth in this region.

The urban pattern of Los Angeles, California, is another example of the constraint that topography can place on urban growth. This city, which has experienced phenomenal population growth in the last half century, exemplifies urban man's determination to subjugate the environment. The population boom seemingly forced the city to expand from the lowlands of the Los Angeles basin onto nearby slopes and hills. The city has become an impressive testomial to man's technological ability to overwhelm his environment—but only if one blindly ignores the risks from landslides, earthquakes, and storms. Even in Los Angeles, though, some steep mountainous areas (those that continue to present physical and economic barriers to urbanization) remain unsettled.

Hills and slopes are not the only kinds of geologic resistance to urban growth. Low, poorly drained areas, such as flood plains, marshes, and tidal flats, also present problems. In the New York metropolitan region, for example, some 15,000 acres (of an original 30,000 acres) of the water-saturated open lands called Hackensack Meadows remain relatively unsettled and unused in the midst of densely populated and industrialized land. Despite dramatic economic pressures, this land is only slowly being drained, filled, and turned to human use. Areas made suitable for some construction by landfilling are not suitable for all types of constructions. Some low-lying fill zones of Manhattan Island, for example, cannot support the skyscrapers that rise on firm bedrock in midtown and lower Manhattan. For many years Chicago was forced to expand horizontally because a means could not be found to soundly support high-rise buildings on the unconsolidated sediments along the lake shore. In recent years this problem has been solved, and vertical growth has become a major feature of Chicago's urbanization.

John D. Ridge, professor of economic geology and mineral resources at the Pennsylvania State University, traces the relationship of mineral deposits to historical affairs from the earliest man, who used a pebble for a weapon, to the impact of the mineral heritage of the United States on the outcome of World War II. The following extract covers a brief part of this span.

MINERALS AND WORLD HISTORY

John D. Ridge

Gold and silver became important for ornamentation and decoration early in the history of mankind, but it was not until much later that a monetary value was attached to them. Perhaps the first real effect one of these precious metals had on the history of the world was made possible by the discovery in 483 B.C. of the rich silver-bearing lead deposits at Lavrion, near the tip of the peninsula that runs southeast from Athens. Although silver had been produced at Lavrion since Solon's time (ca. 550 B.C.), the amounts mined had been small until the discoveries of 483. The sub-surface rights to minerals in ancient Greece belonged to the state, and the Athenian government leased mining claims to private operators for a rental of about 5,000 ounces of silver per year, plus 1/25th of the silver produced. In those days, of course, 5,000 ounces of silver would buy many times what it would today; thus the return of the Athenian government was enormous. Aeschylus spoke of Lavrion as a "fountain of running silver," and the decision of the Athenians as to what to do with the income from the mines was fundamental to the future not only of Europe but of the world.

In 500 B.C., some of the Greek cities in Asia Minor, then under the domination of Darius, Emperor of Persia, revolted. By 494, the Persians had subdued the revolt, but Darius, concluding that future trouble could be prevented only by the conquest of Greece, began to prepare to invade that country. In 491, he attacked Greece, landing 200,000 men from 600 ships on the coast of Euboea. From there, the Persians crossed the strait to Attica and pitched their camp on the plains of Marathon. The Athenians (freemen and freed slaves) under Miltiades, reinforced by most of the men of Platea, moved by forced marches to Marathon. There no more than 20,000 Greeks (perhaps as few as 10,000) faced at least 100,000 Persians. On September 12 in the year 490 B.C., the Greeks under Miltiades routed the Persians in one of the decisive battles of the world. However, it was not until the sea battle of Salamis in 480 and the land battle of Platea in 479 that the Persian threat finally was ended.

Darius, after his defeat at Marathon, did not give up the idea of conquering Greece and began to prepare for another invasion. He died, however, in 486, and his son Xerxes, after some difficulty with rival claimants to the throne, continued the preparation for war with the Greeks. For four years, he collected troops and supplies from the wide reaches of his empire. When he set forth on his westward journey, in 481 B.C., he had what certainly was the largest military force ever assembled on the face of the earth. His army was, of course, a highly heterogenous body of men, but they were brave and well-trained by the standards of the Persians if not by those of the Greeks. In the Spring of 480, this huge force reached the shores of the Hellespont, where two bridges of boats were built across that body of water. For seven days and nights the Persians marched across the bridges, and these wonders of ancient engineering held throughout their passage. With the Persian Army on Grecian soil for the first time in 10 years, it is necessary to see what preparations the Greeks, principally the Athenians, had made to meet their invaders.

After silver began to flow from Lavrion in 483, the Greeks gave thought to the use they should make of the purchasing power it gave to the Grecian state. There were those among the Athenians who wished to use the silver further to beautify the city or to buy more grain from Egypt, but Themistocles convinced his fellow citizens that they should spend the money to increase their fleet so that it might compete successfully with the Persians for the mastery of the Aegean. The outcome of his three years of effort was in the 180 triremes and 9 pentaconters, 60 per cent of the total Greek naval force, that the Athenians contributed to the Greek fleet in 480.

The first clash between the Greek and Persian fleets at Atremisium resulted in a drawn battle. The two armies, the Greeks under the Spartan king, Leonidas, met at the pass of Thermopylae. There, on the second day of Atremisium, largely through treachery and panic on the part of certain of the Greek soldiers, their main forces were overwhelmed from flank and rear, and 4,000 men, including Leonidas himself, were killed by the Persians. The land route through Attica lay open, and the Persians poured down the road toward Athens, taking that city quickly and the Acropolis only after fierce fighting, thus clearing the way to the Isthmus of Corinth. When word of this disaster reached the Greek fleet, the decision was made to retreat south to the narrow waters between the island of Salamis and the Greek mainland northwest of the port of Piraeus. In these narrow waters, under the eyes of Xerxes, who watched from Mount Aeguleos, the Greeks defeated the Persians. In a tactical sense the battle was not decisive, but it cut so deeply into Xerxes' confidence in himself and his troops that he returned to Asia Minor leaving his troops, under Mardonius, to suffer a crushing defeat at Platea (near Thebes) on what probably was August 27, 479 B.C. In that same summer, the Greek troops landed from their fleet on the beach south of Mycale in Asia Minor, marched overland to that port, took the town, and burned the beached Persian fleet. That winter the European terminus of the bridges of boats over the Hellespont was taken by the Greeks, and the bridges were destroyed. The Golden Age of Greece had begun.

Credit for the defeat of the Persians has been given variously to the Greek commanders, such as Themistocles, Aristeides, and Pausanias, as well as to the Greek armor, the Greek organization and discipline, and, above all, to the Greek fleet. But the greatest commanders and the discipline of the Greeks would have been useless had the Greeks not possessed ships, arms, and armor. And none of these would have been available had not the silver of Lavrion paid for them. In short, it was the flowing fountain of silver, even more than the Hellespont or the passes of Thermopylae or the strait of Salamis or the Isthmus of Corinth, that barred the Persians from Greece. Had not the silver of Lavrion belonged to the Athenians, the world of that day after the year 480 would have been under the power of the Persian autocrats and not of the Greek democrats. Whatever the world is today, had the Persians conquered the Greeks, it would be far different from what it is now. Consequently, our present civilization is based on the ships of Salamis and the arms of Platea, and these, in turn, on the silver of Lavrion, and the sound sense of those—foremost among whom was Themistocles—who chose the wooden walls of their ships against the Pentelicon marble walls of buildings the Acropolis was never to know.

Defeated though it was in Greece, the Persian Empire was far from dead, and only the military genius of Alexander the Great finally brought it down at Arbela (Gaugamela) in 331 B.C. Although much has been made of the Greek use of silver to build up its fleet, the Persians did not lack ships at Atremisium and Salamis. What they lacked were Persian ships (instead of ships with crews of faint-hearted allies) manned by well-trained, well-armed, and dedicated sailors and soldiers of the king. The wherewithal to build such ships and to arm and train their crews was available, for the Persian emperor had stocks of gold large enough to make the dreams of Midas seem as idle fancies. But this gold was not used to promote commerce and industry; instead, it was used for ornaments of every variety that the imagination of man could conceive or the desires of kings command. Much remained in the imperial treasure vaults, buried like the single talent of the New Testament steward, in what must have been the most outlandish example of a napkin economy known to human history. When this golden hoard (which might have been turned into a golden horde for the defense of the empire) was captured by the Greeks, it became the sole property of Alexander. But after Arbela, only eight years of life remained to the Macedonian conqueror, and upon his death his successors fought over his gold as energetically as they did over his dominions.

Alexander left his empire "to the strongest," but, since no one was clearly dominant over another, the five most aggressive divided it into five parts. These divisions were: (1) Macedonia and Greece; (2) Thrace; (3) Asia Minor;

(4) Babylonia; and (5) Egypt. Asia Minor was soon joined to Babylonia and, later, much of Babylonia (the Seleucid Empire) was lost to Egypt.

Despite constant boundary changes, the alternation of good and bad kings, and the growing threat of the Romans to the west, commerce and industry flourished in the Near East on the base of the gold of the Persian Empire, while the Hellenistic economy which developed allowed great fortunes to be amassed, huge cities to be built, and a livelihood to be provided for a large fraction of the peoples of these lands. It was through such commerce that the Hellenization of western Asia took place. In turn, the extension of Greek civilization to western Asia paved the way for the advent of the Romans, who took over—one by one—these peoples whose thoughts and ideas were not far removed from those of the Romans themselves. Had not these regions been intellectually prepared for Roman law and Roman peace by the descendants of Alexander's generals, the borders of the Roman Empire would have been drawn much farther west than they eventually were, and the downfall of Rome under the attacks of the barbarians would have occurred far sooner than it did. Had the Greco-Roman civilization of Augustus and Marcus Aurelius, of Trajan and Hadrian fallen hundreds of years before it did, it is doubtful if the end of the far longer Dark Ages would have found a base of Greek civilization—modified though it was by the Romans—on which the Renaissance could be built. Thus, the gold of the Persian Empire, by promoting the growth of Greek industry and commerce and the concomitant development of Greek thought and art, made possible the foundation on which the Renaissance was to grow.

That the possession or lack of mineral raw materials had an overwhelming effect on the course of European history during the millennium between the fall of the Roman Empire and the discovery of America would be difficult to argue. The Roman invasion of Britain well may have, in part, been caused by the Roman need for the tin, copper, and lead that, in those days, were so abundant in the British Isles. Certainly, the struggle of the Romans and the Carthaginians over Spain had its origins in the desire of each of these nations to control tin, gold, copper, and silver that came from that country in such (for that time) profusion. But, it was not until the discovery of America that possession of mineral raw materials played as vital a role in the history of mankind as it had done in the golden age of Greece or in the years after Alexander.

In the latter years of the fifteenth century, the supplies of the precious metals in Europe were much too small to carry the monetary burden imposed on them by the growing population and increasing commercial activity of the time. Then, with the addition of the silver (and to a lesser extent the gold) of the Americas south of the Rio Grande, the monetary climate of Europe was suddenly changed. Had politics not entered into the problem, it well might have been that prosperity on an unprecedented scale might have been the gift of the New World to the Old. But the influx of so much silver (between 1503 and 1600 the silver supplies of Europe trebled, while those of gold increased only by one-fifth) did much to upset the political balance of Europe.

At the beginning of the sixteenth century, the Holy Roman Emperor, a Hapsburg, Charles V (Charles I of Spain), controlled much of Europe and all of the New World that had been seized up to that time. It would seem, at first thought, that this influx of silver might have solidified the hegemony of the Hapsburgs over Europe and the Americas for centuries. What it did do, however, was prevent the domination of the world from passing from the Hapsburgs to the Bourbons of France.

With Henry IV (King of France from 1589 to 1610) the centralization of France was well begun, a centralization that was carried to completion under the next two monarchs of France (Louis XIII and Louis XIV), both of whom were served by the incomparable minister and diplomat, Armand, Cardinal Richelieu. Had the Hapsburg opponents of France been forced to rely on the resources of their European domains, it is almost certain that the French would have conquered as much of Europe as they did in Napoleon's time, and they might actually have been in a far better position to hold their gains than was the Corsican soldier. Only one thing stood between the French and the domination of Europe: as the British fleet stood between Napoleon and the dominion of the world (in Mahan's phrase), so the Spanish infantry stood between France and the dominion of Europe, if not of the entire world. It is usual to think of the defeat of the Spanish Armada as marking the downfall of Spanish (and Hapsburg) power; instead, the military power of that nation, developed in the Moorish wars which ended in the same year that Columbus discovered America and financed by American silver, endured until the battle of Rocroi, where the Great Condé defeated the Spaniards in 1643.

Had the Spanish monarchs made effective use of their almost inexhaustible stores of silver, it is possible that they, instead of the French, might have threatened to take over control of Europe. But the Spanish so reduced the value of the bullion they imported through price inflation at home and debts abroad that only the death of Richelieu and the rise of British power and influence on the continent and in the New World barred the French from complete victory. In short, Spanish ineptitude opposed the unattainable ambitions of Louis XIV until the British were able to cast their weight onto the scales to secure a balance of power in Europe that lasted, with few set-backs, into the twentieth century. Thus, the silver of the Americas gave just enough power to the Hapsburgs and to Spain to prevent France from controlling Europe, while the inefficiency of Spanish administration prevented the incomparable Spanish infantry of the sixteenth century from bringing all Europe under a single dynasty. In the end, the British profited most from this stalemate, with the result that the United States grew to nationhood under a British system of law and justice and with a British heritage instead of a French or Spanish one. Thus, the silver of the Americas, practically none of which came from within the borders of the present United States, was the ultimate determinant of European history in the fifteenth and sixteenth centuries.

By and large, the United States east of the Mississippi was a disappointing area for the mineral prospector. Of all the mines in that portion of the country, only the Cornwall mine in Pennsylvania has been in almost continuous operation since before the Revolution. Though the eastern states contain many large mines, they generally were not of sufficiently high grade or did not contain the proper ores to attract the attention of the prospector of the pre-Revolutionary days. Instead, iron was supplied from residual deposits of minor size; copper, until the discovery of the Keweenaw Point deposits in northern Michigan, was essentially lacking zinc was a useless material; and lead was found in only a few places and usually in small amounts. Even silver and gold generally were of minor importance in the eastern states, and whatever else this portion of the United States may have been, it was not a cornucopia pouring out an inexhaustible supply of wanted and needed minerals.

With the discovery of gold in California at Sutter's Mill on January 24, 1848, the status of minerals in the United States and their effect on its economy underwent a drastic change. Most of the gold produced from the placers and from the Mother Lode of California was directed to the northeastern and middle Atlantic states, where it provided the credit base necessary for the industrial expansion that took place in the 1850's. Without the modern plant and trained workers which the gold of California provided the northern states, the Civil War would have been fought on a far more equal footing than it was. Had the northern industrial base been no larger and effective than that of the South, it is likely that the War between the States would have ended in a draw or might even have resulted in a southern victory, for the imports to the South from the European nations which favored the cause of the Confederacy might well have furnished the weight needed to shift the balance of power south of the Mason-Dixon line.

Even with the greater industrial base of the North, the war might have ground to a halt had the North been forced to finance its war effort with such rapidly depreciating paper currency as the South was forced to use. That the North was, in large measure, able to avoid this fate was primarily due to the discovery of the huge silver deposits of the Comstock Lode in Nevada. The flow of silver from this mine had much the same effect on the northern armies and navy as did the silver from Lavrion for the Greeks in the decade from 490 to 480 B.C. If the North had not been able to keep its superior industrial plant running with no more than a moderate amount of paper money inflation and had that paper money not had the backing of the silver of the Comstock Lode, the north might have been little better able to supply its troops with arms and ammunition and its navy with ships than was the Confederacy. During the two decades from 1859 to 1879, the Comstock Lode produced over 500 million dollars in silver (and much less gold), a tremendous sum for that time. After the war, the discovery of the Comstock's Big Bonanza in 1873 did much to make possible the return of the Federal treasury to specie payments and the gold standard and the recovery from the panic of 1873.

The decline of the Comstock Lode and the failure to find silver or gold deposits of equal value in the United States played a large part in the tightening of credit in the later 1880's, with the result that capital investments in the United States exceeded the possibility of their immediate profitable use. This situation, together with an inflexible banking system, forced prices downward and ruined many small farmers and business men. After the Federal gold reserve fell below the accepted minimum of $100,000,000 in April, 1893, nearly 600 banks in the west and south were closed, at least temporarily, and commercial businesses failed in great numbers. The depression continued through the winter of 1893–1894, and conditions were not substantially improved until 1897, when poor European crops vastly increased exports of American agricultural products. The gold provided by Europe to pay for these crops did much to force prices up and to restore prosperity. Had the crisis of 1893 occurred ten years before it did, however, the European nations might have been unable to pay for American grain in gold, and the imports that, ten years later, revived the American economy might not have been made. This opportune timing in the availability of gold resulted from the discovery of the mines of the Witwatersrand in South Africa in 1884. In 1889, the Main Reef, which has provided over half of the world's gold, was found there. Even by the mid-1890's so much gold had been extracted from this source that the necessary credit had been produced to bring an end to the panic of 1893.

That the climate is cooling and winters are colder is evident to many scientific observers, particularly climatologists and geologists. Whether this trend foreshadows another major glacial advance, or perhaps represents the beginning of an interval of cooling for a few hundred years— another Little Ice Age—are two of the questions posed and answered in this selection. James D. Hays, oceanographer-paleontologist at Columbia University, also discusses some historical developments associated with the Little Ice Age of the later 15th through early 19th centuries, and then relates techniques that can be applied to unravel past long-term global climatic changes.

THE ICE AGE COMETH

James D. Hays

This January the temperature in the country's heartland seemed colder than anyone could recall, aggravating an already chronic fuel shortage. Schools closed in Denver, factories shut down in the Midwest, and thousands of workers were laid off as fuel supplies ran out. The suspicion that winters simply are getting colder is no longer merely a suspicion among climatologists. Over the last 30 years permanent snow on Baffin Island has expanded. Pack ice around Iceland in the winter is increasing and becoming a serious hazard to navigation. Warmth-loving armadillos that migrated northward into the Midwest in the first half of this century are now retreating southward toward Texas and Oklahoma. Russian crop failures are on the increase.

These are just some effects of a global cooling that has been recorded by the world network of weather stations. If all indications are correct, worse is yet to come. Judging by what has happened in the past, it may very well get cold enough to allow great glaciers thousands of feet thick to cover North America as far south as Long Island, burying the highest peaks of the White Mountains and Adirondacks.

The last such glacial advance—the last great Ice Age—ended only 17,000 years ago, when the ice sheets in the mid-latitudes retreated, leaving only some smaller mountain glaciers as they crept back to the poles (where now all that remains are the ice packs on Antarctica and Greenland). As the ice melted, the earth then rapidly warmed until sometime between 3,000 and 5,000 B.C., when it was probably warmer than today—in fact, warmer than it had been for 100,000 years. But these balmy conditions were not long lasting, and the climate gradually cooled until about 900 B.C., during the archaeological Iron Age. It then rebounded again, as evidenced by records of flourishing vineyards in England from 1000 to 1200 A.D. Thus, the great Viking conquests were favored by slightly warmer conditions than today's. This time of mild climate and flourishing agriculture saw the spread of

Christianity over northern and eastern Europe and the most active period of cathedral and abbey construction.

The climate again cooled from the late fifteenth through early nineteenth centuries. This interval, the so-called Little Ice Age, was a refrigeration of considerable intensity, although still a far cry from a full ice age. It saw the greatest advance of mountain glaciers, and probably of ice on the polar seas, since the last great Ice Age. In some degree it touched every aspect of human life.

The effects of the Little Ice Age have commonly been overlooked, presumably because they were overshadowed by the devastation caused by the Black Plague. The late 1500s and the 1600s were probably the coldest time of the four-century period. The 1690s saw the last serious famine anywhere in Great Britain, the result of an eight-year run of harvest failures in the upland parishes of Scotland, where the proportion of the population that perished from hunger rivaled the toll caused by the Black Plague. The consequent weakening of the Scottish nation may have made union with England inevitable.

In England the fifteenth through nineteenth centuries also were significantly colder than now. The Thames River at London had frozen only occasionally prior to the sixteenth century; however, in that century, and during the two that followed, the river froze on numerous occasions. In the late 1600s the river froze repeatedly, and for weeks on end in the winter of 1683–84, Charles II, his court, and just about everybody else in London were able to tread clear across the ice-encrusted Thames. The river froze on only one occasion in the nineteenth century, but what prevented it from freezing more often probably were man-made changes of the river banks and subsequent industrial thermal pollution, rather than climatic warming. The Little Ice Age also affected the Scandinavian colony on Iceland, where grains introduced by the Vikings refused to grow.

On this side of the Atlantic, good historical records don't go back as far, but there is some striking evidence from the eighteenth and early nineteenth centuries that conditions were significantly colder in eastern North America than they are today. In the winters of this period New York Harbor repeatedly froze, allowing Staten Islanders to walk across the narrows to Brooklyn. In 1816 there was frost every month in northern New York and New England, and for this reason it was dubbed "the year without a summer." From the sixth to the nineteenth of June subfreezing temperatures were recorded in upstate New York and New England, and a foot of snow fell on Quebec City. The crop losses that occurred because of this late-spring cold and the following late-September black frost posed serious problems for Americans recovering from the ravages of the War of 1812. Admittedly, 1816 was unusual, even for the Little Ice Age, but the possibility of the recurrence of such anomalous years increases as the world's mean climate cools.

Our knowledge of climatic conditions before the late 1600s, when the thermometer was invented by Gabriel Fahrenheit, is based mainly on observations included in diaries and historical accounts. For the next 150 years we have a temperature record for western Europe and, during the latter part of this period, for North America. By the end of the last century a sufficient number of weather stations had been set up around the world to provide a good indication of global climatic trends.

This climatic record clearly shows the end of the Little Ice Age, with temperatures rising from the late 1800s until about 1940. Since then the trend has reversed, and a steep decline is now clearly underway. Already the climate has reverted to the levels of the 1920s, and the cooling continues.

The effects of the present cooling trend will be most strongly felt in temperate climatic zones, while the tropics will be least affected. Not only will the temperate regions become cooler, but we may also expect an increase in "anomalously" cold years—years with late-spring frosts or early-fall frosts—and a general increase in storminess. Thus, the grain fields of the Canadian prairies and the Russian steppes are vulnerable to a shortening of the growing season; however, the sugar cane harvest of Cuba or the banana crops of Guatemala would probably survive even a fully glacial climate in the north.

It is the vast industrial and grain-producing areas that will undergo the greatest change. The main grain (primarily wheat) exporting nations—the United States, Canada, Argentina, and Australia—all lie in temperate latitudes. A large number of nations, most notably the Soviet Union and China, rely upon these exports. Any reduction of grain production could cause an immediate problem, for the large grain surpluses of some years ago have now been vastly reduced, and most of the world's production is consumed within a year of the harvest. The continuing population growth may put a strain on our food resources even without a marked modification of climate, but a shortening of the growing season in temperate latitudes could easily aggravate this situation.

As climate cools, the demand for more heat will put an ever-increasing strain on our energy resources, causing natural gas, oil, and electricity prices to rise. The present "energy crisis," though partially caused by archaic regulatory procedures, points out how narrow our tolerances are and how closely we regulate our energy supplies to the level of so-called normal winters. However, what was considered an abnormally cold winter is becoming normal, and it is expected that winters will be getting even colder. Some scientists, such as the Czechoslovakian climatologist Jiri Kukla, think we may be on the verge of another Great Ice Age. They point to the fact that previous interglacial mild intervals, which are comparable in warmth to the present one, lasted only about 10,000 years. The present interglacial period already has lasted 10,000 years, so if it is truly similar to earlier ones—and if man's activities do not alter the natural trends—it should be nearly over.

If we indeed have reached the critical point in the climatic cycle where an interglacial changes to a glacial, there is some difference of opinion as to how long this interim period will linger. Recent evidence suggests a shorter time for the transition than was previously thought. Pollen grains preserved in the peat bogs of Macedonia, Greece, reveal that the change from the last interglacial forest to glacial grassland probably took no more than a few hundred years. A comparable length of time was required for the temperate forests of England, the Netherlands, and Denmark to be replaced by subarctic tundra. However, the magnitude of the change from an interglacial to a glacial is such that noticeable changes, even greater than those observed in the Little Ice Age, would probably occur in fewer than 100 years.

Other scientists, such as the British climatologist Hubert Lamb, propose that the present climatic trend may lead us only into another Little Ice Age. They argue that climate appears to oscillate periods of about 200 and 400 years. The coldest part of the Little Ice Age was in the late sixteenth and seventeenth centuries. Another cold spell, though less severe, came in the early nineteenth century. If the system continues to cycle as it has in the past, we can expect substantial cooling in the late twentieth and early twenty-first centuries, perhaps to a degree comparable with the Little Ice Age.

The possibility of a future significant climatic shift, with its consequent impact on our food and energy resources, emphasizes the pressing need to understand the causes and mechanism of climatic change. Like many other environmental problems, climate is international, and the adjustments that must be made in order to deal with a changing climate will affect many nations. The manner in which we will adjust requires long-term planning. Perhaps this planning should logically fall under the jurisdiction of the new United Nations program for safeguarding the human environment headed by the Canadian industrialist Maurice Strong.

In any case, this past winter, when America shipped millions of tons of wheat to Russia and lowered its thermostats to conserve dwindling fuel reserves, should remind us, not only that we are living close to the limits of our resources under present climatic conditions, but also that our climate is changing—and could change substantially.

The eruption in April, 1815 of the volcano Tambora in what was then the Dutch East Indies, and the asking price in shillings for a sack of flour in the commodities market of London in December, 1817 seem to be entirely unrelated events. Yet, in this contribution, Kenneth E. F. Watt, systems ecologist at the University of California, Davis, documents climatic cooling caused by the addition of solid particles of volcanic origin to the atmosphere and demonstrates the limiting effects on food production. He concludes his short thesis with sobering data on the rate at which man is increasing particulate loading of the atmosphere and what the complications might be if an eruption comparable to Tambora or Krakatau were to occur in the next 50 years.

TAMBORA AND KRAKATAU: VOLCANOES AND THE COOLING OF THE WORLD

Kenneth E. F. Watt

You have every reason to be confused as to the possible effects of pollution on the weather. Some scientists have predicted that increasing global air pollution will reduce the amount of solar energy penetrating the atmosphere. This, they say, will cause a decline in the global average air temperature. Other scientists say the earth's weather will warm up. This, it is explained, will come about because of the increase in carbon dioxide concentration in the atmosphere. Since carbon dioxide is a strong absorber of infrared radiation, it will act like the glass roof that prevents loss of heat energy from inside a greenhouse—the infamous "greenhouse effect"—and trap radiation in the atmosphere.

Three types of arguments can be put forth to settle the matter as to which mechanism will be most important. One approach is to argue from mathematical theories of climate determination, using computer-simulation studies. This line of argument is not completely trustworthy, because of the embryonic state of such theories. A second line of argument, pursued by Reid Bryson and Wayne Wendland at the University of Wisconsin, is to discover by statistical analysis the relative strengths of the two processes during recent decades. They found that the impact of the temperature-lowering mechanism will override the impact of the temperature-raising mechanism. Since the temperature has in fact been dropping for more than two decades, their argument is compelling.

However, a third line of argument opens up a most fascinating area of research, both in interpretation of history and in climatological prediction. This argument asserts that the most realistic means of assessing the effect of air pollution is to find some gigantic natural event in the past that operated in a way analogous to modern pollution. Fortunately, we are provided with such historical "experiments" by records of a few particularly gigantic and explosive volcanic eruptions.

A close reading of old newspapers gives evidence that people in the early nineteenth century were not aware of any relationship between unusual weather and volcanoes. To my knowledge, most ancient civilizations thought that the earth's weather was almost totally determined by astronomical phenomena, and this belief explains the spectacular development of observational astronomy in those civilizations. The first man known to have recognized that volcanoes could affect the weather was Benjamin Franklin in 1784, following a series of enormous eruptions in Japan and Iceland the previous year. During the twentieth century many climatologists and geophysicists have come to suspect that volcanoes influenced weather and climate and, more recently, that they had effects similar to the global increase in atmospheric pollution.

There are two obvious sources of information on the impact volcanic eruptions could have on weather and hence on crop production. One is the tables of past weather data recently constructed by historical climatologists such as Gordon Manley and H. H. Lamb. The other is old newspapers, which yield detailed information on local weather, crop growing, planting and harvesting conditions, and agricultural commodity prices.

Consider the effect of the eruption of Tambora in Sumbawa, the Dutch East Indies, in April of 1815. This was the largest known volcanic eruption in recorded history: over the period 1811 to 1818 an estimated 220 million metric tons of fine ash were ejected into the stratosphere. Of this, 150 million metric tons were added to the stratospheric load in 1815 alone, mostly in April. Krakatau, which is much better known because it erupted more recently (1883), ejected only 50 million metric tons of ash into the global stratosphere. It is now known that the ash from certain of these very large and explosive volcanic eruptions spreads worldwide in a few weeks and does not sink out of the atmosphere totally until a few years have elapsed. During all of this time the ash is back-scattering incoming solar radiation outward into space and consequently chilling the earth.

The magnitude of the chilling can be ascertained from Manley's tables for central England. The Tambora volcano erupted most explosively in April of 1815; by November the average temperature of central England had dropped 4.5° F. The following twenty-four months was one of the coldest times in English history. Specifically, 1816 was one of the four coldest years in the period 1698 to 1957; the coldest July in the 259-year period was in 1816; October of 1817 was the second coldest October; May of 1817 was the third coldest May. However, these figures convey little sense of the impact of such chilling on society. For that we must turn to the newspapers of the time. All of the following quotations are from *Evans and Ruffy's Farmers' Journal and Agricultural Advertiser*.

"We had fine mild weather until about the 20th, when it set in cold, with winds at East and North-East, with partial frosts; these together have greatly retarded the operations in Agriculture, and very many cannot purchase seed corn, so that thousands of acres will pass over untilled, and sales of farming stock, and other processes in law, drive many of this useful class in society into a state of despondency. . . . The wheats, late sown . . . have been partially injured by the frosty mornings. . . . Sheep and lambs have suffered from the severity and variableness of the weather. . . . The doing away the Income Tax, and the war duty on Malt, will afford some relief, but are wholly insufficient in themselves to restore this country to its former state of happiness and prosperity." (April 8, 1816)

"From about the 9th or 10th of this month, we have never had a day without rain more or less, sometimes two or three days of successive rain with thunder storms. The hay is very much injured; a considerable part of it must have laid on the ground upwards of a fortnight. . . . Wheat is looking as well as can be expected, considering the deficiency of plants in the ground, and those very weakly . . . but still far short of an average crop." (August 12, 1816)

"Throughout the whole month the air has been extremely cold; there has not been more than two or three warm days, being at other times rather cloudy and dark, and the sun seldom seen. The Oats . . . on high situated ground . . . are the most backward and miserable crop ever seen . . . for the greater part of the Wheat, where the mildew did not strike, has been very much affected by the rust or canker in the head. . . . There have been many seizures for rent this month, and many a farmer brought to nothing, and we hear of very few gentlemen who are inclined to lower their farms as yet; it seems they are determined to see the end." (September 9, 1816)

The preceding quotations all refer to agricultural conditions in England. To indicate that this was a worldwide rather than a local phenomenon, the following quotation, from a letter printed in the same newspaper, suggests the state of affairs in other countries.

"Last year was an uncommon one, both in America and Europe: We had frosts in Pennsylvania every month the year through, a circumstance altogether without example. The crops were generally scant, the Indian Corn particularly bad, and frost bitten; the crops, in the fall and in the spring, greatly injured by a grub, called the cutworm. . . ." (November 10, 1817).

Thus, in the case of a volcano, which puts an immense load of pollution into the atmosphere suddenly—unlike modern pollution, which builds up gradually—we have a gigantic experiment, the effects of which can be clearly traced throughout all social and economic systems. For example, the volcano of 1815 had a clear-cut influence on world agricultural-commodity markets, as one would expect from the preceding descriptions of the consequences of weather deterioration on crops. Many measures of market conditions could be used to make this point, but the one I have selected is the highest asking price for best-quality flour within a month of trading on the London commodities market. The following table shows how this price changed, from the period before the volcano, to the peak price in June of 1817, to the normal price, which was finally reached again by the end of December 1818.

HIGHEST ASKING PRICE FOR A SACK OF FLOUR (IN SHILLINGS)		
Year	June	December
1814	65	65
1815	65	58
1816	75	105
1817	120	80
1818	70	70*
		(*65 by end of month)

The table indicates how volcanic eruptions can serve as the basis for interdisciplinary research, in which a pulse due to a physical event can be tracked through biological, social, economic, and political phenomena.

For example, an interesting feature of this table is the long lag between the time the pollution was introduced into the upper atmosphere and the time that the price elevation was at its peak (April 1815 to June 1817—twenty-six months). These time lags are characteristic of complex systems and indicate why cause and effect are often not connected in peoples' minds: by the time the effect has occurred, everyone has forgotten the cause.

Could the gradual increase in worldwide air pollution concentrations become serious enough to bring crop production to a halt in high latitudes? To answer this question, we must consider the rate of build-up in pollution now and the likelihood that political power to enforce adequate pollution control will materialize.

The worldwide stratospheric particulate loading due to Krakatau and Tambora was 50 and 220 million metric tons respectively. If worldwide man-caused stratospheric particulate loading continues to build up at recently prevailing rates, by about 2018, the permanent particulate load in the stratosphere would be equal to that produced temporarily by Krakatau, and by about 2039 the permanent load would be equal to that produced by Tambora.

Will they continue? The reader must make his own judgments. There is ample printed evidence testifying to the difficulty of requiring automobile manufacturers to conform to the limits required by the Clean Air Act of 1975. Also, a casual reading of industry journals does not suggest that industry would like to arrest the rate of increase in sales of oil, gas, coal, or other polluting substances. Computer-simulation studies of trends in the use of fossil fuels do not indicate any significant lessening of the rate of increase in pollution prior to about 2004 unless stringent controls are introduced. Thus, without a really massive political and social change of a type that does not seem likely, judging from present attitudes, the quotations from English farm newspapers in 1816 and 1817 may well be read as a scenario for the future.

There are further complications if these are not enough. What if man continues to build up the pollution load in the atmosphere, and then a volcano of the order of Tambora adds still more pollution to the atmosphere? How likely is this to happen? The answer is: very likely. The period since 1835 has been remarkably free of major volcanic eruptions of the explosive, Vesuvian type. But over the historical record an average of five very large volcanoes has occurred each century. Luckily for us, the modern period of great technological activity has been free from major volcanic eruptions, to an almost historically unique extent. But at some time our luck may run out.

Another complication is what is going on in the minds of people. The world population of 1816 had no idea that there was any relationship between the phenomenally bad weather they were experiencing and air pollution. Will we be any wiser? Can we change in time? Will we be able to take the necessary action to ensure a brighter end to the scenario for civilization? Simple extrapolation of present trends does not lead to an encouraging answer, but history teaches us that sudden, surprising changes in political and social attitudes do occur. Perhaps it will occur once more—in time.

The Uniformity of Nature

We now propose to examine those changes which still take place on our globe, investigating the causes which continue to operate on its surface.... This portion of the history of the earth is so much the more important, as it has been long considered possible to explain the more ancient revolutions on its surface by means of these still existing causes.... But we shall presently see that unfortunately this is not the case in physical history; the thread of operations is here broken, the march of nature is changed, and none of the agents that she now employs were sufficient for the production of her ancient works.

—George Cuvier

CHAPTER 14

Geopolitics

A common misconception is that science is pure and able to operate above and beyond the rough and tumble of the political arena. Unfortunately, political factors may outweigh scientific ones in funding of research, in land-use applications, in questions of conservation and in provision for National Parks. The geological sciences, particularly in the era of large-scale government funding of research, have come to be involved in the relatively new game of geopolitics, as the three articles of this section portray.

At the time this article was written, the author was on the staff of Science, *one of the most prestigious American scientific journals. The journalistic beat of Daniel Greenberg centered about the political aspects of science in the United States. Of his many reports, the best is perhaps the essay on "Project Mohole." Imagine a hole to be drilled through the crust of the Earth and to the upper mantle. This was an exciting project, one of unquestioned scientific merit and one that would advance geology in the eyes of the public, scientists and politicians. Yet "Project Mohole" ended disastrously and became the only research project ever terminated by act of Congress. How this happened is detailed in the following selection.*

MOHOLE: GEOPOLITICAL FIASCO

Daniel S. Greenberg

Mohole is not a pleasant story; in fact, American scientists still recoil at the mention of it, in large part, it may be, because the Mohole episode marked the beginning of the end of the post-war honeymoon between American science and government. But it is in many respects a uniquely instructive episode, simply because the battles that arose from it became so intense at times that they spilled from the closed committee rooms where science policy is normally formulated, out into public view. Mohole revealed little about the inner composition of the Earth, but it did reveal a great deal about the inner workings of science and government in the United States.

In 1957, when the Mohole idea was conceived, NSF was a mere seven years old. It was growing rapidly, but its annual budget, all of it from government, had to stretch over all the scientific disciplines. Most of its funds were in small packages, and, it is critically important to note, NSF had accumulated virtually no experience in the management of big engineering programmes. Mohole, which was clearly destined to be a multimillion dollar venture, was, at the time, the biggest proposal ever formulated in the Earth sciences. It would be a vast engineering undertaking, involving construction of drilling equipment based on radically new concepts, as well as many years of uninterrupted drilling. From its director, the late Alan T. Waterman—a physicist who had been a research administrator since the beginning of World War II—down, NSF was a deskbound organization, staffed by academic scientists, assisted by others who periodically served as part-time consultants. NSF did no research; it simply gave money, on the basis of expert opinion, to working scientists seeking government finance for their research. NSF was no more qualified to supervise the drilling of a hole through the Earth's crust than it was to operate an international airline. With that bit of background in mind, let us return to the scientists who were the parents of Mohole.

Having NSF's bankroll in mind, they realized that some sort of organization—"institutional base" is the formal term—would be needed to apply formally for a research grant and to administer the project. It occurred to them that there already existed an organization that might fill this role. It was a whimsical organization, really nothing but a very "in" joke among geophysicists: the American Miscellaneous Society. AMSOC, as it was known, had existed for a number of years solely as a harmless caricature of conventional scientific organizations and their ponderous ways. It did nothing, held no meetings, had no membership rolls, by-laws, officers, or publications. A get-together for drinks at a club in Washington now and then was the sole manifestation of its existence. But many prominent geophysicists had joined in the joke and proudly asserted their association with AMSOC. Let AMSOC at last do something useful, was the decision of Mohole's progenitors. And so, a number of AMSOC "members," meeting at a "wine breakfast in California, constituted themselves into a special group—the "drilling committee"—and wrote out an application asking NSF for $30,500 to finance detailed planning of the project.

Under the conservative hand of Director Waterman, who had to account for his activities to an often critical Congress, NSF was not disposed to turn over 30,000 of its scarce dollars to such an organizational peculiarity. It advised the group to get itself moored to an established organization before seeking government funds; so AMSOC got itself attached to one of the oldest, stablest and organizationally impeccable institutions in the United States— the National Academy of Sciences.

In its formal application to NSF, AMSOC solicited funds "for support of the study of the feasibility of drilling *a hole to the Moho discontinuity."* Reference was later made in the application to various geophysical studies that could be made in *"the hole."* It was stated that sedimentary materials might be sampled at the upper levels, but on the face of it, it appeared that AMSOC's ambitions were directed toward a single, crust-piercing hole. Hindsight shows the absurdity of suggesting the construction of a multi-million dollar apparatus simply to drill one hole; not unlike constructing an immensely powerful microscope simply to make a single observation. But the creators of Mohole, eager for a project that would have dramatic as well as scientific appeal—Earth scientists were feeling a bit envious of the money and glamour then attaching to the space programme—formulated their proposal with an eye toward public relations impact. And NSF, with its tight budget and small, inexperienced staff, took up the proposal with the under-standing that Mohole was a one-shot affair. The ultimate cost could not be reckoned until detailed studies had been completed. But in 1958, two key figures in AMSOC published an article in which the cost was estimated at $5 million, spread over several years, and it may be assumed that NSF was thinking of a figure in that vicinity when it decided to commit its money and reputation to the support of Mohole.

By 1965, estimates had risen to at least $125 million. But, long before that, a remarkable figure came into the affairs of Mohole, a young ocean-ographic engineer named Willard Bascom. One of the few live wires on the staff of the venerable Academy, Bascom naturally gravitated to this exciting project, and, in short order, the university-based scientists who made up AMSOC, and who came to Washington only now and then to look into it among other matters, asked Bascom to become the first full-time employee of Mohole. Bascom enthusiastically accepted, becoming executive secretary of the project, with instructions to turn the idea into a reality. As plans for Mohole crystallized in Bascom's mind, he published an article in which he rejected the one-shot concept. "The ultimate objective is to reach the mantle," he wrote, but "an intermediate step is likely to yield equally valuable and interesting results. . . . No one site or hole," concluded Bascom, "will satisfy the requirements of the Mohole project." Noting reports that the Russians had set up a committee to look into very deep drilling, Bascom speculated, "Perhaps there will be a race to the mantle." Bascom then proceeded to propose a technical strategy for the project, one that involved a seaborne approach to the mantle. It would be difficult, of course, to drill through thousands of meters of rock from a vessel riding on thousands of meters of water. But the Earth's crust was thinnest under the deep sea, and Mohole's chief planner felt certain that when all the difficulties were calculated, it would be faster, cheaper and easier to drill a shorter distance in deep water than a far greater distance on dry land. But since deep drilling at sea was then in its infancy, Bascom suggested that the necessary experience be ac-quired by starting out with a relatively modest waterborne project. For this purpose, he proposed that AMSOC hire an oil-drilling barge, modify and re-equip it for Mohole's specialized needs, and undertake a preliminary drilling exercise, preparatory to designing and building a special vessel for going all the way to the mantle. Meanwhile, NSF provided $80,700 to finance studies of promising sites for drilling. This was shortly followed by another $80,500 to pay the salaries of the growing AMSOC staff.

The time was now June 1959, and planning was sufficiently advanced for the AMSOC executive committee to give NSF some idea of the costs that would ultimately be incurred. AMSOC had no special funds of its own; NSF was footing all the bills. The committee assembled at the site of the now-famous "wine breakfast" where Mohole was hatched two years earlier, and came up with a figure of $14 million "for the entire project." Only a few months earlier, the cost had been placed at $5 million. Why had it risen nearly three-fold? The answer, quite simply, was that since the project had never been clearly defined, the costs could not possibly be estimated with any reasonable accuracy. Furthermore, AMSOC consisted mainly of scientists, not engineers, and they had little or no experience in the intricacies of estimating the costs of a unique engineering system. Finally, and perhaps most important, the Cold War and the early days of the space age had virtually obliterated financial conservatism in scientists' dealings with government. Money in unprecedented quantities was flowing into many fields of science and technology. As a White House science adviser later explained, "I took a quick look at the project and decided, 'Why not? It's only going to run about the cost of one space shot'". AMSOC did not require the $14 million all at once. It informed NSF that $1,250,000 would suffice for the first stage of the drilling scheme proposed by Bascom. (Shortly afterwards, a club of geologists in Washington put on its annual show. The title that year was *Mo-Ho-Ho and a Barrel of Funds*.)

Employing a barge equipped with engines and a navigation system designed to keep it nearly motionless at any point on the high sea, Mohole's Phase I was a stunning success. Brilliantly directed by Bascom, at a site off the California coast, it set a world record for deep drilling by boring 197 meters into the ocean bottom at a water depth of nearly three km. This remarkable feat drew wide public acclaim, and was even hailed in a message from President John F. Kennedy, who, to the delight of American scientists, had brought an enthusiastic appreciation of science into the White House. But Phase I, successful as it was, constituted no more than a mere scratch on the ocean bottom compared to the objective of drilling through the Earth's crust. That feat called for securing a vessel in nearly five km of water, and drilling about 5,000 meters into the ocean floor, most of the way through rock so dense that two years of nearly continuous drilling would be required. High temperatures, corrosion, the crushing weight of mile upon mile of piping, weather uncertainties—all together posed engineering problems far beyond the state of the art. At that point, spirits were high. What was not realized was that though Mohole was so far a great technical success, its administrative and, ultimately, its political underpinnings were beginning to crumble.

With Phase I successfully completed, the Academy began to feel uncomfortable about having an operational research programme headquartered on its premises. As a statutory adviser to government, the Academy had always remained aloof from any matter on which its advice might be sought, and clearly Mohole, with all of its technical and financial complexities, was the very sort of matter on which the Academy's counsel might be solicited. So, the Academy wanted to get out; but then so did AMSOC.

The two organizations agreed that it would be desirable to establish a new administrative base, though, of course, they hoped that NSF would continue to foot the bills for the continuation of the project. (These were estimated at \$4.2 million for the fiscal year beginning 1 July 1961). But what was to be done with Bascom and the staff that had so successfully carried out Phase I? It was concluded that they should go into the employ of whatever organization was selected for the next phase. In response to the suggestion that it hire a new organization to run the project, NSF invited public bids from various academic and industrial organizations. But bids for what? ". . . the drilling of a series of holes in the deep ocean floor, *one of which* will completely penetrate the Earth's crust." That is how the invitations for bidders read. In response to this and related specfications, 11 individual and combined organizations submitted bids. The initial evaluation, by a group of NSF officials, gave first place to a consortium headed by the Socony Mobil petroleum company. Fifth place was given to a late-comer that just barely met the deadline for bid, a vast engineering combine from Houston, Texas, Brown and Root, whose business fortunes, mainly in the form of government contracts, were alleged to have been closely linked to the Democratic Party, to which it regularly contributed campaign funds. Houston was also the home base of a publicity-shy but immensely powerful member of Congress, the late Albert Thomas, the chairman of the very sub-committee that annually decided how much money would be voted for NSF—the sole source of finance for Project Mohole. Brown and Root got the contract, touching off angry allegations of a political fix.

Brown and Root was now responsible for Phase II of Project Mohole. In accord with the expressed desire of AMSOC and the Academy, it duly hired Bascom and his crew to help in the design of the vessel needed to carry out the project. Two months later, Bascom and his colleagues resigned, asserting that Brown and Root was unqualified for the task. Meanwhile, an important change occurred within AMSOC. A tough-minded geologist, Hollis Hedberg, a vice-president of Gulf Oil and a part-time professor of geology at Princeton University, became chairman of the committee. Hedberg decided that before the project proceeded any further, all uncertainties should be resolved about methods and objectives. The one-shot concept, he declared, was publicity-seeking nonsense. "The project . . . should in no way be considered merely a stunt in deep drilling; it is a most serious search for fundamental information concerning the rock character and the rock record of the Earth on which we live. And," continued Hedberg, "the scope of the project should be such as to take advantage of opportunities for contributions to this geological end wherever they may be found—water or land, deep or shallow." Hedberg then came out for a cautious approach to Phase II, one that would involve construction of an "intermediate" drilling vessel, as proposed by Bascom. However, Brown and Root, the Texas firm that had been awarded the contract for Phase II, was proceeding with the understanding—as written into the contract—that its task was to design a ship that could drill a hole through the crust. Much time had been lost through Brown and Root's inability to assemble a staff, and several members of the U.S. Congress, aroused by the political allegations surrounding the contract, had taken to sniping at NSF and its contractual choice. Finally, early in 1963, some 13 months after the award of the contract, Brown and Root unveiled its plans. These called for building a gigantic, self-propelled sea-going platform, 76 by 82 meters, resting on two submarine-shaped hulls, each 120 meters long. The cost of construction and drilling was estimated at $67,700,000.

When this figure was stated, the White House baulked and directed NSF to suspend the project pending clarification. NSF set up a special study committee; meanwhile, several congressional committees began to probe into the tangled affairs of Mohole. Hedberg went before one of these committees and declared that, though Mohole had unfortunately started out as a gimmick, there was time to salvage and build upon its true scientific merit. Whereupon the President of the National Academy of Sciences—AMSOC was still affiliated with the Academy—demanded that Hedberg clear all future public statements with him. Hedberg promptly resigned, and thus both he and Bascom, perhaps the two most effective men ever associated with Mohole, were out of the picture.

In the meantime, Waterman, then aged 71, had retired as director of NSF, and was succeeded by Leland J. Haworth, a veteran research administrator long associated with the U.S. Atomic Energy Commission. Working closely with the Academy, Haworth finally persuaded the White House to permit the project to proceed, though NSF officials now conceded that the costs might ultimately be as high as $125 million. For a time, the prospects looked good, but then in January 1966, Mohole's chief congressional guardian, Representative Albert Thomas of Texas, died. One of the first steps taken by his successor as chairman of the sub-committee that controlled Mohole's funds was to cut off the project. There was a brief chance of a reprieve by the Senate. Then it was revealed that a few days after the Senate had been asked to attempt to salvage the project, members of the Brown and Root family gave $23,000 to a Democratic Party fund-raising organization.

Politically that was the last straw. In August 1966, a little less than a decade after a group of scientists in Washington had proposed drilling a hole through the Earth's crust, the U.S. Congress voted to eliminate all further funds for *Project Mohole*.

*The Florissant fossil beds of Colorado are unique for the abundance of
delicate plant and insect fossils they contain. That the area should be
preserved for study and enjoyment for future generations as a National
Park or National Monument seems obvious. Yet to do so proved a long
and arduous task. Initial attempts were made as early as 1921, and it was
only in 1971 that partial success finally was attained. The political procedures and maneuverings, and the fragile beauty of Florissant fossils are
all part of the story in Berton Roueche's contribution,* A Window on the
Oligocene.

A WINDOW ON THE OLIGOCENE

Berton Roueché

There are few more beautiful natural parks than the mountain valley that
encloses the village of Florissant (pop. 102), in Teller County, Colorado,
about a hundred miles southwest of Denver, on the western slope of the
Front Range of the Rockies. This valley has been variously known as South
Park and Castello's Ranch, but it now takes its name from the village. It is
ten miles long and a mile or two wide, and it follows a gently winding
course, with many sudden, serpentine arms. The valley floor is a rolling
meadow. It flows, deep in grass, around an occasional thrust of rust-red
granite outcrop, and it is framed on the east and west by low hills crowned
with open groves of ponderosa pine. Florissant is high country—almost nine
thousand feet—but above its highest hills, encircling the infinite distance,
are the peaks of higher mountains. There are antelope in the Florissant valley,
and deer and elk and bear and badgers and coyotes and porcupines, and in
July the meadows are a garden of wild flowers—bluebells, yellow clover,
purple locoweed, pink trumpet phlox, alpine daisies, brown-eyed Susans,
flaming Indian paintbrush. The beauty of Florissant, however, is not its only
natural glory. It has another, a rarer and a richer one. Beneath the flowering
Florissant meadows lies a treasury of paleontological history. Florissant is
one of the richest repositories of plant and insect fossils in the world.

The Florissant fossil beds were formed by a series of volcanic eruptions
that began some thirty-eight million years ago. This places the beds, by
geologic time, in the Oligocene epoch of the Cenozoic era. Florissant was
then, as now, a valley, but it was a highland rather than a mountain valley,
and it had a mild, humid, probably subtropical climate. A wandering stream
flowed south through the valley, and the banks of the stream and the slopes
above it were covered with willow and oak and pine and beech and giant
sequoia. The first, progenitive volcanic eruption changed that distant scene.
Its convulsions deposited a levee of volcanic mud at the mouth of the valley,
blocking the outward flow of the stream, and the stream thus dammed became

a pond, and the pond became a lake. Fish flourished there, and insects swarmed, and trees and shrubs grew dense along the shore. The volcano presently came to life again, and for perhaps three million years there were frequent wild eruptions. With every eruption, a cloud of fine volcanic ash blew into Florissant. The ash fell thickly on the lake, and as it settled it carried with it to the bottom a multitude of leaves and insects and tiny fish and stored them there in a thin, impressionable casing. Millennia passed, and the layered deposits of silt and ashy shale accumulated.

The first Florissant fossils to be properly studied were those collected by Scudder in 1877. That visit, though brief, was prophetically productive. In the mere five days that Scudder spent in the beds, he and three companions (two fellow-scientists and a local rancher Adam Hill) gathered a total of around five thousand specimens. The quality of the material also was prophetically high. "The shales of Florissant," Lesquereux enthusiastically noted, "have preserved the most delicate organisms—feathers, insects, small flies, petals, even anthers and stamens of flowers."

The paleontological importance of Florissant that Scudder first proclaimed has been ardently confirmed by many other paleontologists. At least a dozen major scientific expeditions have worked the Florissant beds and assembled large and comprehensive collections. The respect of the owners of Florissant for science has undoubtedly been a factor in the preservation of its fossil beds for continued scientific study. It has not, however, been an important one. Florissant has been largely preserved by the accident of nature that made it cattle country.

It has long been felt that the fossil riches of Florissant are too fragile to be so lightly guarded. The desirability of some defense against the human compulsion to rearrange nature which would be stronger than a long winter and a short summer was briefly considered as early as 1921. In that year, a group of interested scientists and others suggested that Florissant be brought into the protective embrace of the National Park Service. The suggestion received a politely sympathetic hearing, but no action was taken, and the interest of even its sponsors faded away and finally vanished under the impact of the Depression and the Second World War. It was revived, in 1953, by Edmund B. Rogers, then (he has since retired) the superintendent of Yellowstone National Park, with a testimonial based on first-hand exploration and entitled "Florissant Fossil Shale Beds, Colorado." Rogers' recommendation excited sufficient influential interest to set in motion the standard salvational procedure. This involved a pilot study, a preliminary report, and a formal proposal, and it proceeded at the standard pace. A formal proposal to preserve six thousand of the twelve thousand acres at Florissant as a Florissant Fossil Beds National Monument was approved by the director of the National Park Service in 1962. The following year, an enabling bill was introduced in the House of Representatives. It died, ignored. A similar bill was introduced in 1965, and it was similarly received. Two years later, in 1967, the House was offered a third Florissant bill. An amendment to it reduced the size of the proposed Monument from six thousand to one thousand acres. This truncated preserve attracted a wave of penny-pinching support, and the bill was approved with little opposition and sent along to the Senate. There it once more died. Another year went by and another Florissant bill was conceived. The new proposal differed from its predecessor in that it restored the cut in acreage and originated in the Senate. It was introduced on February 4, 1969. A few days later, an identical bill was introduced in the House. The winter passed, and also the spring. Then, on June 20th, the Senate approved its Florissant bill. The House bill remained in committee. Summer came on.

Two weeks went by. I watched the New York papers, but there was no mention of the Florissant affair. Then, around the middle of August, I got a letter from a friend in Denver. The letter enclosed a clipping from the Washington bureau of the *Rocky Mountain News*. It read, "The House Monday approved on a voice vote a bill creating the 6,000-acre Florissant Fossil Beds National Monument 35 miles west of Colorado Springs. The bill now goes to the Senate, where approval of technical amendments is expected. The Senate earlier passed a measure identical in substance to the House version. Following anticipated Senate passage, the measure will go to President Nixon for his signature. . . ." The story had no dateline, but the date of issue was printed at the top of the clipping. It was August 5th. That had left a week for the unleashing of the bulldozer. Too bad, but not irreparable. On August 20th, President Nixon signed the bill bringing Florissant under the protection of the National Park Service. A few days later, the bulldozer had vanished and the culvert liner had been hauled away.

If political decisions are made that affect geology, and it is evident that they are, then should not geologists enter into politics and attempt to influence or control such decisions? Dr. Peter Flawn, president of the University of Texas, believes that geologists, particularly those whose work is people-oriented, can no longer afford to shun the political aspects of their science. He is forceful in his assertion that geologists must develop political muscle and take action through their professional associations to modify or prevent uninformed political decisions that relate to the earth sciences.

THE ENVIRONMENTAL GEOLOGIST AND THE BODY POLITIC

Peter T. Flawn

I use the term environmental geologist to include those geologists directly concerned with society's use of the Earth, with the limitations and capacities of the Earth, with Earth resources and processes. In more traditional terms these are engineering geologists, economic geologists, hydrologists, and some marine geologists—but I am restricting the term to those in this group who are *people* oriented—the man interested only in the composition of ore-forming fluids, for example, is not an economic geologist. He is a geochemist. The *urban geologist* is commonly singled out with a special name, but only because Earth hazards and resource problems are more acute in urban areas.

Interest in environmental geology is accelerating. The U.S. Geological Survey under the leadership of W. T. Pecora has focused efforts on applications of geology to social problems. In state government, interest in the area is at a high level. At Virginia Polytechnic Institute there has been established an Institute of Environmental Geology; at Portland (Ore.) State College, Project Auger contemplates creation of an urban geology center. The Kansas and Illinois Geological Surveys have operating environmental geology sections. In California, Senate Bill 1401 recently passed by the State Legislature requires any city or county with a master plan to include a "natural resources element." The intent is to insure consideration of resources such as sand and gravel. Florida is actively studying sinkhole development in urban areas. The Maryland Geological Survey has been given responsibility for adminsitration of new strip-mining legislation. In New Hampshire, geologic maps are being prepared for town and city planners. West Virginia is emphasizing environmental geology for land-use planning. A committee to review the program of the Pennsylvania Bureau of Topographic & Geologic Survey recommended expansion of urban and environmental studies.

Knowledge about Earth processes, Earth resources, and behavior of Earth materials benefits few people unless it is put to work.

When an environmental geologic report has been prepared it must be sold. It must be sold to planning agencies and engineering departments. Circulation of reports by mail to the heads of agencies is not enough. I admit to differences of opinion on this point. Does a scientific and engineering fact-finding department really have a "selling" mission? Does not its job stop after the data are developed and turned over to authorities responsible for planning and policy? Perhaps if the data were non-technical and their implication readily apparent to non-technical people, it would be enough to develop the information and distribute it. However, considering the nature of the geological information, the very least the data-producing agency can do is to explain the report in conference with appropriate government officials charged with responsibility for planning, engineering, and management.

The most vigorous opposition to land-use plans and recommendations commonly comes from landowners, individually or in associations, whose right to use their land as they see fit is abridged, or who believe the value of their property has been damaged by geologic appraisal as unsuited for certain uses or certain kinds of construction. Federal and state geological surveys have been attacked politically because they incurred the wrath of property owners by appraising their land as hazardous for residential or commercial development. As a result, they have been careful not to condemn property for particular uses but simply to present the data which should indicate to engineers the hazard, or, for example, that the area of potential flood hazard is below such and such a contour with a frequency of once every 10 years.

It is difficult for the geologist who is employed by a public agency to take action individually beyond that sanctioned by his department head; such action may result in trouble for the department and a sudden change in employment for the crusader. Other geologists in the community, however, may be in a position to take individual action as citizens when it becomes clear that dangerous and costly mistakes are being made in the face of sound and documented conclusions in a professional report. This may take the form of standing up in a city council meeting and publicly fixing the responsibility for ignoring the report. A public statement from a professional man in good standing with facts to support his position carries weight with politicians, at least at the local level. A permit for a subdivision in an area subject to potential flooding would certainly be reviewed if a geologist with the data on runoff, flood-frequency, and maps showing areas of hazard were to make a public protest. Responsibility for the possible death by drowning of a young child is then put into the record.

It would be more desirable, however, if unhappy situations could be changed without public geological demonstrations in the seat of the government.

Recognizing that voluntary consultants are seldom welcome, and that the actions of individuals within public agency departments are limited by political realities, the question is—how can a small group of scientists and engineers effectively work for the community good?

I think the answer is the professional association.

Responsibility is a burden that comes with knowledge and education. Professional men who have observed poor engineering practices in their communities or who can evaluate the causes of engineering failures have an opportunity to initiate corrective action through a professional organization. The geologist driving to work who sees a contractor using a montmorillonite clay as a stable fill is professionally obligated to do more than shake his head in dismay.

I suggested the geological society as a mechanism for community action. Let us now look at the geological society and what is happening to it. Until recently, the major national geological societies proclaimed without exception that "We are scientific!" Because of registration and accreditation traditions, engineering societies are concerned with these areas—but for *engineers* and not their maverick geologist members. "Register as a geological engineer" is always the recommendation. But until the last few years the national geological scientific societies have shunned the political arena. Perhaps this was because of the lack of any real need for geological lobbying—there, I've said it—*lobbying* instead of liaison, public education, or some other euphemism—perhaps it was because of the danger of losing their tax-exempt status under the Internal Revenue Code, perhaps it was because of the feeling that politics is a dirty business and beneath the dignity of a professional man, perhaps it was because they did not know how to go about it, or perhaps it was because of what happened to Herbert Hoover. Local geological societies from time to time have taken political action on urgent local issues but such cases have not really established a trend.

Things are different now. There is growing concern about professionalism. There is concern about the effects of uninformed political decisions. The proof is in the corpus of the Association of Engineering Geologists, in the recently organized American Institute of Professional Geologists, in the certification efforts of the American Association of Petroleum Geologists, in the existence of the Society of Independent Professional Earth Scientists, and in the new look of the Association of American State Geologists, which has established a liaison committee to keep in touch with Federal agencies. But geologists have come late to the political game. There is the big league in Washington, the minor league in the statehouse, and the sandlot in the city. The last is most important for the geologist. These leagues play by different rules. There is some feeling that if a society issues a handsome certificate and seal certifying its member as competent, the job is done. This is naive. Any certifying society has got to declare its professionalism, forget about tax exemption, and be willing to jump into a legislative committee or a courtroom in defense of the profession.

And at the very top of the heap in the big Washington league where the big money pie is being cut, where the top-level committees, councils, and boards are appointed, and where the science and engineering advisers to high officials are, what do we see? We see an influential and prosperous American Chemical Society, an influential and prosperous American Institute of Physics, and a chronically impoverished American Geological Institute that only recently has been recognized by Government as the voice of geology. AGI, a federation of geological societies, is scarcely able to field a team because of the reluctance of some member societies to relinquish any sovereignty for unity in geology.

Only the compelling reason that too many people are getting into our business has forced the modest shift toward professional action that we have seen so far:

(1) Too many politicians are writing laws to control geological processes and create mineral resources.

(2) Too many economists are proposing answers to resource problems without knowing the facts of mineral occurrence.

(3) Too many planners with flatland concepts are deciding how the land should be used without knowing anything about the properties of the land itself.

(4) Yes, and too many soil scientists are playing around with rocks, and too many civil engineers are writing geological reports.

The geologist must develop political muscle for very altruistic reasons. On this crowded Earth the margin for error, the slack in the rope, the surpluses, are disappearing too fast for us to continue making costly mistakes or to operate inefficiently. We are dealing with too many irreversible environmental reactions.

CHAPTER 15

Geology in the Service of Humankind

We inhabit a benevolent planet. "Surely there is a vein of silver and a place of gold. Iron is taken out of the Earth and copper is smelted from the ore. Men put an end to darkness and search out the opening in a valley away from places where men live," we read in the book of Job. A familiarity with the Earth for early man bred not only fear, but ultimately an understanding and a growing utilization of the planet on which he lived. The history of man is, to some extent, a history of successive exploitation of his tiny planet: the use of accidentally shaped stones for weapons and tools, the discovery of fire, the later crafting of selected stones as weapons; the discovery of the usefulness of certain metals, copper, bronze, and then iron.

We find this progression beginning about 5000 B.C., represented by the first copper needles around the Nile Valley. About 1,500 years later, men learned to mix a little tin with the copper, to produce bronze. 1,500 years later, the same cultural sequence can be traced in Europe. One way to divide human history is by such inventive stages. The use of fire, the use of different metals, the growing use of language and writing, the domestication of animals and plants, the development of communities, the invention of the wheel—all these were developed as long as 6,000 years ago in parts of Europe and the Middle East. And yet, ancient as are the origins of these foundations of our current prosperity, the remarkable thing is that the provision of this bountiful planet has been such that there is a greater difference between our present life style and that of the early nineteenth century, than there was between that era and the period of the life of Christ. The reason for this is the increasingly successful exploitation of the crust of the planet on which we live. Life expectancy as recorded in one of the older censuses based on Sweden in the middle of the 18th century was 19 years in towns and villages. In those same places, it is now 70 years, and the difference that has arisen in that brief time span results from four particular discoveries and applications: medical care, an adequate diet, clean water supplies, and an effective system of sewage disposal. And all those things ultimately depend upon materials provided by the surface of the planet on which we live.

The development during the 17th and 18th centuries of cheap forms of power and the invention of the blast furnace to produce inexpensive iron and steel have revolutionized the world in which we live. It is somewhat difficult to realize the totality of our dependence on the Earth. The buildings in which we live are built of material manufactured from the Earth's crust. Heating, cooling, and lighting systems involve the use of energy derived from materials in the crust, energy stored from the sun by plants and animals millions of years ago. Most of the clothing we wear, the food we eat, gasoline, cars, radios, TV—all these are triumphs of extraction of crude materials from the crust of the earth: dull, unpromising rocklike materials, discovered, extracted, concentrated, refined, and then manufactured into something of immediate value.

This is not to deny the overriding importance of man's intellectual and moral development. Yet even here, there is a measure of interdependence. The supremacy of Greece was established on the discovery around 1000 B.C. of the rich silver and lead ores of Laurium, southeast of Athens. Containing up to 120 ounces of silver per ton, these ores were worked by slaves (who had a life expectancy of only four years in the mines) who sunk 2,000 shafts, some 400 feet deep, and produced about two million tons of ore. For three centuries the glory of Athens, the basis of her commerce, and the power of her armies, were financed by this silver.

And so, behind the rise of man in the intellectual realm, there has also been the increasingly successful exploitation of the crust of the planet on which we live. Each of us today, has at our elbow the equivalent of the service of about two hundred workers in the form of the energy supplies that we use. And without that energy, all derived from the surface of the Earth, life would be a very different proposition.

There has, however, been a price for this progress. With this transformation has gone a change in attitude. Early Man worshipped the Earth. He attempted to placate it. In a literal sense, the Earth was the mother of Man. But now, we have changed our attitude towards the Earth and we look upon it, not so much as a thing to be worshipped, but as a thing to be exploited. Rene Dubos recently declared that the problem with modern technology is that we use it as a magic instead of a religion. And he follows Malinorsky in defining religion as referring to fundamental issues of human existence; whereas, magic turns around specific concrete and practical problems. We have used technology as magic, and have debased it by failing to relate it to the ultimate issues of human existence. With the increasing success of our exploitation of this bountiful planet, we have changed our attitude towards the planet itself.

For some of our most essential commodities—oil, gas, mercury, tin, silver, amongst them—the end of exploitation looms. Reserves dwindle, yet demand increases. Our nonreplenishable resources, our burgeoning population, our unequal patterns of prosperity and our careless technology—all these present both threat and challenge to our survival.

Water

Most alarming of all, vast underground reserves of water, deposited over thousands of years, have been seriously depleted in a matter of decades. All water comes as rain from the sky, but 92 per cent of it either evaporates immediately or runs off, unused, to the oceans. One-quarter of the water that irrigates, powers and bathes America is taken from an ancient network of underground aquifers. In 1950, the nation took 12 trillion gallons of water out of the ground; by 1980, the figure had more than doubled, and each day, 21 billion more gallons flow out than seep in. . . . The nation's water outlook now is frighteningly reminiscent of the oil outlook a decade ago—affording, cynics say, a matchless opportunity to make the same mistakes again. And in this pinch, Saudi Arabia is not going to be of much help.

Just as Americans have discovered the hidden energy costs in a multitude of products—in refrigerating a steak, for example, on its way to the butcher—they are about to discover the hidden water costs. Beginning with the water that irrigated the corn that was fed to the steer, the steak may have accounted for 3,500 gallons. The water that goes into a 1,000-pound steer would float a destroyer. It takes 14,935 gallons of water to grow a bushel of wheat, 60,000 gallons to produce a ton of steel, 120 gallons to put a single egg on the breakfast table.

—*Newsweek* "The Browning of America"

Lord Ritchie-Calder, author, journalist, Senior Fellow, Center for the Study of Democratic Institutions, and formerly Professor of International Relations at Edinburgh University, has been widely involved as a member of many international scientific and technological commissions. Author of some 30 books on science and social problems, one of his most recent is How Long Have We Got?

Calder argues that the power and carelessness of modern technology pose a threat to the terrestrial homestead we have inherited. We now use the biosphere as a giant laboratory for experiments whose effects cannot be predicted (the atomic bombs in 1945, atomic waste disposal, pollution on a Torrey Canyon scale—for which the cure was worse than the complaint—the detonation of an atomic bomb in the Van Allen Belt and the increase in atmospheric CO_2 content are examples). Scientific superspecialism leads to a lack of both consultation and comprehension. The extract we quote concerns an irrigation scheme in Pakistan, and illustrates that major mistakes can result, even from the best of intentions. Those and other hazards are compounded by the continuing population explosion.

MORTGAGING THE OLD HOMESTEAD

Lord Ritchie-Calder

Past civilizations are buried in the graveyards of their own mistakes, but as each died of its greed, its carelessness or its effeteness another took its place. That was because such civilizations took their character from a locality or region. Today ours is a global civilization; it is not bounded by the Tigris and the Euphrates nor even the Hellespont and the Indus; it is the whole world. Its planet has shrunk to a neighborhood round which a man-made satellite can patrol sixteen times a day, riding the gravitational fences of Man's family estate. It is a community so interdependent that our mistakes are exaggerated on a world scale.

For the first time in history, Man has the power of veto over the evolution of his own species through a nuclear holocaust. The overkill is enough to wipe out every man, woman and child on earth, together with our fellow lodgers, the animals, the birds and the insects, and to reduce our planet to a radioactive wilderness. Or the Doomsday Machine could be replaced by the Doomsday Bug. By gene-manipulation and man-made mutations, it is possible to produce, or generate, a disease against which there would be no natural immunity; by "generate" is meant that even if the perpetrators inoculated themselves protectively, the disease in spreading round the world could assume a virulence of its own and involve them too. When a British bacteriologist died of the bug he had invented, a distinguished scientist said,

"Thank God he didn't sneeze; he could have started a pandemic against which there would have been no immunity."

Modern Man can outboast the Ancients, who in the arrogance of their material achievements built pyramids as the gravestones of their civilizations. We can blast our pyramids into space to orbit through all eternity round a planet which perished by our neglect.

A hundred years ago Claude Bernard, the famous French physiologist, enjoined his colleagues, "True science teaches us to doubt and in ignorance to refrain." What he meant was that the scientist must proceed from one tested foothold to the next (like going into a mine-field with a mine-detector). Today we are using the biosphere, the living space, as an experimental laboratory. When the mad scientist of fiction blows himself and his laboratory skyhigh, that is all right, but when scientists and decision-makers act out of ignorance and pretend that it is knowledge, they are putting the whole world in hazard. Anyway, science at best is not wisdom; it is knowledge, while wisdom is knowledge tempered with judgment. Because of overspecialization, most scientists are disabled from exercising judgments beyond their own sphere.

In the Indus Valley in West Pakistan, the population is increasing at the rate of ten more mouths to be fed every five minutes. In that same five minutes in that same place, an acre of land is being lost through water-logging and salinity. This is the largest irrigated region in the world. Twenty-three million acres are artificially watered by canals. The Indus and its tributaries, the Jhelum, the Chenab, the Ravi, the Reas and the Sutlej, created the alluvial plains of the Punjab and the Sind. In the nineteenth century, the British began a big program of farm development in lands which were fertile but had low rainfall. Barrages and distribution canals were constructed. One thing which, for economy's sake, was not done was to line the canals. In the early days, this genuinely did not matter. The water was being spread from the Indus into a thirsty plain and if it soaked in so much the better. The system also depended on what is called "inland delta drainage," that is to say, the water spreads out like a delta and then drains itself back into the river. After independence, Pakistan, with external aid, started vigorously to extend the Indus irrigation. The experts all said the soil was good and would produce abundantly once it got the distributed water. There were plenty of experts, but they all overlooked one thing—the hydrological imperatives. The incline from Lahore to the Rann of Kutch—700 miles—is a foot a mile, a quite inadequate drainage gradient. So as more and more barrages and more and more lateral canals were built, the water was not draining back into the Indus. Some 40 percent of the water in the unlined canals seeped underground, and in a network of 40,000 miles of canals that is a lot of water. The result was that the watertable rose. Low-lying areas became waterlogged, drowning the roots of the crops. In other areas the water crept upwards, leaching salts which accumulated in the surface layers, poisoning the crops. At the same time the irrigation régime, which used just 1½ inches of water a year in the fields, did not sluice out those salts but added, through evaporation, its own salts. The result was tragically spectacular. In flying over large tracts of this area one would imagine that it was an Arctic landscape because the white crust of salt glistens like snow.

The situation was deteriorating so rapidly that President Ayub appealed in person to President Kennedy, who sent out a high-powered mission which encompassed twenty disciplines. This was backed by the computers at Harvard. The answers were pretty grim. It would take twenty years and $2 billion to repair the damage—more than it cost to create the installations that did the damage. It would mean using vertical drainage to bring up the water and use it for irrigation, and also to sluice out the salt in the surface soil. If those twenty scientific disciplines had been brought together in the first instance it would not have happened.

Always and everywhere we come back to the problem of population— more people to make more mistakes, more people to be the victims of the mistakes of others, more people to suffer Hell upon Earth. It is appalling to hear people complacently talking about the population explosion as though it belonged to the future, or world hunger as though it were threatening, when hundreds of millions can testify that it is already here—swear it with panting breath.

Professor Kingsley Davis of the University of California at Berkeley, the authority on urban development, has presented a hair-raising picture from his survey of the world's cities. He has shown that 38 percent of the world's population is already living in what are defined as urban places. Over one-fifth of the world's population is living in cities of 100,000 or more. And over one-tenth of the world's population is now living in cities of a million or more inhabitants. In 1968, 375 million people were living in million-and-over cities. The proportions are changing so quickly that on present trends it would take only 16 years for half the world's population to be living in cities and only 55 years for it to reach 100 percent.

Within the lifetime of a child born today, Kingsley Davis foresees, on present trends of population-increase, 15 billion people to be fed and housed —nearly five times as many as now. The whole human species would be living in cities of a million-and-over inhabitants, and—wait for it!—the biggest city would have 1.3 billion inhabitants. That means 186 times as many as there are in Greater London.

For years the Greek architect Doxiadis has been warning us about such prospects. In his Ecumenopolis—World City—one urban area like confluent ulcers would ooze into the next. The East Side of World City would have as its High Street the Eurasian Highway stretching from Glasgow to Bangkok, with the Channel Tunnel as its subway and a built-up area all the way. On the West Side of World City, divided not by the tracks but by the Atlantic, the pattern is already emerging, or rather, merging. Americans already talk about Boswash, the urban development of a built-up area stretching from Boston to Washington; and on the West Coast, apart from Los Angeles, sprawling into the desert, the realtors are already slurring one city into another all along the Pacific Coast from the Mexican Border to San Francisco. We don't need a crystal ball to foresee what Davis and Doxiadis are predicting; we can already see it through smog-covered spectacles; a blind man can smell what is coming.

The danger of prediction is that experts and men of affairs are likely to plan for the predicted trends and confirm these trends. "Prognosis" is something different from "prediction." An intelligent doctor having diagnosed your symptoms and examined your condition does not say (except in novelettes), "You have six months to live." An intelligent doctor says, "Frankly, your condition is serious. Unless you do so-and-so, and I do so-and-so, it is bound to deteriorate." The operative phrase is "do so-and-so." We don't have to plan for trends; if they are socially undesirable our duty is to plan away from them; to treat the symptoms before they become malignant.

We have to do this on the local, the national and the international scale, through intergovernmental action, because there are no frontiers in present-day pollution and destruction of the biosphere. Mankind shares a common habitat. We have mortgaged the old homestead and nature is liable to foreclose.

Richard Jahns, former Dean of the School of Earth Sciences at Stanford University, discusses the need for recognition of the geologic jeopardy in which society still stands. Trained at California Institute of Technology and Northwestern, Jahns has also held appointments at Pennsylvania State University and the U. S. Geological Survey. In spite of man's technological accomplishments and extractive skills, burgeoning population, endless wastes, and planning blunders make man more, rather than less, vulnerable to the effects of natural hazards. Jahns analyzes the various geologic hazards of life in California, including floods and landslides, as well as earthquakes, from which section of this article our extract is taken. Though informed legislation can reduce dangers from most of these hazards—such as more stringent grading ordinances and building codes, which have averted many disasters—far more comprehensive planning and prediction are required, and Jahns argues for a new level of public stewardship.

GEOLOGIC JEOPARDY

Richard H. Jahns

Not long ago, man was often inclined to reflect with unqualified satisfaction upon his growing record of accomplishment in competing with nature. But as he has continued to reshape the terrain, to modify much of its drainage, to extract useful materials from the subsurface, and to control various elements of his environment on larger and larger scales, it has become less and less clear that so pleasant a view is justified by the record. Today man is being more widely recognized as the kind of schizoid competitor he really is—imaginative, ingenious, resourceful, and remarkably courageous, but with distressing capacities for vastly increasing his own numbers, for enveloping himself with wastes of many kinds, and for making serious mistakes in dealing with his natural surroundings.

As human population has burgeoned and clustered during recent decades, unpleasant confrontations with geologic reality have become more frequent and more challenging. Among the so-called geologic hazards, or risks, those most commonly encountered are related to floods, earthquakes, and various kinds of unstable ground, and as such they are normal and widespread manifestations of natural processes operating upon and within the earth's crust. From time to time some of the risks are translated into disasters, either unavoidably or through the active cooperation of man; the nature, location, and even the magnitude of such disasters often can be predicted well in advance of their occurrence, but the advance dating of these occurrences is another matter. Intervals between successive natural catastrophes of the same kind ordinarily are so great that studies of the geologic and historic records rarely lead to forecasts sharply enough focused to be useful without supplementary information of other kinds.

A brief sampling of well-documented geologic risks and disasters from the state of California may provide some notion of the diversity of problems that have been recognized. By no means does California have a corner on such problems, but its great variety of geologic materials and features, combined with major concentrations of population, establishes it as an excellent testing ground for the basic wisdom of its occupants. And in few other places are large numbers of residents confronted by such stimulating assemblages of natural hazards as those in the Los Angeles and San Francisco regions, where contrasts in topography, climate, and geology are especially prominent.

From California's subsurface come those recurring violent actions known as earthquakes, several of which have dealt rather harshly with man and his works during the period of historic record. These shocks have originated along faults, or breaks in the earth's crust that represent repeated slippage over very long spans of time. Thousands of faults are known within the state, and many of them can be classed as large in terms of their total displacements. Many of them also are geologically active in the sense of having moved within the past 10,000 years, and more than a few have been active in historic time.

The release of energy in the form of a large earthquake can be assumed to begin with sudden fault movement initiated at some depth in the earth's crust, and to continue as this slippage is rapidly propagated in all directions along the fault. Under appropriate conditions, the displacement may reach the earth's surface and appear as horizontal, vertical, or oblique offsets along the trace of the fault. Such surface faulting also may accompany a relatively small earthquake if the focus, or point of original rupture, is sufficiently shallow.

During the San Francisco earthquake of 1906, predominantly horizontal surface displacement occurred along perhaps as much as 270 miles of the San Andreas fault, California's widely known master break. Maximum observed offset of reference features such as roads and fences was nearly twenty feet. Displacements of similar magnitude may well have occurred farther south along the same fault during the great Fort Tejon earthquake of 1857. Among earthquakes originating in California and nearby parts of Nevada during the past century, at least twenty are known to have been attended by measurable surface faulting. The largest offsets observed in relatively recent years were nineteen feet (horizontal) with the Imperial Valley earthquake of 1940, and twelve feet (horizontal component) and fourteen feet (vertical component) with the Fairview Peak, Nevada, earthquake of 1954.

Although surface rupture is neither common nor widespread over periods of human generations, there is obvious risk in erecting homes or other structures athwart the most recent traces of past movements along active faults. Yet this is precisely what has been done, especially along the San Andreas fault southward from San Francisco and along the San Andreas and San Jacinto faults in the San Bernardino area of southern California. Moreover, this kind of "gamble-in-residence" is being taken more and more often, and in areas where the positions of active faults are well known; it is particularly distressing to note the number of schools represented among installations that some day might serve as reference features for large-scale shear. The gamble might seem safe enough in terms of the odds against losing, but any loss in this ill-advised game could be a major disaster. Fortunately, the nature of such risk is being increasingly noted and appraised for aqueducts, dams, and other special engineering works, if not for housing developments.

A more immediate threat to some installations that extend across fault traces is the slow creep, or progressive slippage in the apparent absence of earthquakes, that is being recognized as currently characteristic of several major breaks in California. These include the Hayward fault on the east side of San Francisco Bay, the Calaveras fault in the vicinity of Hollister, and parts of the San Andreas fault for a distance of about two hundred miles southeastward from San Francisco. Tunnels, railroad tracks, roads, fences, culverts, and buildings are being deformed and displaced where they straddle the narrow strands of creep, with cumulative offsets generally measured in inches over periods of years or decades. At the Almaden-Cienega winery southeast of Hollister, horizontal slippage along the San Andreas fault averages more than half an inch per year; the main building, which evidently lies astride the active fault trace, has been aptly described as experiencing "obvious structural distress."

Returning now to large earthquakes, it should be emphasized that by far their most widespread effects upon man are results of severe ground shaking that accompanies the sudden rupturing along faults. For a given moment of energy released, the severity of shaking can vary greatly from one locality to another according to the interplay of many factors. Perhaps most important among these is the nature of the rocks or other foundation materials at the site in question, as demonstrated by marked variations in the distribution of damage from individual historic shocks. In general, shaking is least intense in hard, firm rocks like granite and gneiss, and most intense where the energy is coupled into relatively soft and loose materials like alluvial silts, sands, and gravels, swamp and lake deposits, and hydraulic fill. The oft-used analogy of a block of jello on a vibrating platter is grossly simplified but nonetheless reasonable, and assuredly a structure that in effect is built upon the jello must be designed to accommodate greater dynamic stresses if it is to survive the shaking as effectively as a comparable structure built upon the platter.

This focuses principally upon California's metropolitan areas, large parts of which are underlain by relatively soft and poorly consolidated materials, and it further points up the need for prudent design of buildings to be erected on reclaimed marsh land or filled portions of lakes and bays. Much of value has been learned during recent decades in the important field of earthquake-resistant design for structures, and a great deal of this knowledge has been wisely applied in actual construction. Unhappily, however, it rarely is possible to offer more than broad generalizations concerning the nature of ground motion to be expected beneath these structures during future earthquakes. Relations between the release of energy at its source and the attendant ground response at a given surface locality are incompletely understood, and pertinent empirical data are still sketchy at best; thus the engineer with capability for earthquake-resistant design cannot be readily supplied with precise design criteria.

A considerable sum of recorded observations indicates that strong shaking during past California earthquakes has found expression in many different ways. In addition to its widely known effects upon man-made features, it has led to cracking, fissuring, warping, lurching, and local elevation or depression of the ground, to triggering of slumps, landslides, avalanches, and debris flows, to shifting of surface drainage and groundwater circulation, and to broad-scale sloshing of lakes and other water bodies. Various combinations of these and other abrupt changes in the normal scene doubtless will accompany future earthquakes that are certain to occur; indeed, the notion that the state is now "overdue" for another great earthquake is not without some foundation.

It seems obvious that man cannot take for granted the ground he occupies, and that responsibility for troubles stemming from a careless attitude cannot be easily fixed upon someone else, legally or otherwise. Nor can a defeatist attitude survive under the growing pressure of population increase, with corollary expansion of settlement into areas where questions of ground stability must be faced and answered. Here some real progress already has been made, especially in the San Francisco and Los Angeles regions, as more geologists, engineers, land developers, and public officials appear to be asking themselves:

> Will posterity participate
> In chaos we create,
> Or will our heirs commemorate
> Mistakes we didn't make?

Granting man's limitations in controlling certain important elements of his geologic environment, in California and elsewhere, at least he is learning that when he imposes improperly upon nature, nature is likely to respond by imposing more seriously upon him. He has been modifying his approach by seeking better to understand natural processes and more effectively to apply this understanding in the primary struggle, which really lies more with himself than with nature. Nature now can be identified less as the antagonist than as the arena in which ever-increasing numbers of people are competing with one another for food, for air and water, for energy, and for space—there is no other readily available arena, hence this one needs a bit more respect and care.

Toward environmental understanding and improvement, geological scientists and engineers are vigorously investigating many kinds of so-called natural hazards. Active and potentially active faults are now being precisely mapped over large areas, and their respective styles of behavior during the geologic past are being deciphered via a remarkable variety of approaches. New data on creep along faults, the accumulating strain in ground adjacent to faults, and the behavior of the ground during recent earthquakes are revealing some instructive and unexpected relationships. Engineering seismology has fully emerged as a highly significant field of study, with major efforts now being devoted to determining the response of bedrock terranes and surficial deposits to earthquake shaking, the behavior of buildings and other structures during earthquakes, and the most satisfactory types of seismic design for many kinds of structures.

Prediction of earthquakes no longer seems to be an objective for some future century; indeed, at least one important break-through in this area can be expected within the coming decade. Soon perhaps we shall have warning of a few minutes to as much as an hour in advance of strong shocks along two or three of this country's most prominent active faults, a contribution of incalculable value to people if not to their property. In the meantime, increasing coordination of empirical and basic data and of observation, experiment, and theoretical analysis can be expected further to improve our dealings with floods, landslides, ground subsidence, and other kinds of natural hazards. May we look forward to the days when all kinds of ground failure can be forestalled or reduced in impact, when existing landslides can be made safe for useful development, when our buildings and utilities can survive the severest earthquake, and when sound programs for disaster insurance can be predicted upon knowledge not yet available.

The human side of these dealings is even more complex than the problems posed by nature. The general public and numerous stewards at all levels of government have been rapidly awakening to the existence and scope of geologic jeopardy in its numerous forms, with reactions ranging from apathy to panic but tending properly to consolidate into deep concern. The growing record of damage and death, especially in some thickly populated areas, has become so compelling that direct responses are now extending beyond temporary reactions to individual disasters. Homeowners and public officials, scientists and engineers, universities and utilities, conservation groups and industrial organizations, government agencies and legislatures, and increasingly large numbers of individual citizens are discovering that they have a common stake in learning how better to live with their physical environment, regardless of their other interests. As more and more of them come to recognize the game we all have been playing in nature's arena for so many centuries, more and more of them will want to know what the score is. Geoscientists and engineers must be continuingly ready with the answer, however constrained or unpalatable it might be at any given place or time.

The Most Exalted Object

"It is interesting to contemplate a tangled bank, clothed with many plants of many kinds, with birds singing on the bushes, with various insects flitting about, and with worms crawling through the damp earth, and to reflect that these elaborately constructed forms, so different from each other, and dependent upon each other in so complex a manner, have all been produced by laws acting around us. These laws, taken in the largest sense, being Growth with Reproduction; Inheritance which is almost implied by reproduction; Variability from the indirect and direct action of the conditions of life, and from use and disuse: a Ratio of Increase so high as to lead to a Struggle for Life, and as a consequence to Natural Selection, entailing Divergence of Character and the Extinction of less-improved forms. Thus, from the war of nature, from famine and death, the most exalted object which we are capable of conceiving, namely, the production of the higher animals, directly follows. There is grandeur in this view of life, with its several powers, having been originally breathed by the Creator into a few forms or into one; and that, whilst this planet has gone cycling on according to the fixed law of gravity, from so simple a beginning endless forms most beautiful and most wonderful have been, and are being evolved."

—Charles Darwin

The Great Lakes, hub of the North American continent, and largest lake body in the world, contain one-fifth of the Earth's fresh water. Gladwin Hill, of the New York Times, *respected and informed writer on environmental topics, illustrates the effects of the continuous pollution of the Great Lakes, and defines the magnitude and the complexity of the problems of their restoration. In spite of recent improvements, the problems described by him are still serious.*

THE GREAT AND DIRTY LAKES

Gladwin Hill

Whatever a honeymoon visit to Niagara Falls was like in the days when Blondin was crossing on his tightwire, it's different now. Something new and unpleasant has been added. Sightseers boarding the famous *Maid of the Mist* excursion boat are likely to find themselves shrouded in a miasma that smells like sewage. That's what it is—coming over the American falls in the Niagara River and gushing out of a great eight-foot culvert beneath the Honeymoon Bridge. As the little boat plows through the swirling currents to a landing on the Canadian side, it has to navigate an expanse of viscous brown foam—paper-mill waste out of the culvert—that collects in a huge eddy across the river.

While aesthetically shocking, this annoyance is hardly factually surprising. The falls in effect are the funnel through which the water from four of the Great Lakes passes on the final leg of its trip to Lake Ontario, the St. Lawrence River, and the sea . . .

The Great Lakes constitute the largest reservoir in the world, containing about 20 per cent of the fresh water on the face of the earth. They are the principal water source for one of the nation's largest concentrations of population and industry. If their waters became corrupted, it would be a calamity of unprecedented magnitude. And it would involve not one but two nations, for the Canadian boundary bisects all the lakes except Lake Michigan.

If you stand on the shore at Duluth, Minnesota, and look out over the endless expanse of Lake Superior, it seems impossible that such a vast body of water could ever become tainted. Yet the flow of industrial pollutants alone from Lake Superior into Lake Huron has been measured by the International Joint Commission in hundreds of millions of gallons a year. The contamination gets rapidly worse as you move eastward along the chain of lakes to Michigan and Erie. . . .

Chicago, at the southern tip of Lake Michigan, has spent more than half a century and billions of dollars developing a good water system. The city draws a billion gallons a day from the lake, to serve 4,400,000 people. Its sewage, treated to remove most of the pollutants, is channeled southward into tributaries of the Mississippi, so that it does not affect the lake.

But around Chicago, extending past the Indiana line only ten miles to the southeast, is a network of small, sluggish waterways—the Grand Calumet River, the Little Calumet River, Wolf Lake, and various canals—that serve as a drainage system for a dense industrial complex sprawling for more than twenty miles along the shore, from Chicago through Hammond, Whiting and Gary, Indiana. The complex includes ten steel mills, five oil refineries, and dozens of other plants ranging from paper mills to soap factories. Six major plants discharge a billion gallons of waste a day that includes 35,000 pounds of ammonia nitrogen, 3,500 pounds of phenols, 3,000 pounds of cyanide, and fifty tons of oil. A good deal of this finds its way into Lake Michigan. There it has spoiled some of Chicago's best beaches, exterminated much aquatic life and recently defied city water officials' best efforts to provide a supply free of objectionable tastes and odors.

Pollution from the western Great Lakes is augmented to a repulsive level as their waters come down through the Detroit River—actually an interlake strait—and flow into Lake Erie.

"These waters," the Public Health Service summarizes, "are polluted bacteriologically, chemically, physically and biologically, and contain excessive coliform (intestinal bacteria) densities as well as excessive quantities of phenols, iron, ammonia, suspended solids, settleable solids, chlorides, nitrogen compounds and phosphates."

Michigan has prided itself on enforcement of its pollution regulations throughout most of its 11,000 lakes and 36,000 miles of streams. But Detroit, with its great industrial complex and a surprising horse-and-buggy municipal sewage system, has long been a stumper.

Apart from relatively minor contributions from Canadian communities just across the strait, the Detroit area dumps twenty million pounds of contaminant materials into Lake Erie every day, in a waste flow totaling 1.6 billion gallons.

About two-thirds of this is from industry and one-third is municipal sewage. Detroit gives the sewage of three million people only "primary" treatment, which means just the settling-out of grosser solids; standard sewage treatment today includes a "secondary" stage of chemical and biological neutralization of up to 90 per cent of the contaminants.

Late in 1961 Michigan asked for federal help in cleaning up the Detroit situation. The Public Health Service spent two years on an exhaustive investigation of pollution sources. The report named some of the nation's leading automobile, chemical, and paper companies as major offenders. A cleanup program similar to the Chicago one was arrived at in hearings this spring.

Michigan waters account for only about 1 per cent of Lake Erie's surface area. Conditions like Detroit's are repeated in various degrees at Toledo, Cleveland, Erie, and Buffalo. No one knows the total amount of pollution pouring into the lake. But there is plenty of other evidence, both visual and scientific, that the aggregate pollution has long since passed the danger point. It has transformed the 240-mile-long lake, Senator Nelson remarked, "from a body of water into a chemical tank."

Two weeks of federal hearings in Cleveland and Buffalo last August laid the groundwork for a cleanup program. It will cost upward of $100 million just to bring Detroit's sewage system up to standard. The outlay confronting industry there for adequate waste treatment facilities is probably several times that. Along the whole lakeshore, comprehensive remedial measures will probably run into the billions.

The disruption of normal processes has been aggravated by a runaway growth of algae, the plant life that in a healthy lake is microscopic; in Lake Erie it grows like seaweed, in great swatches up to 50 feet long. The dominant type is *cladaphora,* a coarse growth that looks as if it had dripped off the Ancient Mariner and smells terrible.

Algae growth is promoted by phosphates, which come particularly from municipal sewage. Detergents are composed up to 70 per cent of phosphates. Millions of pounds of waste detergent pour into the lake every year. Each pound will propagate 700 pounds of algae. The algae absorbs more oxygen. When it dies, it sinks to the bottom as silt and releases the phosphate to grow another crop of algae. Copper sulfate will kill algae in a small volume of water, such as a pool. But biologists know of no chemical that can exterminate it on a scale such as is found in Lake Erie.

Even if the flow of pollution into Lake Erie were stopped entirely tomorrow, the reversal of this accelerating deteriorative cycle would be problematical. "All we can do," the scientists say, "is try."

Lake Erie contracted this disease rapidly because it is quite shallow. Its maximum depth of 210 feet is only one-eighth that of Superior, the deepest of the lakes. But Erie's fate obviously is simply a preview of what awaits all the lakes if pollution is not stemmed.

Michel Halbouty, consulting geologist in Houston and former President of the American Association of Petroleum Geologists, discusses the role of geology in ministering to human needs, ranging from the burgeoning demand for raw materials and fuels on the one hand, to water supply and environmental health on the other. Analyzing the growing responsibilities of the geologist in meeting human needs, Halbouty invites enlistment in the army of scientists who constitute one of the most vital and essential of the sciences.

GEOLOGY—FOR HUMAN NEEDS

Michel T. Halbouty

Every man who has entered the field of science did so in order to serve his fellow man. The science of geology especially serves mankind, which is the reason that the title "Geology—For Human Needs" was chosen for this article.

Our science, which is based on the ancient studies of astronomy, biology, chemistry, mathematics, and physics, deals with a variety of subjects. Among other things, geology deals with the distribution of natural resources such as ores, rocks and gems, weather and climates of the past, the origin and evolution of animals and plants, history of the rocks, and even the origin of man. Geology is a curious and fascinating mixture of many sciences, yet it is distinct from all others.

It has been the responsibility of geologists to explore for all types of minerals, metals, precious stones, and gems and to examine and evaluate the waters and soils of the earth. In the field of energy supply our profession is responsible for discoveries of raw energy fuels which are so important to the progress and prosperity of this nation and to the remainder of the world. A new age of human progress has been opened in nuclear energy alone, based on results of geological exploration for fissionable materials, such as uranium.

Petroleum geology employs the largest number of geologists, but before a geologist is competent to lead in the hydrocarbon exploration effort he first must be well-grounded in the broad spectrum of science. In other words, a person is a *geologist* first, and a specialist in petroleum geology, second. The petroleum geologist is called upon continually to explore in new provinces, new regions, and new countries. There is no limit to the scientific effort of the petroleum geologist in searching for world-wide deposits of oil and gas to supply energy for the world's needs.

As the space program progresses we are becoming more involved in studies of other planets of the universe. Recently astronaut-geologists have been selected to become members of the space exploration teams of the future.

A new branch of geology called astrogeology has developed as a result of our interest in space.

As communications improve, civilization spreads and creates innumerable contacts between once isolated communities. If the American standard of living should spread to only half the world in the next century, it is certain that known supplies of raw materials would be used up far more rapidly than new deposits could be found. All of the scientific potential now available for the exploration for the required new mineral reserves could hardly satisfy the demand.

To meet the growing demand for raw materials we are slowly but surely turning to the seas. The seas make up one of the planet's richest ecological units and they have reservoirs of resources scarcely touched at present. The oceans will absorb an increasing proportion of man's research and development energies for generations to come. Minerals and life of the sea are already challenging to the geologists, biologists, and oceanographers. Marine geology is an important area of specialization for new geologists.

Along with engineers, geologists have been responsible for the development of many major building and construction programs. They have changed the courses of rivers and located dams, harbors, high-rise structures, housing developments, railroads, highways and sites for new cities. However, many large industrial sites, new cities, housing developments, and other types of construction are built without geological advice. But construction undertaken without an understanding of the geology of an area frequently has had tragic consequences.

Tremendous losses have been suffered where homes have fallen into ravines; reservoirs have broken and washed away entire communities, taking the lives of innocent home dwellers. Such tragedies should be a warning to the public to seek geological advice *before* construction, not after a catastrophe.

Geologists and seismologists currently are involved in studying past earthquakes and seeking criteria for the possible predictions of future earthquakes. These studies include investigations of areas vulnerable to earthquakes and could result in recommendations for the removal of major cities, or portions of them, to locations where the likelihood of earthquakes is negligible.

Today a world-wide water shortage demands much of our attention. Lack of sufficient water limits economic growth, undermines and lowers living standards, jeopardizes the security of nations, and above all, and most important, endangers people's health. A supply of water is one of our profession's most serious challenges in meeting human needs for the future. Answers must be found for this baffling and seemingly impossible problem or the world's multiplying population slowly but surely will perish without this vital natural resource.

Modern geology has penetrated even the field of medicine. Recently, a symposium on the subject, "Medical Geology and Geography" was sponsored by the American Association for the Advancement of Science and The Geochemical Society. The symposium revealed that a relation between trace elements and human health may be more than a fascinating hypothesis. Research along this line undoubtedly will enable our science to add biomedical geology to its many fields of endeavor.

Although geology has abandoned none of its old goals, the science is required to meet new challenges and to participate in new projects to help meet the ever expanding needs of mankind. Demands on geologists will continue to increase as human needs increase. An immediate challenge to the profession today is the need to see that enough geologists are trained to meet the growing demands of society. New geologists must be found, trained and put to work.

With world population doubling every 35 years and with per capita consumption of almost every raw material constantly mounting with the progress of civilization, nations of the world are faced with a problem of great magnitude in finding and providing additional mineral reserves. This presents a challenge to all scientists and technologists everywhere, but primarily to geologists and all other earth scientists.

In the more civilized areas of the earth, geologists already have tapped the most obvious sources of raw materials. For example, the oil and gas which were easiest to find have been discovered. The same is true of all other minerals which are now being mined and produced. Now we must find the more difficult and less obvious deposits. We will find these hidden reserves only by employing bold geological deductions, imagination, and relentless enthusiasm: Geologists continuously must plan for the world's increasing multitude of people and therefore the science of geology is important to the future welfare of the world's people.

Challenges to geologists today are greater than ever before but so are the opportunities. Soon people of all nations will demand the type of civilization that we enjoy in America today. Petroleum geologists will be needed to explore for an estimated 700 billion barrels of petroleum which is now waiting to be found along the continental shelves of the world. However, this oil will be only a fraction of that needed to meet demands that seem to have no ending.

Solving the mysteries of space, and finding hidden treasures of the sea will demand the services of hundreds of new geologists. Geologists also are needed in the fields of waste disposal, pollution control, land development, and dam construction aimed at providing new sources of water supply. Growth of the techniques of geologic aerial photographic interpretation has resulted in new importance to this field. Micropaleontology and sedimentology are finding new applications.

Thousands of new geologists will be needed to fill the many assignments in diversified areas of scientific endeavors that fall within the general field of geology. Yet the enrollment of students taking geology in our universities throughout the nation is far less than the demand—not only for the present, but the next ten years.

Teachers who have scientifically inclined or highly imaginative students should give every consideration to leading them into the study of geology. The author belongs to a family of geologists all of whom agree that there is no finer or more rewarding profession. Ours is one of the most vital and essential sciences on earth. It is also one of the most attractive and fascinating.

The profession is not a closed union as many are today. Instead geologists constitute an army of scientists reaching far out for knowledge. We need the help of every brilliant, inquisitive mind that we can attract. The more we learn, and the better the men we attract to geology, the more qualified we will be to meet the future challenges of human needs through geology.

The Meaning of Evolution

It is another unique quality of man that he, for the first time in the history of life, has increasing power to choose his course and to influence his own future evolution. The possibility of choice can be shown to exist. This makes rational the hope that choice may sometime lead to what is good and right for man. Responsibility for defining and for seeking that end belongs to all of us.

—George Gaylord Simpson

The short, but cogent, article on the need for a national energy policy was written by Dewey F. Bartlett, petroleum geologist and former governor of Oklahoma. In precise and unemotional terms it traces the origin of the present energy crisis, showing the complex interaction of geologic, technologic, economic, and legislative factors.

NATIONAL ENERGY POLICY

Dewey F. Bartlett

This country currently is facing an energy shortage of major proportions. The gap between the energy needs of the United States and its ability to supply these needs is expanding rapidly. However, this is not the only problem which we face. The problems of the oil and gas industry are closely intertwined with those of other fuels industries, and together, all of the energy problems become extremely complicated. The solution is to develop a coherent national energy policy.

This country is relying on foreign imports of oil and gas in increasing amounts. Any significant disruption in the foreign oil supply will result in immediate fuel oil, natural gas, and/or gasoline rationing in this country.

The Gross National Product (GNP) is a measure of work accomplished. The consumption of energy has a close and direct relationship with the gross national product. To keep up with the anticipated 4 percent annual growth in the GNP, the consumption of energy is anticipated to double by 1985—energy to heat our homes, run our factories, power our transportation systems, and in general to keep this highly industrialized country running.

Oil contributes 44 percent of our total energy supply. Our current rate of consumption is a staggering 15 million bbl of oil per day. Until 1948, this country was a net exporter of oil. After 1948, we began to import more and more oil for economic reasons. Since 1967, our domestic oil industry has not produced our full requirements of petroleum. By 1969, our net imports of oil exceeded by approximately 1 million bbl/day the amount of reserve capacity in the United States that could be used to replace imports. This deficit increased greatly during 1970.

The difference between productive capacity and actual production of oil is the amount of reserve capacity that can be used when needed. Large reserve capacities of oil in the past have served well our national security and our day-to-day needs during World War II, the Korean War, the Suez crisis of 1956 and 1957, the Arab-Israeli War of 1967, and the disruptions of last summer caused by the shutting down of the Trans-Arabian pipeline and by production cutbacks ordered by the government of Libya.

The reserve capacities of oil and gas have been created by state regulatory bodies limiting or prorating oil and gas production from certain wells to be within the maximum efficient producing rates (MER) and the posted demand for oil and gas production. This reserve capacity has proved to be an expensive extra cost to oil and gas producers.

The reserve capacity based on IPAA reserve figures will disappear by 1974 or 1975 because we are producing more oil than we are finding and current economic conditions do not justify carrying extra reserves.

However, from information which I have assembled on the demand and supply of oil and gas, the reserve capacity of this country is practically gone now and should be gone long before 1974 or 1975.

The East Coast now depends on foreign imports for 93 percent of its residual fuel supply. This dependence on foreign imports without sufficient reserves has severe national security implications.

The current limited reserve capacity of oil and gas means that Americans soon are in for a new experience. Instead of opening the valves of some domestic oil and gas wells in order to make up for a loss in foreign oil imports when an interruption occurs, we shall feel the loss directly and will have to ration fuel oil, gas, and/or gasoline.

The federal government has encouraged the depletion of our domestic oil-reserve capacity by publicly criticizing the action of state regulatory bodies in limiting production below the maximum efficient rate (MER)—which thereby creates producible reserves.

The delays in approving the 800-mi Trans-Alaskan pipeline make it unlikely that oil from the North Slope will reach this country before 1975 at the earliest. During 1970, oil imports represented approximately 23 percent of our domestic consumption. If 500,000 bbl/day come through the pipeline from Alaska in 1975, the North Slope production will reduce imports only from 28 to 25 percent. By 1980, imports without North Slope oil are projected to be 44 percent; with 2 million bbl/day of North Slope oil, imports will be 35 percent.

The reduction in the federal depletion allowance from 27½ percent to 22 percent, the reduction of some state depletion allowances, federal and state tax increases, increased costs of steel and labor, the low price established by the federal government for natural gas, and the threat of a tariff on foreign oil for the purpose of reducing the price of domestic crude oil have reduced profits and discouraged exploratory, research, and drilling activities within the United States.

Thus, the absence of a domestic oil and gas reserve producing capacity in the near future is a sobering thought.

Natural gas demand during the last 20 years has grown about 6 percent per year, or just about 50 percent more than the growth in energy consumption. By 1969, gas was contributing about a third of our energy supply. This period of rapid growth is at an end. The success rate of discovering new gas reserves in the last 3 years is down. The decline in exploratory activity occasioned by leaner profits resulting from controlled prices has existed for more than 10 years.

In the early 1950s, more than 1.5 bbl of oil was found for each barrel taken out of the ground. About 2 cu ft of gas was then discovered for each one produced. We are using up our reserves of both oil and gas faster than we are replacing them.

The northeastern part of this country could become dependent on foreign natural gas for as much as half of its supply by 1980. The increased consumption of natural gas, just as with oil, means a greater reliance on foreign sources.

The demand for natural gas skyrocketed mainly because it was priced so low by federal controls. The abnormally low prices in turn have discouraged drilling and have led to the present gas shortage.

In 1969, nuclear generated electric power supplied two tenths of 1 percent of the energy consumed in this country. In 1980, it is projected to be 10 percent. Nuclear power will provide a partial solution to our increasing reliance on foreign energy, if pollution problems relating to disposal of extremely high temperature waters can be solved.

We have a coal supply for hundreds of years in discovered and recoverable reserves in the United States. In 1910, coal supplied more than three fourths of our energy requirements; today, it provides a fifth, whereas three fourths of our energy now comes from oil and gas.

Coal consumption has been reduced sharply by the advent of nuclear power and air quality standards that have taken much high-sulfur coal off the market.

Our energy laws have specious results. The pollution laws reduce the consumption of high-sulfur coal but its side effect is to replace American coal with foreign residual fuel oil. Price control on gas was established so house owners could enjoy a premium fuel at a low price; but, the price was set so low that much of it was used where coal could have served as well, and now the prospects of a gas shortage are imminent. The concern over the pollution of the polar ice cap is appreciated by everyone, but the delay in Trans-Alaskan pipeline is delaying much needed energy for the security and well-being of this country.

The interrelations of the various fuels and the short supply of domestic energy lead to the following conclusion: a national energy policy must be established by our federal government. This policy may be stated as follows.

It is the public policy of the United States to have available at all times and in the future a supply of energy from secure sources to meet adequately the essential day to day and emergency needs of its citizens and to safeguard our national security.

In implementing such a policy, the following should be considered:

1. Ways of eliminating waste and inefficient use of various fuels should be determined.

2. Federal and state governments should take steps immediately to create an economic climate favorable for strengthening domestic reserves for oil and gas, and the establishment of a reserve producing capacity to be available in case of national emergency or disruption in foreign sources of supply as a safeguard for our national security.

3. The use of coal, uranium, and/or hydroelectric power as reserves of energy for oil and gas should be developed.

4. Additional research in exploratory techniques for finding oil, gas, and uranium should be encouraged.

5. Additional research to develop an economic method of extracting oil from oil shales should be encouraged.

6. Additional research to develop an economic method of extracting oil from tar sands should be encouraged.

7. Additional research to develop improved secondary and tertiary methods of oil recovery should be encouraged.

8. Research to develop economic methods of reducing the high sulfur content in certain coals should be encouraged.

9. Technical guarantees to eliminate the threat to the environment should be encouraged so that the construction of the Alaskan pipeline may start soon.

10. Research to control extremely high water temperatures in connection with nuclear generated electric power should be encouraged.

11. Research to construct a deuterium fusion reactor capable of containing a heat of at least 45 million degrees Celsius should be encouraged.

One of the most influential books of the late 1960s was Affluence in Jeopardy. *Charles Park, the author, Professor Emeritus of Geology at Stanford, argues that exploding world population, growing aspirations of the citizens of the less industrialized nations, and dwindling and non-renewable resources converge to threaten present levels and styles of living. In the present extract, Park reviews the situation, assesses the probabilities of obtaining substitutes and the cost factors, and calls for a threefold program of population control, national mineral policies, and new trading patterns.*

MINERALS, PEOPLE, AND THE FUTURE

Charles F. Park, Jr.

"Never look back," advised the philosopher Satchel Paige, also known in the realm of baseball, "something may be gaining on you."

But this philosophical pitcher did not mean that he would ignore the scoring threat of a runner behind him on second base. To the contrary, his record shows that he worked on the future danger in the batter's box with his strategy founded in part on what had already occurred.

As citizens of today face the problems (such as wars, civil unrest, and air pollution) that are lined up in front of us, claiming our immediate attention, we have another problem that is catching up on us from behind, so to speak, because it is largely unobserved and unheeded—the peril of mineral shortages. Through our past treatment of our mineral resources, we have let this problem get at least to second base. If we ask ourselves whether or not there is likely to be a score against us, we are back at this book's starting point.

Can the earth afford the world's affluence if the philosophy and technology of our society are extended? Can the earth even continue to afford the affluence that exists today? May these questions be dismissed with the assurance that substitutes will be found as they are required, or that the oceans will supply enough minerals for the future, so that our economies may continue to expand? Or must we realize that there are limits to the supplies of minerals that form the basis of a modern industrial civilization? Is affluence in jeopardy?

A few mineral commodities, iron and aluminum for example, exist in sufficient quantities to maintain the industries of the world and to permit considerable expansion. A few other commodities, such as mercury, tin, and silver, are already pressed to maintain their current production and markets; their prices to consumers are rising and, as a consequence, substitutes are being sought (and not always being found).

Between these two extremes lie the great majority of mineral products. They are present in the earth's crust in amounts sufficient to satisfy current demands and to allow for moderate expansion. They cannot, however, over a long period of time keep abreast of the rapid increase in world population. This is especially true of the fossil fuels that provide so much of the energy required by our civilization.

While some substitution will be possible, general substitution will not be—substitutes cannot fill all our needs of essential minerals.

And, after all, any substitution must come from the earth's mineral supplies.

The mineral industries have made remarkable progress in meeting their needs for the cheaper extraction of lower-grade ores. Still, we cannot look forward to the indefinite utilization of lower and lower grades of any ore. The iron and steel industry, for example, has received the benefit of technological development in the form of pelletizing, which has made economic the mining of lower grades of iron ore. In addition, iron is a common mineral—anyone may take a ton of soil or rock from his back yard and find iron in it. It is doubtful, however, that anyone could afford to extract a pound of iron from a ton of heterogeneous backyard mixture, and to process it, for 7½ cents. If a ton of material contains too little iron, not even the use of pellets can make the extraction of the iron economic at that price.

Although technology is continually improving, the discovery of new deposits and the mining of ever lower grades of ore, both of which must in the future turn to greater depths underground or to the oceans, will reach a point where they are uneconomic. Shortages of essential minerals can come, not only from the ultimate physical exhaustion of the minerals, but also from their unavailability at reasonable cost.

When we speak of costs, we talk in terms of money—of dollars and cents if we are in the United States. But such terms are only symbols of the real cost—that of the energy used. Money is energy, and energy, as has been demonstrated, is one more commodity that is limited in its supply.

Mineral resources and cheap energy are not available in unlimited quantities. Since there are limits to the supplies of minerals that form the basis of a modern industrial civilization, the affluence of our society is in jeopardy. If we are to preserve it, we must answer two questions: How can sufficient quantitites of raw materials be obtained over long periods of time? How can they be equitably distributed? In trying to answer these two questions, we must ask a third: How can the emerging nations, which produce large amounts of the world's minerals and energy, be encouraged to develop and to improve their standards of living?

The future of our affluent civilization depends upon the answers to all three of those questions. Since they are so important, their answers cannot come only from a few educated politicians (and one shrinks from thinking of the answers that would come from *uneducated* politicians); it is a knowledgeable public that must everywhere provide the answers. Political leaders must indeed, for the future welfare of the world, be informed about minerals, and so must economists—in fact, the impossibility of divorcing minerals from the political economy has been shown by the number and the wide range of subjects relating to politics and economics that this book has had to mention, however generally, in order to provide background for understanding the mineral industries. The more all citizens know about minerals, however, and the better they recognize their dependence on minerals, the greater the hope that people will in the future have the minerals they need.

This book has three recommendations to make in attempting to solve the problem of mineral shortages in the future.

First, the rate of population growth must be controlled. Without such control, all efforts to improve the lot of peoples in the underdeveloped nations, and even to maintain the status quo, can at best achieve only temporary and partial success. Shortages of fertilizers and foods, of energy, and of mineral products of all kinds are inevitable; poverty and starvation are bound to increase. With such conditions at their sources of supply, how will the industrialized nations, with growing needs and growing problems from their own overpopulation, obtain the raw materials that are the basis of their affluence?

Second, every nation must formulate a carefully thought out and clearly stated mineral policy that is suited to its particular needs and that recognizes the necessity for international cooperation. Such a policy must not be entrusted to an entrenched, unenlightened, and arbitrary bureau. Such a policy must be under constant review, since the place of minerals in the world economy is seldom static.

The emerging nations need help from the industrialized nations to develop the mineral resources which can raise their standards of living. They need foreign capital and foreign personnel to provide the money and the skills required for exploration and for the building and maintaining of mines, mills, and smelters. If they permit foreign capital to earn a fair return, the result will be mutual advantage. If, on the other hand, they overtax foreign investment or resort to shortsighted nationalistic measures of expropriation, they will drive away what they most need—money and trained personnel—to develop their mineral resources and thus to develop their civilization.

Third, the United States should explore the possibilities of establishing—and joining—a common market that might include all peoples in the Western Hemisphere. This would assure the Latin American nations of a market for their raw materials and of a supply of the manufactured articles they need and give them an opportunity to develop the industries for which they are best suited, while this would assure the United States of a source of raw materials for the future and an expanding market for many of its finished products.

John Fuller's article, The Geological Attitude, *analyzes the relationships between geological activity, industrial and scientific developments, and social pressures. Fuller, geologist with Amoco Canada Petroleum Co., draws on the history of geology, both early and modern, to clarify the "consumption-pollution and exploitation-injury" dilemma facing our society. The present extract from his article analyzes the problems facing the petroleum industry in the closing years of the 20th century.*

THE GEOLOGICAL ATTITUDE

John G. C. M. Fuller

By "the geological attitude," I mean that particular point of view which has been a characteristic of geologists since they first tracked mud and heresy into the schools of learning some 200 years ago. To explain this attitude we must know something about the nature of geological activity itself, and something about the social environment in which it is found. I begin with two statements to cover these points.

First, geological activity takes place mainly in response to industrial and social pressures.

Second, past geological reaction to these pressures profoundly altered popular conceptions of time, the Church, man, and the balance of nature.

Examples both of pressure and reaction range through the whole of geology; and although the emphasis is always shifting, according to the particular demands of the time, present-day circumstances are not essentially different from those of the past, as I hope to show.

A third point about geological activity is that for at least 100 years its practical style has depended on technology. Peaks of activity cluster obviously on the introduction from time to time of new instrumental capabilities, and not infrequently such activity is testing concepts or relationships which had been perceived as long as a century beforehand.

Modern geology began in style. In 1785, James Hutton, a 58-year-old naturalist and farmer, presented to the Royal Society of Edinburgh his thoughts "concerning the system of the earth, its duration and stability." The paper dealt with the lands and the seas in a succession of former worlds, and the power of water to wear away rocks. It painted a scene of immeasurable antiquity, constant in change, and evidently endless. "With respect to human observations," he wrote, "this world has neither a beginning nor an end." Hutton's words, as later events were to prove, struck a mortal blow to the ruling notion of secular time.

And of what importance was that? When Isaac Newton and the astronomers a century before had expanded the universe in space, beyond any constraint other than natural law, the established and popular view of Creation suffered no harm. When Hutton and the geologists, however, expanded the created world in time, they seemed to be challenging the Divine word. The reason was this: the English-speaking peoples had for two centuries resisted the authority of their original Church, preferring to substitute the authority of the English Bible. Divine authority rested in the Scriptures, and to question them seemed to question Providence itself. The creation of the world, said the Bible, was accomplished in six days: Hutton and the geologists were saying that even six millennia could not meet the facts. Suddenly, the geological world view became a religious issue, and a conflict of appropriate ferocity commenced.

The battle of time was fought mainly on the evidence of stratigraphy, which was then new to academic geology, although it had been in use for 100 years or more as traditional working rules and empirical observations among colliers and quarrymen. In its primitive state stratigraphy had been of little or no philosophical interest to scholars, whose chief geological concerns were focused on cosmogony.

The change came with the Industrial Revolution in the second half of the 18th century, when the demand for coal and minerals, and the resulting acceleration of mining and quarrying operations, brought together practical geology and industrial interests. By this means, a working knowledge of the strata—their order, regularity, and distinguishing features—came to the notice of naturalists and scientists. One can argue, in a sense, that organized stratigraphy as a branch of geology was a product of 18th-century industrial pressure. By the beginning of the 19th century stratigraphy was reacting on popular beliefs.

Hutton's world view, and much of the geological doctrine of uniformity that followed from it (Hooykaas, 1963) can be summarized in a single phrase: the inevitability of gradualness. But the stratal succession was not gradual: it was manifestly serial and punctuated by relics of past turmoil. At many places the upper-most layer of sand and clay was so clearly a deposit such as might have resulted from a sudden inundation of the landscape, that stratigraphy seemed to prove with certitude the fact of the Flood; and thus proved, only the foolhardy and tough-minded would doubt Moses' account of it. These glacial deposits were mapped for years as "Diluvium" or deluge-stuff, long after the Flood peak had passed.

The apparently catastrophic nature of the event that left the diluvium, and the nearly universal extinction of life which it was believed to have caused, together with the relics of upheaval in the stratigraphic succession, became the chief elements in a new doctrine of successive creations which held the field for 50 years. Decayed fragments of it can still be found in sensational literature.

It was a popular theory, for it incorporated an old belief that the major divisions of secular time, the six celestial days of Creation, were indeed millennia. Shakespeare knew in 1599 that the poor world was almost 6,000 years old, and said so; and at Cambridge in 1659 John Lightfoot fixed the exact date.

Geology so stirred up public imagination during the middle years of the last century that people responded in thousands. The story of the rocks was news. In 1835, at Boston, 1,200 people turned out day after day for six weeks to hear Benjamin Silliman, Professor at Yale College, lecturing on geology. "Clergymen thank me warmly," he wrote, "for the manner in which they say that delicate points are treated." No less than 4,500 ticket applications were made for Charles Lyell's lectures in the same city in 1841. Henry D. Rogers, whose name reappears in connection with petroleum geology, drew an audience of 1,000 in the comparatively small town of Portsmouth, New Hampshire, in 1845. In England, Adam Sedgwick, founder of the Cambrian System, made a public sensation in 1838 with a speech to some 3,000–4,000 people on the seashore at Tynemouth. Hugh Miller in 1854 gave a geological lecture in London at which there was an attendance of 5,000.

Another spectacular display of popular enthusiasm took place near Birmingham in 1849, when Sir Roderick Murchison, celebrated author of the Silurian System, led a field meeting on the Silurian outcrops at Dudley. H. B. Woodward recorded that the Bishop of Birmingham, taking a gigantic speaking-trumpet, called on those present to offer their salutations to Murchison, repeating after him the words "Hail, King of Siluria." This they did three times; and 25,000 people visited the place that day. This kind of thing was not new to Murchison. When he was appointed Director of the Geological Survey of Great Britain, members of the House of Commons rose to their feet and cheered on hearing the Prime Minister's announcement.

But during all this time, while stratigraphy piled system upon system, and the fossils multiplied, the conviction was growing among some geologists that a continuous process governed the evolution of living forms, a process in which the hand of Providence was scarcely to be seen, and which had produced man almost incidentally. When Charles Darwin published these convictions in 1859 he detonated an explosion that scorched the very floorboards of society; and the burning fallout can still be seen.

This confrontation with the established order affected everybody. Darwin's book, *On the Origin of Species,* quickly went through three editions. All the scientific doubts of the preceding years seemed suddenly to be public issues. "The rational cat was among the mystical pigeons," wrote Nicolas Bentley, "but with beak and claw and with celestial vision dimmed by tears of outrage, the pigeons fought back." The Bishop of Oxford enquired of Thomas Huxley whether he claimed descent from an ape on his father's or his mother's side, and Huxley retorted that he would rather have an ape for a grandfather than a man who misused his gifts to bring ridicule and religious prejudice into a grave scientific discussion.

Huxley followed Darwin's book with *Evidence as to Man's Place in Nature,* and this subject, after lying dormant for a long time, has become the central issue of our day. It is, in fact, the key to the social impact of geological activity past and present.

Going back to Darwin for a moment, one finds that he was actually following a well-trodden path, which his grandfather, Erasmus Darwin, had traveled before him. And among other well-known precursors it is something of a curiosity—a quite astonishing one in view of later events—to find John Wesley, founder of the Methodist Church. Nearly 100 years before the *Origin of Species,* Wesley was putting his name to questions which asked: "By what degrees does nature raise herself up to man? How change these paws into flexible arms? What method will she make use of to transform these crooked feet into supple and skilful hands?"

This only goes to show that a search for the originator of an idea or the first time it was recorded really does not shed much light on the social pressures and industrial demands which bring that idea into prominence, and cause it to function as the focal point of new activity. Geological ideas grow and spread when social conditions are right for them; and the social pressures, working on the technology at hand, determine the style and scope of the geological activity which results.

Consider, for example, the history of American petroleum geology. Stated simply, drilling for petroleum in Pennsylvania during the 1860s was an outcome of overhunting the whales whose oil served as lamp oil during the first half of the century. Kerosene provided a substitute for whale oil. And although petroleum at that time was already known to yield kerosene, a second competitive answer to whale-oil shortage was found in oil distilled from cannel coal and bituminous shale. The technology of oil-well drilling and casing moreover was also in existence during the first half of the century, having been developed in response to the demand for brine suitable for salt production in the territory west of the Alleghenies.

Petroleum geology thus presents a peculiar feature: the industry possessed a working technology before the geological facts of its commodity were known, and as a consequence its operational and theoretical conceptions diverged wildly.

Forty or fifty years passed in Pennsylvania and the Mid-Continent area before geological principles relating to the origin and accumulation of oil—in particular the "anticlinal" theory—found their way into industrial practice. The reason was quite simply that science was not needed to make a profit from sweet oil lying in huge quantity near the surface of the ground. Drilling mania and greed sustained the momentum until the early 1900s when, by coincidence, wildcat prospecting which had spread southward into Oklahoma encountered proof of mappable geological control over the sites of the oil fields. From that point, between 1913 and 1915, the formal organization of geology in the major American oil companies began. That is one reason why the headquarters of this Association of petroleum geologists, founded in 1917, is in Oklahoma, at Tulsa, and not in Pennsylvania, at Titusville, where the oil industry originated.

Yet it is a fact that an integrated geological conception of the origin of petroleum in black shales and its concentration into oil pools, and of the link between coal rank and indigenous petroleum, went back 50 years to Henry D. Rogers, the first State Geologist of Pennsylvania. Rogers published his findings in 1863, only four years after Drake's discovery well was drilled at Titusville.

> That the volatile hydrocarbons were distilled, as it were, from out the low-lying carbonaceous strata, into the pores and fissures of the over-resting ones, receives strong confirmation from the fact that the elsewhere bituminous shales of the Silurian and Devonian ages, deep under the coal, are altogether as much desiccated and debituminized everywhere in the districts contiguous to the anthracites as the coal beds themselves.

As he wrote this passage did Rogers see in his mind's eye the workings of a refiner's retort, distilling lamp oil from a charge of heated shale? Within a few years such models for metamorphism of organic matter in sedimentary rocks had disappeared from America.

Geology in the petroleum industry serves one function only: to provide eyes in the ground. And all geological activity that is of any use whatever to the industry is directed toward this function. Consequently, petroleum geology clusters round the inventions and innovations which, quite literally, see oil.

The man of action in organized petroleum geology in the United States after the First World War undoubtedly was Everette DeGolyer. What the plane table could do on the surface, a remote sensor could do underground. He succeeded first in Gulf Coast salt-dome terrane, in 1922, with Eötvös torsion balances. Three years later, by converting the principles of German acoustical-seismic apparatus to electrical instrumentation, he assembled what was to be the most effective physical sensing device ever used in geology. Its effect on structural prospecting was immediate and explosive.

The Schlumbergers' invention of "electrical coring" and its introduction to the United States in 1929, later provided the means of detecting by downhole physical attributes alone the existence of unsuspected closure to potential reservoirs. From this sprang a vogue for "stratigraphic traps."

But the main problem in exploration is now no longer one of simply locating favorable structure or favorable reservoirs, for the fact that there exists only a finite number of pools in a finite amount of continental rock increasingly involves statistical conceptions of hydrocarbon potential and sedimentary volume in the exploratory process.

What about recent technological developments? The major concentrations of activity are related to new instrumentation. Consider for example the scanning electron microscope, which is essentially a microtopographic plotting machine. In petroleum geology it is transforming rotary-drill and core chips into storehouses of submicroscopic fossil remains, and to judge by its performance since it was introduced about 7 years ago, it promises to be truly a magic lantern in the dusty passageways of stratigraphy. Its capability of revealing microtopography is being applied also to inorganic forms and it is already a leading instrument in the study of mineral diagenesis and reservoir porosity. Another leader, also operating on the same scale, is the electron microprobe, a microanalytical instrument of unparalleled discrimination.

Concepts generally thought of as belonging to "conservation" are now finding their way into exploration, particularly on the continental shelves, for geologists and the public alike are beginning to visualize the total inventory of prospective rock that extends under the sea to the edge of the continents as final sources of supply, and questions about the meaning of "ownership" are in the newspapers. This is to be expected because social pressure on the industry exceeds anything that it has experienced before.

But the fact remains that the industry faces an immediate necessity of finding and producing unprecedented amounts of petroleum to meet the foreseeable demand: and it is at this point that the petroleum geologist can be confused by the options presented in a pollution-conservation dilemma, of which the essential issues are not at all clear. One can treat the two main options separately.

First, petroleum accounts for something close to three quarters of world energy demand, most of which is concentrated in specific areas of very high consumption. Waste products of this high consumption visibly pollute the environment. Each one of us, for example, produces about 20 times as much pollution by consumption of nonrenewable resources as a so-called underdeveloped person. Problems of this kind involve the petroleum geologist to the extent that he is part of a community which is consuming more energy in its lifetime than the sum total consumed by all preceding generations. Pressure is on the geologist to find oil and gas to satisfy the predicted rates of future consumption, but at the same time he must express his concern for the maintenance of the quality of life in his surroundings, and act on his concerns if he sees any deterioration.

The second part of the dilemma is this: petroleum exploration has now reached areas of the continental shelf and the remote Arctic that were formerly closed through lack of a technology that could operate in those places. The marine and Arctic environments, particularly the latter, are in their own ways peculiarly prone to damage by operations which elsewhere might seem quite normal and inoffensive. The petroleum geologist, if he has any mind for conservation principles at all, is involved to the degree that the natural environment of the ground in which he hopes to find petroleum is itself a resource, the use of which cannot be simply regarded as a matter of capture, or even of prorated exhaustion.

These alternatives, which we can summarize as consumption-pollution and exploitation-injury are to my mind the horns of the dilemma. The geologist can escape from the dilemma if he wishes by saying that the problems are mainly of recovery, carriage, and use, rather than problems of exploration. But if he does this without first analyzing the whole geological scene he can find himself confused in aim, and very likely suffering from what K. H. Crandall called "motivation mortality." I think many of us suffer bouts of it from time to time, and it is something, as I mentioned before, that the educational and informational resources of the Association can help to correct. Quite a lot depends on how far we align ourselves with the trend that Linn Hoover noted "toward a more humanistic view of the geosciences," because this trend, if it continues for the next few years, he wrote, "may have as much impact as a new scientific concept." That really is the crux of the pollution-conservation problem. It is a lighted fuse.

Summarizing briefly, I began by linking geological activity to industrial demand, and suggested that geology in turn had profoundly altered some popular concepts of man and nature. That is still true. I also claimed that geological activity was dependent on technology, and this I illustrated by the accelerated activity which takes place when new instrumentation is applied to old concepts.

Lastly, I endeavored to say that awareness of man's place in nature, which is a fundamental perception of geology, governs the geological attitude.

In this Association let us use the earth for the benefit of all. This world is one world. Perhaps the biggest achievement of geologists in this century will once again turn out to be a general principle, a worldwide concept, that reveals beyond its strictly geological meaning the essential unity of nature.

In a host of ways, each individual undergoes a ceaseless chemical and physical interchange with the Earth and its atmosphere. Respiration, nutrition, shelter and sight—all these and more reflect this interaction. Yet environmental health is still imperfectly understood. The importance of fluorine in water supplies, selenium in soils and carbon in the atmosphere are recognized largely because of the conspicuous, if different, effects they produce. But little is known of the effects of trace elements in soil on food and health, even though some, such as molybdenum, are vital to the metabolism of many organisms. Harry Warren and Robert DeLarault, of the Geology Department of the University of British Columbia, examine the need for more knowledge of medical geology.

MEDICAL GEOLOGY

Harry V. Warren and Robert DeLarault

At first thought one might be tempted to assume that there can be little to relate the sciences of geology and medicine. However, a few moments of reflection should be enough to permit the most skeptical to realize that an association between geology and medicine is a natural one. After all, most people are prepared to accept, at least as a hypothesis, the possibility that man's health is determined, to some extent, by the food he eats. The quality of food reflects the makeup of the soil which, in turn, is determined, in part, by the chemistry of the rocks and vegetal life on the Earth's crust, which, between them, contribute much to the makeup of soil.

It is not difficult to find medical men who can point out how different communities have different ailments. Some of these ailments probably have little or nothing to do with geology, but a few of them may. For generations, the more fortunate members of our society have enjoyed a change of air, food, and water by going each year for a holiday. For some of those who are not too sophisticated, a change of food may come as a delight. Perhaps a change of trace elements is involved in a holiday. At all events, as society gets more and more removed from close contact with the soil, and more and more closely tied to the supermarkets of today, the ties that surely do exist between man and the earth's crust will become more and more tenuous. Even today, it is difficult to find communities suitable for simple epidemiological studies in which trace elements may be involved. However, with care it is still possible to find some communities suitable for trace element epidemiological studies: such communities do exist in some of the undeveloped countries, but they can also be found in conservative and comparatively long-established settlements in Western Europe and North America.

It has long been known that an iodine deficiency can affect whole communities, although we now realize that the problem of goiter is much more involved than was originally realized. Human beings can be affected by poisonous materials in soils when those soils contribute to a large proportion of the diet of a community. In parts of the Dakotas, not only cattle but humans were affected by the selenium that was found in some specific soils. Drinking water may also be the vehicle by which undesirable elements may be introduced into a diet: arsenic and fluorine are two elements that may be present in harmful amounts in local water supplies.

To most geologists, lead is merely lead when it is reported in food or drinking water. Actually, there is one isotope of lead, namely lead 210, derived from uranium by way of radon 222, which is a gas slightly soluble in water. Lead 210 is radioactive with a half-life of some 22 years. However, lead 210 itself breaks down and yields polonium 210, which is a strongly radioactive and poisonous element, particularly dangerous because it finds a resting place in the soft tissues of human bodies.

There may be some good reason, but superficially it seems strange that agriculturists have not done more than they have done to draw doctors and geologists together. It is literally an article of faith with agronomists that soils are related on the one hand to their parent material, and, on the other, to the quality of the food they produce. Nevertheless, doctors and nutritionists alike still publish books that contain tables showing the amounts of various elements, such as copper and iron, that are to be found in different foodstuffs. One may search these medical journals, and all too rarely find any evidence that the significance of geology in relation to food is appreciated. It is equally true that one can search geological literature for a long time before finding any analyses of foodstuffs in relation to their geological background. It is easy to forecast that all this will change before long. Already, in Scandinavia, in Australia, in Great Britain, and in Canada, foresters are realizing more and more how the productivity of their forests is governed, in some areas to a significant extent, by the bedrock involved. It may surprise some geologists to learn that there are areas in the world where foresters would not dream of planting new trees before consulting with their geological confreres.

What makes biogeochemistry such a fascinating subject for a geologist is that appropriate quantities of such elements as copper, zinc, and molybdenum are now known to be just as vital to the health of some plants and animals as they are to the success of mining operations; the concentrations of a specific element needed for the fulfillment of these different needs naturally varies considerably. However, zinc, in appropriate concentrations, is as essential to a diabetic patient as it is for the farmer trying to produce a maximum poundage of pig from each pound of food, and for a forester who is attempting to cultivate a forest. Humans, pigs, and trees would all die if no zinc were available, and equally they can all be adversely affected if they are given too much zinc. The problem yet to be solved in most instances is that of knowing just what is the optimum concentration of zinc needed by a human, a pig, or a tree, as the case may be.

Unfortunately, it is much more difficult to solve equations where there are two variables instead of one, and with each additional unknown element, the complexity of the inquiry increases. A comparatively short time ago only some dozen elements were known to be essential to plant and animal health, but today it may be said that nearer 40 than 30 are thought to play some part in nutrition. Some elements that are always present in plants and animals are not yet known to serve any useful purpose: in the fullness of time it may be found that even these elements, such as lead, mercury, and silver, which are now considered as inhibitors or poisoners of enzyme systems, may actually, if present in appropriate concentrations, act as useful moderators, or regulators, of specific metabolic processes. A word coming more and more into use by biogeochemists is "imbalance." For maximum health in both plants and animals, it is desirable not only that the elements should be present in appropriate concentrations, but also that there should be harmonious relationships between the concentrations of the different elements. It has taken hundreds of millions of years for life to evolve as we now have it on our planet. Surely it is reasonable to assume that life today represents a complex synthesis of the energy and the elements present on the crust of the earth? Thus we might expect to have an infinite variety of living cells representing responses to the different facies extant on the earth's crust. Some forms of life are able to adapt to a sharply different environment, but others perish. Some species of grass have been found to adapt in two generations to a high lead content in soil, but sheep eating this grass could not adapt themselves so promptly: fortunately, it seems that the shepherds in the area involved with this lead soon learned just how long their sheep could tolerate this leaded herbage. Interestingly enough, sheepherders for generations seasonally have moved their flocks. Sometimes this was done merely to ensure a supply of food, but occasionally these moves were dictated by the necessity of avoiding poisonous concentrations in herbage of such elements as molybdenum and selenium.

If it be accepted that variety is the spice of life, then, indeed, biogeochemistry should attract many geologists in the years ahead. However, the number of subjects with which a biogeochemist may become involved is large. Biology, botany, agriculture, bacteriology, virology of the life sciences, and physics and chemistry of the physical sciences all have much to offer to the biogeochemist who must sample them with prudence and wisdom if he is not to be overwhelmed with a mental flatulence likely to be neither congenial nor stimulating but merely frustrating.

At a meeting of the AAAS in Montreal, there was a symposium on Medical Geology and Geography. The participants included a geologist, a pharmacognosist, a biochemist, a general practitioner, and a geological engineer. It might be a moot question as to whether each one of these persons appreciated fully all the finer points raised by their fellow participants in the panel. Nevertheless, the participants, themselves, seemed to find the interdisciplinary exercise well worthwhile and, judging from the reactions of the press and the audience, cross-fertilization experiments of a similar nature may be expected to attract increasing attention in the future. Biogeochemistry is, to many, an uninspiring name: perhaps we should change the name to "Geological ecology," which would relate this field of study to that of life in relation to its geological environment.

EPILOGUE

It is always necessary to close a lecture on Geology in humility. On the ship *Earth* which bears us into immensity toward an end which God alone knows, we are steerage passengers. We are emigrants who know only their own misfortune. The least ignorant among us, the most daring, the most restless, ask ourselves questions; we demand when the voyage of humanity began, how long it will last, how the ship goes, why do its decks and hull vibrate, why do sounds sometimes come up from the hold and go out by the hatchway; we ask what secrets do the depths of the strange vessel conceal and we suffer from never knowing the secrets. . . . You and I are of the group of restless and daring ones who would like to know and who are never satisfied with any response. We hold ourselves together on the prow of the ship, attentive to all the indications that come from the mysterious interior, or the monotonous sea, or the still monotonous sky. We console each other by speaking of the shore toward which we devoutly believe we sail, where we shall indeed arrive, where we shall go ashore tomorrow, perhaps. This shore not one of us has ever seen, but all would recognize it without hesitation were it to appear on the horizon. For it is the shore of the country of our dreams, where the air is so pure there is no death, the country of our desires, and its name is "truth."

—Pierre Termier

Sources

Chapter 1

THRILLS IN FOSSIL HUNTING: George F. Sternberg, *The Aerend*, Volume 1, No. 3, Summer 1930, Kansas State Teachers' College, Hays, Kansas, pp. 139–153. Reprinted in the *Journal of Geological Education*, 1967.

SHIP'S WAKE: From Hans Cloos, *Gespräch mit der Erde*. Copyright © R. Piper & Co. Verlag, München, 1947.

KING'S FORMATIVE YEARS: R.A. Bartlett, 1962, from *Great Surveys of the American West*, copyright 1962 by the University of Oklahoma Press, Norman, Oklahoma, pp. 127–140.

THE MAKING OF A VOLCANOLOGIST: Haroun Tazieff, 1952, *Craters of Fire*, Harper and Brothers, New York, pp. 24–29. Copyright 1952 by Haroun Tazieff. Reprinted with permission of Harper & Row Publishers, Inc.

WITH SHACKLETON IN THE ANTARTIC: Mary Edgeworth David, 1937, *Professor David, The Life of Sir Edgeworth David*, Edward Arnold & Company, London, pp. 129–177.

LIFE, TIME, AND DARWIN: F.H.T. Rhodes, 1958, *Life, Time and Darwin*, University College of Swansea, pp. 3–7.

THE VOYAGE OF THE BEAGLE: Charles Darwin, 62nd printing, 1969, *The Voyage of the Beagle*, P.F. Collier & Son Corporation, New York, pp. 380–384.

A LONG LIFE'S WORK: Sir Archibald Geikie, 1924, *A Long Life's Work*, Macmillan and Company, London, pp. 55–58.

THE OLD RED SANDSTONE: Hugh Miller, 1922, *The Old Red Sandstone*, J.M. Dent and Sons, Ltd., London, pp. 35–44, First edition 1841.

LIFE AT FREIBERG: Raphael Pumpelly, 1918, *My Reminiscences*, Vol. I, Henry Holt and Company, New York, pp. 121–125.

Chapter 2

THE TURTLE MOUNTAIN SLIDE: R.G. McConnell and R.W. Brock, 1904, Report on the Great Landslide at Frank, Alberta: *Canada Department of the Interior: Annual Report, 1902–1903*, Pt. VIII, pp. 6–7.

THE NIGHT THE MOUNTAIN FELL: Gordon Gaskill, 1965, "The Night the Mountain Fell," Reprinted with Permission from the May 1965 Reader's Digest. Copyright©1965 by the Reader's Digest Association, Inc.

CANDIDE: From *The Portable Voltaire* edited by Ben Ray Redman, Copyright 1949,©1968 by The Viking Press Inc., pp. 239–243. Reprinted by Permission of Viking Penguin Inc.

THE LISBON EARTHQUAKE: from James R. Newman, 1970, *Science and Sensibility*, Doubleday and Company, Inc., Garden City, New York, pp. 56–66. Original source *Scientific American*, July 1957 pp. 164–170, Reprinted with permission.

NO MORE WOODEN TOWERS FOR SAN FRANCISCO, 1906: Mary Austin, "The Temblor: A Personal Narration" in David Starr Jordan's *The California Earthquake of 1906*, A.M. Robertson, San Francisco, 1907, pp. 341–360.

TSUNAMI: from *Our Changing Coastlines* by F.P. Shepard and H.R. Wanless. Copyright 1971. Used with permission of McGraw Hill Book Company.

NOT A VERY SENSIBLE PLACE FOR A STROLL: Haroun Tazieff, 1952, *Craters of Fire*, Harper and Brothers, New York, pp. 11–21. Reprinted by permission of Harper & Row, Publishers, Inc.

LAST DAYS OF ST. PIERRE: Fairfax Downey, *Disaster Fighters*, G.P. Putnam's and Sons, 1938. Reprinted with permission from the author.

BEACONS ON THE PASSAGE OUT: Hans Cloos, 1953, *Conversation With the Earth*, Alfred A. Knopf, New York, pp. 14–21. Reprinted with the permission of Random House Inc. The original was substantially abridged for this edition.

ERUPTION OF THE ORAEFAJÖKULL, 1727: Sir George Mackenzie, 1812, *Travels in Iceland*, Constable, Edinburgh, Scotland, pp. 81–85.

Chapter 3

FOOTPRINTS IN RED SANDSTONE: Willy Ley, *Dragons in Amber*, The Viking Press, New York, 1951, pp. 53–68. Reprinted with Permission.

APES, ANGELS, AND VICTORIANS: William Irvine, *Apes, Angels, and Victorians*, Weidenfeld and Nicolson, London, 1955, pp. 3–7. Reprinted with permission of McGraw-Hill Book Company.

FOSSILS AND FREE ENTERPRISERS: Howard S. Miller, *Dollars for Research*, 1970, University of Washington Press, Seattle, pp. 133–143. Reprinted with permission.

THE FOUNDERS OF GEOLOGY: Archibald Geikie, *The Founders of Geology*, 1962, Dover Publications, Inc., New York, pp. 144–152.

Chapter 4

CONCERNING THE SYSTEM OF THE EARTH: James Hutton, 1785, "Abstract of a Dissertation Read in the Royal Society of Edinburgh, upon the Seventh of March, and Fourth of April, M,DCC,LXXXV, Concerning the System of the Earth, Its Duration, and Stability", pp. 3–30. Published in *Illustrations of Huttonian Theory of the Earth*, University of Illinois Press, 1956. Reprinted with Permission.

THE METHOD OF MULTIPLE WORKING HYPOTHESES: from the reprinted article by T.C. Chamberlin, 1965, "The Method of Multiple Working Hypotheses," *Science*, Volume 142, pp. 754–759. Copyright 1965 by the American Association for the Advancement of Science.

THE DISCRIMINATION OF SHORE FEATURES: G.K. Gilbert, *Lake Bonneville*, 1890, United States Geological Survey, Monograph 1, U.S. Government Printing Office, pp. 74–89.

HISTORICAL SCIENCE: George Gaylord Simpson, "Historical Science", from *The Fabric of Geology*, C.A. Albretton (ed.) 1963, Freeman, Cooper & Company, Stanford, Cal., pp. 24–48. Reprinted with Permission of Addison-Wesley Publishing Company from a Geological Society of America Publication. Original was substantially abridged for this edition.

Chapter 5

EARTH AND MAN: Frank H.T. Rhodes, "Earth and Man", *Wooster Alumni Magazine*, October, 1972, The College of Wooster, Ohio, pp. 5–7.

THE FLOWERING EARTH: *The Flowering Earth* by Donald C. Peattie, 1939, Viking Press, New York, pp. 337–339, 366–370. Reprinted by Permission of Noel R. Peattie and his Agent James Brown Associates, Inc. Copyright 1939, by Donald Culross Peattie.

HABITS AND HABITATS: R. Claiborne, *Climate, Man, and History*, 1970, W.W. Norton and Company, Inc., New York, pp. 170–178. Reprinted with Permission. Copyright 1970 by Robert Claiborne.

Chapter 6

Chapter 7

Chapter 8

TO A TRILOBITE: T.A. Conrad, quoted by P.E. Raymond, 1939, *Prehistoric Life*, Harvard University Press, p. 15.

THE LISBON EARTHQUAKE: E. R. Dumont, Ferney Edition: *The Works of Voltaire*, 1901, Volume XXXVI, Paris, London, New York, pp. 10–13.

A SHROPSHIRE LAD: from "A Shropshire Lad"—Authorized Edition—from *The Collected Poems of A.E. Housman*, Copyright 1939, 1940,©1965 by Holt, Rhinehart and Winston. Copyright©1967, 1968 by Robert E. Symons. Reprinted by permission of Holt, Rhinehart and Winston, Publishers.

WHERE SHALL WISDOM BE FOUND?: The Book of Job, Chapter 28, Verses 1–28; Chapter 38, verses 1–38, *The New Oxford Annotated Bible with the Apocrypha*, New York, Oxford University Press, 1973, pp. 407–408, 414–415.

MENTE ET MALLEO: Frances E. Vaughn, 1970, *Andrew C. Lawson, Scientist, Teacher, Philosopher*, Arthur H. Clark Company, Glendale, California, pp. 356.

PARADISE LOST: John Milton, *Paradise Lost Book VII*, Thomas Nelson & Sons, Ltd., London & Edinburgh, pp. 158–162, 1900.

LYELL'S HYPOTHESIS AGAIN: Kenneth Rexroth, 1967, *The Collected Shorter Poems*, New Directions, New York, pp. 180–181. Copyright 1949 by Kenneth Rexroth. Reprinted with permission.

Chapter 9

LANDSCAPE AND LITERATURE: Sir Archibald Geikie, 1905, *Landscape in History and Other Essays*, Macmillan & Company, Ltd., London, pp. 76–77, 124–127.

ROUGHING IT: Mark Twain, 1903, "Roughing It", *Writings of Mark Twain*, v. VII, The American Printing Company, Hartford, Connecticut, pp. 294–307.

MONO LAKE—AURORA—SONORA PASS: William H. Brewer, 1930, *Up and Down California, 1860–1864*, Yale University Press, New Haven, pp. 415–418. Reprinted with permission of University of California Press.

SEVEN PILLARS OF WISDOM: T.E. Lawrence, 1938, *Seven Pillars of Wisdom*, Garden City Publishing Company, New York, pp. 183–186. Reprinted with permission of Jonathon Cape, Ltd., and by Doubleday and Company Inc. for the USA.

TRIP TO THE MIDDLE AND NORTH FORKS OF SAN JOAQUIN RIVER: John Muir, 1912, *The Yosemite*, The Century Company, New York, pp. 153–158.

SEA OF CORTEZ: John Steinbeck and E.F. Ricketts, 1941, *Sea of Cortez: A Leisurely Journal of Travel and Research*, The Viking Press, New York, pp. 224–228. Copyright

1941 by John Steinbeck and E.F. Ricketts Jr. Copyright renewed 1969 by John Steinbeck and E.F. Ricketts Jr. Reprinted with Permission of McIntosh & Otis Inc., and Viking Penguin Inc. in North America.

WIND, SAND AND STARS: Excerpted from *Wind, Sand, and Stars* by Antoine de Saint-Exupéry, copyright 1939, 1940 by Antoine de Saint-Exupéry; copyright 1967 by Lewis Galantiere; copyright 1968 by Harcourt Brace Jovanovich, Inc. Reprinted by permission of the publisher, and in the United Kingdom and British commonwealth by William Heinemann, Ltd.

GREEN HILLS OF AFRICA: Ernest Hemingway, 1935, *The Green Hills of Africa*, Charles Scribner's Sons, New York, pp. 250–251. Used by permission of Charles Scribner's Sons, from *The Green Hills of Africa* by Ernest Hemingway. Copyright 1935 Charles Scribner's Sons; copyright renewal©1963 Mary Hemingway, and Jonathon Cape Ltd. for the British Commonwealth.

THE FRENCH LIEUTENANT'S WOMAN: John Fowles, 1970, *The French Lieutenant's Woman*, Signet Books, New American Library, Times-Mirror, New York, pp. 111–114. Reprinted with permission of Little Brown & Company, and Anthony Sheil Associates Ltd. in the British Commonwealth.

LETTERS FROM SWITZERLAND AND TRAVELS IN ITALY: A.J.W. Morrison, 1882, *Goethe's Letters from Switzerland and Travels in Italy*, S.E. Cassino, Publishers, Boston, pp. 161–163, 240–243, 261–262, 324–326.

THE LOST WORLD: A. Conan Doyle, 1912, *The Lost World*, John Murray, London, pp. 173–176.

Chapter 10

A LAND: Jacquetta Hawkes, 1951, *A Land*, Random House, New York, pp. 100–140. Reprinted with permission of David & Charles.

BEYOND MODERN SCULPTURE: Jack Burnham, 1968, *Beyond Modern Sculpture*, George Braziller, New York, pp. 94–100. Reprinted by permission of the publisher, George Braziller, Inc.

THOMAS MORAN: AMERICAN LANDSCAPE PAINTER: R.A. Bartlett, 1962, *Great Surveys of the American West*, U. of Oklahoma Press, Norman, Okla., pp. 41–42. Reprinted with permission.

GREEK MARBLES: DETERMINATION OF PROVENANCE BY ISOTOPIC ANALYSIS: Harmon and Valerie Craig, 1972, "Greek Marbles: Determination of Provenance by Isotopic Analysis," *Science*, V. 176, pp. 401–403. Copyright 1972 by the American Association for the Advancement of Science. Reprinted with permission.

Chapter 11

THE GREAT DIAMOND HOAX: From *Exploration and Empire*, by William H. Goetzmann. Copyright © 1966 by William H. Goetzmann. Reprinted by permission of Alfred A. Knopf, Inc.

THE GREAT PILTDOWN HOAX: William L. Straus, Jr., 1954, "The Great Piltdown Hoax," *Science*, V. 119, pp. 265–269, 26 February 1954. Reprinted with permission.

Chapter 12

MEGALONYX, MAMMOTH, AND MOTHER EARTH: B.T. Martin, 1953, *Thomas Jefferson: Scientist*, Abelard-Schuman, Ltd., New York, pp. 107–130. Copyright 1952 by Harper & Row, Publishers, Inc. Reprinted with permission. Original was substantially abridged for this edition.

THREE SHORT, HAPPY MONTHS: William A. Stanley, January 1968, *Three Short, Happy Months*, ESSA World, U.S. Dept. of Commerce/Env. Sci. Serv. Admn.

STANFORD UNIVERSITY, 1891–1895: Herbert C. Hoover, 1951, *The Memoirs of Herbert Hoover: Years of Adventure, 1874–1820*. MacMillan & Co., New York, pp. 16–24.

MOUNTAIN-WORSHIP: W.G. Collingwood, *The Life of John Ruskin*, 1905, Methuen and Co., London, pp. 41, 60–61, 246–247.

MINERALOGY, GEOLOGY, METEOROLOGY: R. Magnus, 1949, *Goethe as a Scientist*, Abelard-Schuman, Ltd., New York, pp. 200–217. Copyright 1949 by Harper & Row, Publishers, Inc. Reprinted with Permission. Original was substantially abridged for this edition.

BENJAMIN FRANKLIN AND THE GULF STREAM: H. Stommel, 1965, *The Gulf Stream*, Univ. of California Press, pp. 4–5. Reprinted with permission.

LEONARDO DA VINCI AS A GEOLOGIST: Thomas Clements, *Leonardo Da Vinci as a Geologist*, Univ. of Southern California, Los Angeles, California.

Chapter 13

TOPOGRAPHY AND STRATEGY IN THE WAR: D.W. Johnson, 1917, *Topography and Strategy in the War*, H. Holt & Co., New York, pp. 1–10.

THE GEOLOGIC AND TOPOGRAPHIC SETTING OF CITIES: by Donald F. Eschman and Melvin G. Marcus from Thomas R. Detwyler and Melvin G. Marcus, 1972,

Urbanization and Environment, Duxbury Press, Belmont, California, pp. 27–34. Copyright 1972 by Wadsworth Inc., Belmont, Ca. 94002. Reprinted with permission of the publisher, Duxbury Press.

MINERALS AND WORLD HISTORY: John D. Ridge, 1967, "Minerals and World History," *Mineral Industries*, The Pennsylvania State University, College of Earth and Mineral Sciences, University Park, Pa., Vol. 36, No. 9, pp. 3–12. Reprinted with permission.

THE ICE AGE COMETH: James D. Hays, 1973, "The Ice Age Cometh", *Saturday Review of the Sciences*, Vol. 1, N.S., April 1973, pp. 29–32.

TAMBORA AND KRAKATAU: Kenneth E.F. Watt, 1972, "Tambora and Krakatau: Volcanoes and the Cooling of the World," *Saturday Review*, December 23, 1972, pp. 43–44. Copyright 1972 by *Saturday Review*. All rights reserved. Reprinted with permission.

Chapter 14

MOHOLE: GEOPOLITICAL FIASCO: Daniel S. Greenberg, 1971, Chapter 25 in *Understanding the Earth*, (copyright©1971, The Open University), M.I.T. Press, Cambridge, Massachusetts, pp. 343–348. Reprinted with permission of The Open University Press.

A WINDOW ON THE OLIGOCENE: Berton Roueché, "A Window on the Oligocene," *New Yorker*, November 13, 1971, pp. 141–148. Copyright 1971, 1978 Berton Roueché. Reprinted with Permission of Harper and Row, Publishers, Inc. for North America and Harold Ober Associates for the British Commonwealth.

THE ENVIRONMENTAL GEOLOGIST AND THE BODY POLITIC: Peter T. Flawn, *Geotimes*, Vol. 13, No. 6, July-August 1968. Reprinted with permission.

Chapter 15

MORTGAGING THE OLD HOMESTEAD: Lord Ritchie-Calder, "Mortgaging the Old Homestead," *Foreign Affairs*, Vol. 48, No. 2, January 1970, pp. 207–220. Reprinted with permission from Foreign Affairs. Copyright 1969 by Council on Foreign Relations. Original was substantially abridged for this edition.

GEOLOGIC JEOPARDY: Richard H. Jahns, "Geologic Jeopardy", *Texas Quarterly*, University of Texas Press, Summer 1968, Vol. XI, No. 2, pp. 69–79. Reprinted with permission.

THE GREAT AND DIRTY LAKES: Gladwin Hill, "The Great and Dirty Lakes", *Saturday Review*, October 23, 1965, pp. 32–34. Copyright Saturday Review, 1965. All rights reserved.

GEOLOGY—FOR HUMAN NEEDS: Michel T. Halbouty, 1967, "Geology—For Human Needs," *Journal of Geological Education*, Vol. 15, pp. 80–82.

NATIONAL ENERGY POLICY: Dewey F. Bartlett, August 1971, "National Energy Policy", *The American Association of Petroleum Geologists Bulletin*, Vol. 55, No. 8, pp. 1132–1134. Reprinted with permission.

MINERALS, PEOPLE, AND THE FUTURE: Charles F. Park, Jr., 1968, *Affluence in Jeopardy, Minerals and the Political Economy*, Freeman, Cooper and Company, San Francisco, pp. 321–337. Reprinted with permission.

THE GEOLOGICAL ATTITUDE: J.G.C.M. Fuller, "The Geological Attitude", *The American Association of Petroleum Geologists Bulletin*, Vol. 55, No. 11, November 1971, pp. 1927–1938. Reprinted with permission.

MEDICAL GEOLOGY: Harry V. Warren and Robert DeLarault, "Medical Geology", *Geotimes*, September 1965, pp. 14–15. Reprinted with permission.

Index

405

About The Editors

Frank H. T. Rhodes was born in England and educated at Solihull School and the University of Birmingham. He studied as a post-doctoral Fulbright Fellow at the University of Illinois and served on the faculty at the University of Durham, the University of Illinois, and the University of Wales, where he was, for 12 years, professor and head of the department of Geology at the University College of Swansea. He served successively as professor of Geology, dean of the College of Literature, Science and the Arts, and vice president for academic affairs at The University of Michigan. He has been President of Cornell University since 1977. Dr. Rhodes is the author of four books, four monographs, and numerous scientific papers in the fields of paleontology, stratigraphy, evolution and geology. He holds honorary degrees from four universities and received the Bigsby Medal from the Geological Society.

Richard O. Stone was born in Los Angeles, California and studied at the Colorado School of Mines and the University of Southern California. He spent the whole of his academic career at the University of Southern California, rising from the position of assistant professor in 1956, to the chairmanship of the department of Geological Sciences. A widely respected teacher, lecturer, and member of the University community, Richard Stone's chief scientific contributions were in the field of geomorphology, with a special reference to the geology of deserts and their morphology and terrain. He also published various eolian studies, as well as papers in sedimentation and quantitative geomorphology. He died in 1977.

ber 1, 1755, was in every sense a colossal seismic disturbance. It was felt over a large area. It shook heavily the whole southwestern corner of Portugal; there was a tremendous upheaval in North Africa in the area of Fez and Meknes, with much loss of life; Spain and France experienced shocks, as did Switzerland and northern Italy; tidal waves were recorded in England and Ireland at 2 P.M. and in the West Indies at 6 P.M. Over almost all Europe, including Scandinavia, the water in rivers, canals, and lakes was suddenly agitated. But while the ravages and astonishing side-phenomena of this cruel day caused widespread alarm, there were other factors that had a special turning effect on men's minds.

Consider, first, how "men of action" responded to the disaster. It must be said that they came off well. Lisbon was, of course, in a state of terror and panic. Corpses were everywhere. Men, women, and children crept out of the rubble bleeding and mangled, and searched frantically for others in their families. Suicidal attempts were made to escape from the upper stories of partly wrecked houses. The air was filled with screams and groans, and with the piteous cries of wounded horses and dogs. As flames roared through the town, it was soon enveloped in an impenetrable sulphurous cloak. And the earthquake continued. Brief aftershocks, some quite violent, although they did little damage, kept hysteria alive. (In the week after November 1, there were nearly thirty earthquakes, and by August, 1756, some five hundred aftershocks had been recorded.) There was reason to believe, as many thought, that God would not desist until the city had been razed.

But the men of action wasted no time on such feckless speculations. Foremost among these energetic leaders was King José's secretary of state, Sebastião José Carvalho e Mello, the future Marquês de Pombal—the name by which he is remembered. Pombal, who was dictator of Portugal throughout the reign of José I, was troublesome, touchy, and aggressive, but a man of exceptional ability. Where others, on learning of the catastrophe, wrung their hands and looked to heaven for succor, he saw the need for firm and immediate terrestrial steps to avert social chaos. When the unhappy King asked despairingly what was to be done, Pombal is said to have replied, "Bury the dead and feed the living."

This he did and much more.

It cannot be said that the response of the men of science, such as they were, compared favorably with that of the men of action. Lisbon needed some plain truths about the earthquake to counteract the monstrous exaggerations. But the very suggestion that the earthquake was a natural phenomenon, like an eclipse or a storm, shocked the devout and enraged their religious instructors. Even honest men deemed it prudent to hedge. A physician, José Alvares da Silva, was one of the first to venture the opinion that while the earthquake might be a judgment of God, it could also be naturally explained. He put forth several possible theories, among them the notion that compressed air was responsible for the quake, and also that electricity—an increasingly fashionable phenomenon—was an important factor. It is our duty, he said, to find out how nature works before looking to supernatural causes; undoubtedly Lisbon is a wicked city, but to compare it with Babylon is absurd, and if God intends to set an example there are much more deserving candidates in other countries. A noteworthy aspect of da Silva's essay is his stress on the spiritual enrichment of the world, which the researches of such men as Descartes and Newton had effected. He proposed that their approach be emulated, and the physical sciences encouraged. God cannot be expected to help man if he is too witless or too wayward to help himself.

The theme of God's anger and of retribution was endlessly repeated in sermons, tracts and poetry throughout Europe. In Lisbon, especially when the exhortations were insanely ferocious, the effect was to make people apathetic and seriously to hamper the work of recovery. Should a man rebuild and start afresh, or should he, as the saintly and crazy Jesuit Gabriel Malagrida passionately urged, "set all this miserable worldly business aside and seek in what might well be his last hours to save his soul"? These extremes could not be reconciled; there was no conceivable compromise here between the men of action and the more fervid men of God. But it was a conflict not confined to Lisbon: its repercussions sounded everywhere.

"One generation passeth away, and another generation cometh: but the earth abideth forever," said the preacher. But now man could not be sure that the earth would abide. We must recall the setting for this upheaval in thought. Looking backward in time from the first half of the eighteenth century, it is astonishing to see the change in outlook that had taken place in two hundred years. Science, philosophy, politics, technology, trade and commerce had transformed society. This life was no longer to be regarded as a mere preparation for the hereafter. Man was to enjoy what he had. He saw himself and the world about him as no medieval thinker could have imagined, as few of even the most ebullient philosophers of the Renaissance would have dared hope. Sir Thomas writes: "It is said that the first half of the eighteenth century with its enlightenment, its optimism, its cult of happiness and its content with the *status quo* was a fortunate age, so much so that it might be preferred to all other times in the past as the one in which a sensible man might elect to live." It was in all things an "age of stability." For the wealthy and educated, Basil Willey has said, it was "the nearest approach to earthly felicity ever known to man"; and even the common man could share in this cozy feeling. Thus the Lisbon earthquake could scarcely have come as a more terrible shock. Suddenly the whole edifice of confidence began to crumble; in ten awful minutes the world's sense of security was swallowed up. There is nothing like an earthquake to make men feel helpless, to remind them of their mortality, of the vanity of all they covet and acquire, of the ridiculous insubstantiality of what they have made and built. If the ground itself will not stand firm, what remains? Even in our time earthquakes strike terror; how much more shattering the experience must have been in an age when earthquakes were not understood as natural events but regarded as occurrences "instinct with deity."

Voltaire was the foremost figure who seized upon the Lisbon earthquake as an opportunity to attack the climate of optimism. The chief target of his attack was what he called the *tout est bien* philosophy expressed twenty years earlier in Alexander Pope's "Essay on Man." We must admit, said Voltaire, that there is injustice and evil in the world, that there is inexcusable suffering, that there are inexplicable calamities. It is stupid and self-deluding to pretend that every misfortune is a benefit in disguise. It is folly to believe that Providence will assure safe-conduct to the virtuous. Man is "weak and helpless, ignorant of his destiny, and exposed to terrible dangers, as all must now see." Optimism must be replaced by realism; at best, by an "apprehensive hope that Providence will lead us through our dangerous world to a happier state."

The poem was widely read and discussed. Voltaire himself was a little apprehensive as to the offense it might give in religious circles and for that reason tinkered with some of the lines. It was scarcely calculated to nourish the belief in a kind and loving God. Nor, on the other hand, did it support a blind faith in a just God whose ways might be hard but whose over-all scheme could not be questioned. Rousseau was one of those who strongly opposed Voltaire's pessimism; Kant, while less cheery, took the position that "the only possible theodicy is a practical act of faith in divine justice."

In 1759 Voltaire published his immortal satire *Candide*, which was far more influential than his poem. It blew the *tout est bien* philosophy to bits.

Optimism never recovered from Lisbon and Candide. "There was no more to be said; the case was finished and the case was lost." Not, of course, that it vanished all at once. "A doctrine," as the noted French literary historian Paul Hazard wrote, "lives on for a long time, even when wounded, even when its soul has fled." But within a few years the wounds proved fatal, and a French poet could say that the age of optimism had degenerated into the Dark Ages.

The second earthquake described in the present book is more familiar than that of Lisbon. It concerns the great San Francisco earthquake of 1906. Mary Austin describes the event with clarity and perception, as well as with an obvious feeling of tenderness and understanding toward those individuals who suffered. Her own impressions convey the intensely personal details that remain after the terror of such a catastrophe has subsided.

NO MORE WOODEN TOWERS FOR SAN FRANCISCO, 1906

Mary Austin

There are some fortunes harder to bear once they are done with than while they are doing, and there are three things that I shall never be able to abide in quietness again—the smell of burning, the creaking of house-beams in the night, and the roar of a great city going past me in the street.

Ours was a quiet neighborhood in the best times; undisturbed except by the hawker's cry or the seldom whistling hum of the wire, and in the two days following April eighteenth, it became a little lane out of Destruction. The first thing I was aware of was being wakened sharply to see my bureau lunging solemnly at me across the width of the room. It got up first on one castor and then on another, like the table at a séance, and wagged its top portentously. It was an antique pattern, tall and marble-topped, and quite heavy enough to seem for the moment sufficient cause for all the uproar. Then I remember standing in the doorway to see the great barred leaves of the entrance on the second floor part quietly as under an unseen hand, and beyond them, in the morning grayness, the rose tree and the palms replacing one another, as in a moving picture, and suddenly an eruption of nightgowned figures crying out that it was only an earthquake, but I had already made this discovery for myself as I recall trying to explain. Nobody having suffered much in our immediate vicinity, we were left free to perceive that the very instant after the quake was tempered by the half-humorous, wholly American appreciation of a thoroughly good job. Half an hour after the temblor people sitting on their doorsteps, in bathrobes and kimonos, were admitting to each other with a half twist of laughter between tremblings that it was a really creditable shake.

The appreciation of calamity widened slowly as water rays on a mantling pond. Mercifully the temblor came at an hour when families had not divided for the day, but live wires sagging across housetops were to outdo the damage of falling walls. Almost before the dust of ruined walls had ceased rising, smoke began to go up against the sun, which, by nine of the clock, showed bloodshot through it as the eye of Disaster.

It is perfectly safe to believe anything any one tells you of personal adventure; the inventive faculty does not exist which could outdo the actuality; little things prick themselves on the attention as the index of the greater horror.

I remember distinctly that in the first considered interval after the temblor, I went about and took all the flowers out of the vases to save the water that was left; and that I went longer without washing my face than I ever expect to again.

I recall the red flare of a potted geranium undisturbed on a window ledge in a wall of which the brickwork dropped outward, while the roof had gone through the flooring; and the cross-section of a lodging house parted cleanly with all the little rooms unaltered, and the halls like burrows, as if it were the home of some superior sort of insect laid open to the microscope.

South of Market, in the district known as the Mission, there were cheap man-traps folded in like pasteboard, and from these, before the rip of the flames blotted out the sound, arose the thin, long scream of mortal agony.

Down on Market Street Wednesday morning, when the smoke from the burning blocks behind began to pour through the windows we saw an Italian woman kneeling on the street corner praying quietly. Her cheap belongings were scattered beside her on the ground and the crowd trampled them; a child lay on a heap of clothes and bedding beside her, covered and very quiet. The woman opened her eyes now and then, looked at the reddening smoke and addressed herself to prayer as one sure of the stroke of fate. It was not until several days later that it occurred to me why the baby lay so quiet, and why the woman prayed instead of flying.

Not far from there, a day-old bride waited while her husband went back to the ruined hotel for some papers he had left, and the cornice fell on him; then a man who had known him, but not that he was married, came by and carried away the body and shipped it out of the city, so that for four days the bride knew not what had become of him.

There was a young man who, seeing a broken and dismantled grocery, meant no more than to save some food, for already the certainty of famine was upon the city—and was shot for looting. Then his women came and carried the body away, mother and betrothed, and laid it on the grass until space could be found for burial. They drew a handkerchief over its face, and sat quietly beside it without bitterness or weeping. It was all like this, broken bits of human tragedy, curiously unrelated, inconsequential, disrupted by the temblor, impossible to this day to gather up and compose into a proper picture.

The largeness of the event had the effect of reducing private sorrow to a mere pin prick and a point of time. Everybody tells you tales like this with more or less detail. It was reported that two blocks from us a man lay all day with a placard on his breast that he was shot for looting, and no one denied the aptness of the warning. The will of the people was toward authority, and everywhere the tread of soldiery brought a relieved sense of things orderly and secure. It was not as if the city had waited for martial law to be declared, but as if it precipitated itself into that state by instinct at its best refuge.

In the parks were the refugees huddled on the damp sod with insufficient bedding and less food and no water. They laughed. They had come out of their homes with scant possessions, often the least serviceable. They had lost business and clientage and tools, and they did not know if their friends had fared worse. Hot, stifling smoke billowed down upon them, cinders pattered like hail—and they laughed—not hysteria, but the laughter of unbroken courage.

That exodus to the park did not begin in our neighborhood until the second day; all the first day was spent in seeing such things as I relate, while confidently expecting the wind to blow the fire another way. Safe to say one-half the loss of household goods might have been averted, had not the residents been too sure of such exemption. It happened not infrequently that when a man had seen his women safe he went out to relief work and returning found smoking ashes—and the family had left no address. We were told of those who had dead in their households who took them up and fled with them to the likeliest place in the hope of burial, but before it had been accomplished were pushed forward by the flames. Yet to have taken part in that agonized race for the open was worth all it cost in goods.

Before the red night paled into murky dawn thousands of people were vomited out of the angry throat of the street far down toward Market. Even the smallest child carried something, or pushed it before him on a rocking chair, or dragged it behind him in a trunk, and the thing he carried was the index of the refugee's strongest bent. All the women saved their best hats and their babies, and, if there were no babies, some of them pushed pianos up the cement pavements.

All the faces were smutched and pallid, all the figures sloped steadily forward toward the cleared places. Behind them the expelling fire bent out over the lines of flight, the writhing smoke stooped and waved, a fine rain of cinders pattered and rustled over all the folks, and charred bits of the burning fled in the heated air and dropped among the goods. There was a strange, hot, sickish smell in the street as if it had become the hollow slot of some fiery breathing snake. I came out and stood in the pale pinkish glow and saw a man I knew hurrying down toward the gutted district, the badge of a relief committee fluttering on his coat. "Bob," I said, "it looks like the day of judgment!" He cast back at me over his shoulder unveiled disgust at the inadequacy of my terms. "Aw!" he said, "it looks like hell!"

No matter how the insurance totals foot up, what landmarks, what treasures of art are evanished, San Francisco, *our* San Francisco is all there yet. Fast as the tall banners of smoke rose up and the flames reddened them, rose up with it something impalpable, like an exhalation. We saw it breaking up in the movements of the refugees, heard it in the tones of their voices, felt it as they wrestled in the teeth of destruction. The sharp sentences by which men called to each other to note the behavior of brick and stone dwellings contained a hint of a warning already accepted for the new building before the old had crumbled. When the heat of conflagration outran the flames and reaching over wide avenues caught high gables and crosses of church steeples, men watching them smoke and blister and crackle into flame, said shortly, "No more wooden towers for San Francisco!" and saved their breath to run with the hose.

What distinguishes the personal experience of the destruction of the great city from all like disasters of record, is the keen appreciation of the death-lessness of the spirit of living.

The Excursion

He who with pocket-hammer smites the edge
Of luckless rock or prominent stone, disguised
In weather-stains or crusted o'er by Nature
With her first growths, detaching by the stroke
A chip or splinter—to resolve his doubts;
And, with that ready answer satisfied,
The substance classes by some barbarous name,
And hurries on; or from the fragments picks
His specimen, if but haply interveined
With sparkling mineral, or should crystal cube
Lurk in its cells—and thinks himself enriched,
Wealthier, and doubtless wiser than before!

—William Wordsworth

Tsunami are giant waves, produced in the oceans by submarine earthquakes, and sometimes by explosive submarine volcanic eruptions. These "tidal waves," which may reach heights in excess of 100 feet, cause great destruction when they strike land. The account of the 1946 Tsunami in Hawaii is based upon an eyewitness description by Frances P. Shepard, who had to escape from the violent effects of the wave that reached the Hawaiian Islands after crossing 2,300 miles of ocean. The tidal wave originated around an earthquake near the Aleutian Island of Unimak, and the waves spread out from that area. They were imperceptible to ships moving across their path, because on the open ocean they had a height of only about a foot, although moving at an average speed of over 500 miles an hour. On coastlines, the waves reached heights of over 30 feet, and in a few restricted areas flung debris over 50 feet above normal beach level. Shepard, Professor Emeritus of Submarine Geology at Scripps Institute of Oceanography at La Jolla, California, and one of the world's leading oceanographers, who was living in a rented cottage on the waterfront near Kahuku Point, on the north shore of Ioahu, describes the 1946 Tsunami.

TSUNAMI

Francis P. Shepard

I was awakened at 6:30 in the morning by a hissing noise that sounded as if hundreds of locomotives were blowing off steam. I looked out in time to see the water lapping up around the edge of our house, rising 14 feet above its normal level. Grabbing my camera instead of clothes, I rushed out to take photographs. My wife and I watched the rapid recession of the water exposing the narrow reef in front of the house and stranding large numbers of fish. A few minutes later, we saw the water build up at the edge of the reef and move shoreward as a great breaking wave. Because the wave looked very threatening, we dashed in back of the house, getting there just in time to hear the water reach the front porch and smash in all of the glass at the front. As the water swept around the house and into the cane field, we saw our refrigerator carried past us and deposited right side up, eggs unbroken.

Water swept down the escape road to our right, leaving us no chance to get out by car. The neighbors' house, which was vacant at the time, had been completely torn apart, but the back portion of our house still stood, thanks probably to the casuarina (ironwood) trees growing along the edge of the berm in front.

As soon as the second wave had started to retreat, we ran along the beach ridge to our left, and then through the cane field by path to the main road, arriving there just ahead of the third wave.

We found quite a group on the road. The house of one family living at the edge of Kawela Bay had been carried bodily into the cane field and dropped without the wave having done much damage. In fact, their breakfast was still cooking when they were set down. Other unoccupied houses did not fare so well. Some of them were swept into a small pond, and others out into the bay by the retreat of the second wave.

We watched three or four more waves come into Kawela Bay at intervals (shown later by tide gauge records) of fifteen minutes. They had steep fronts, looking very much like the tidal bore that comes up the Bay of Fundy. Just before the eighth wave, I decided that the excitement was abating and ran back to the house to try to rescue some effects, particularly necessary as we were in pajamas and raincoats. Just as I arrived, a wave that must have been the largest of the set came roaring in, and I had to run for a tree and climb it as the water surged beneath me. I hung on swaying back and forth as the water roared by into the cane field.

That was the end of the adventure, and we gradually got what we could rescue of our possessions, but had to leave behind quantities of notes and the beginning of a new book, *Submarine Geology*, which had been scattered in the cane field. We were given wonderfully hospitable treatment by new and old friends whom we shall never forget. After a few days, I got in touch with Gordon Macdonald, of the U.S. Geological Survey, and Doak Cox, the geologist of Hawaiian Planters Association, and together we started a program of investigating the tsunami effects around the different islands.

In Memoriam

There rolls the deep where grew the tree.
O earth, what changes hast thou seen!
There where the long street roars, hath been
The stillness of the central sea.

The hills are shadows, and they flow
From form to form, and nothing stands;
They melt like mist, the solid lands,
Like clouds they shape themselves and go.

—Alfred Lord Tennyson

Volcanoes provide the most dramatic of all illustrations of the cauldron-like energy of the earth's interior. Haroun Tazieff, a contemporary volcanologist, contributes two descriptions of volcanoes, the "very substance of the earth itself," as he describes them. Both these accounts convey something of the noise, the fumes, the molten lava and the dangers of the volcanic eruptions that, on the pages of our textbooks, seem so ordinary. Tazieff, awakened one morning by a strange noise "like a herd of antelope galloping through the bush" provides not only a superb account of two volcanoes, but also describes the curious mixture of delight and desperation, of determination and despair, that represents the tension between personal discovery and personal danger. His account of his climb around the rim of an active crater of a "growling cone," is as good a piece of adventure writing, as it is a scientific description. In the process, Tazieff, like the rest of us, learned something about himself. "The calm of this fiery pool . . . spoke to me in enigmatic terms of a mighty and mysterious power. I was spellbound, and literally had to wrench myself out of the fear-laden ecstasy . . ."

NOT A VERY SENSIBLE PLACE FOR A STROLL

Haroun Tazieff

Standing on the summit of the growling cone, even before I got my breath back after the stiff climb, I peered down into the crater.

I was astonished. Two days previously the red lava had been boiling up to the level of the gigantic lip; now the funnel seemed to be empty. All that incandescent magma had disappeared, drawn back into the depths by the reflux of some mysterious ebb and flow, a sort of breathing. But there, about fifty feet below where I was standing, was the glow and the almost animate fury of the great throat which volcanologists call the conduit or chimney. It was quite a while before I could tear my eyes away from that lurid, fiery centre, that weird palpitation of the abyss. At intervals of about a minute, heralded each time by a dry clacking, bursts of projectiles were flung up, running away up into the air, spreading out fan-wise, all aglare, and then falling back, whistling, on the outer sides of the cone. I was rather tense, ready to leap aside at any moment, as I watched these showers, with their menacing trajectories.

Each outburst of rage was followed by a short lull. Then heavy rolls of brown and bluish fumes came puffing out, while a muffled grumbling, rather like that of some monstrous watch-dog, set the whole bulk of the volcano quivering. There was not much chance for one's nerves to relax, so swiftly did each follow on the other—the sudden tremor, the burst, the momentary intensification of the incandescence, and the outbreak of a fresh salvo. The bombs went roaring up, the cone of fire opening out overhead, while I hung in suspense. Then came the hissing and sizzling, increasing in speed and intensity, each 'whoosh' ending up in a muffled thud as the bomb fell. On their black bed of scoriae, the clots of molten magma lay with the fire slowly dying out of them, one after the other growing dark and cold.

Some minutes of observation were all I needed. I noted that today, apart from three narrow zones to the west, north, and north-east, the edges of the crater had scarcely been damaged at all by the barrage from underground. The southern point where I stood was a mound rising some twelve or fifteen feet above the general level of the rim, that narrow, crumbling lip of scoriae nearer to the fire, where I had never risked setting foot. I looked at this rather alarming ledge all round the crater, and gradually felt an increasing desire to do something about it . . . It became irresistible. After all, as the level of the column of lava had dropped to such an exceptional degree, was this not the moment to try what I was so tempted to do and go right round the crater?

Still, I hesitated. This great maw, these jaws sending out heat that was like the heavy breathing of some living creature, thoroughly frightened me. Leaning forward over that hideous glow, I was no longer a geologist in search of information, but a terrified savage.

'If I lose my grip,' I said aloud, 'I shall simply run for it.'

The sound of my own voice restored me to normal awareness of myself. I got back my critical sense and began to think about what I could reasonably risk trying. 'De l'audace, encore de l'audace. . . .' That was all very well, of course, but one must also be careful. Past experience whispered a warning not to rush into anything blindly. Getting the upper hand of both anxiety and impatience, I spent several minutes considering, with the greatest of care, the monster's manner of behaving. Solitude has got me into the habit of talking to myself, and so it was more or less aloud that I gave myself permission to go ahead.

'Right, then. It can be done.'

I turned up my collar and buttoned my canvas jacket tight at the throat— I didn't want a sly cinder down the back of my neck! Then I tucked what was left of my hair under an old felt hat that did service for a helmet. And now for it!

Very cautiously indeed, I approach the few yards of pretty steep slope separating the peak from the rim I am going to explore. I cross, in a gingerly manner, a first incandescent crevasse. It is intense orange in colour and quivering with heat, as though opening straight into a mass of glowing embers. The fraction of a second it takes me to cross it is just long enough for it to scorch the thick cord of my breeches. I get a strong whiff of burnt wool.

A promising start, I must say!

Here comes a second break in the ground. Damn it, a wide one, too! I can't just stride across this one: I'll have to jump it. The incline makes me thoughtful. Standing there, I consider the unstable slope of scoriae that will have to serve me for a landing-ground. If I don't manage to pull up . . . if I go rolling along down this funnel with the flames lurking at the bottom of it . . . My little expedition all at once strikes me as thoroughly rash, and I stay where I am, hesitating. But the heat under my feet is becoming unbearable. I can't endure it except by shifting around. It only needs ten seconds of standing still on this enemy territory, with the burning gases slowly and steadily seeping through it, and the soles of my feet are already baking hot. From second to second the alternative becomes increasingly urgent: I must jump for it or retreat.

Here I am! I have landed some way down the fissure. The ashes slide underfoot, but I stop without too much trouble. As so often happens, the anxiety caused by the obstacle made me over-estimate its importance.

Step by step, I set out on my way along the wide wall of slag-like debris that forms a sort of fortification all round the precipice. The explosions are still going on at regular intervals of between sixty and eighty seconds. So far no projectile had come down on this side, and this cheered me up considerably. With marked satisfaction I note that it is pretty rare for two bombs of the same salvo to fall less than three yards apart: the average distance between them seems to be one of several paces. This is encouraging. One of the great advantages of this sort of bombardment, compared with one by artillery, lies in the relative slowness with which the projectiles fall, the eye being able to follow them quite easily. Furthermore, these shells don't burst. But what an uproar, what an enormous, prolonged bellowing accompanies their being hurled out of the bowels of the earth!

I make use of a brief respite in order to get quickly across the ticklish north-eastern sector. Then I stop for a few seconds, just long enough to see yet another burst gush up and come showering down a little ahead of me, after which I start out for the conquest of the northern sector. Here the crest narrows down so much that it becomes a mere ridge, where walking is so difficult and balancing so precarious that I find myself forced to go on along the outer slope, very slightly lower down. Little by little, as I advance through all this tumult, a feeling of enthusiasm is overtaking me. The immediate imperative necessity for action has driven panic far into the background. And under the hot, dry skin of my face, taut on forehead and cheekbones, I can feel my compressed lips parting, of their own accord, in a smile of sheer delight. But look out!

A sudden intensification of the light warns me that I am approaching a point right in the prolongation of the fiery chimney. In fact, the chimney is not vertical, but slightly inclined in a north-westerly direction, and from here one can look straight down into it. These tellurian entrails, brilliantly yellow, seem to be surging with heat. The sight is so utterly amazing that I stand there, transfixed.

Suddenly, before I can make any move, the dazzling yellow changes to white, and in the same instant I feel a muffled tremor all through my body and there is a thunderous uproar in my ears. The burst of incandescent blocks is already in full swing. My throat tightens as, motionless, I follow with my gaze the clusters of red lumps rising in slow, perfect curves. There is an instant of uncertainty. And then down comes the hail of fire.

Suddenly I hurl myself backwards. The flight of projectiles has whizzed past my face. Hunched up again, instinctively trying to make as small a target of myself as I can, I once more go through the horrors that I am beginning to know. I am in the thick of this hair's-breadth game of anticipation and dodging.

And now it's all over; I take a last glance into the marvellous and terrible abyss, and am just getting ready to start off on the last stage of this burning circumnavigation, all two hundred yards of it, when I get a sudden sharp blow in the back. A delayed-action bomb! With all the breath knocked out of me, I stand rigid.

A moment passes. I wonder why I am not dead. But nothing seems to have happened to me—no pain, no change of any sort. Slowly I risk turning my head, and at my feet I see a sort of huge red loaf with the glow dying out of it.

I stretch my arms and wriggle my back. Nothing hurts. Everything seems to be in its proper place. Later on, examining my jacket, I discovered a brownish scorchmark with slightly charred edges, about the size of my hand, and I drew from it a conclusion of immense value to me in future explorations: so long as one is not straight in the line of fire, volcanic bombs, which fall in a still pasty state, but already covered with a kind of very thin elastic skin, graze one without having time to cause a deep burn.

I set off at a run, as lightly as my 165 pounds allow, for I must be as quick as I can in crossing this part of the crater-edge, which is one of the most heavily bombarded. But I am assailed by an unexpected blast of suffocating fumes. My eyes close, full of smarting tears. I am caught in a cloud of gas forced down by the wind. I fight for breath. It feels as if I were swallowing lumps of dry, corrosive cotton-wool. My head swims, but I urge myself at all costs to get the upper hand. The main thing is not to breathe this poisoned air. Groping, I fumble in a pocket. Damn, not this one. How about this other one, then? No. At last I get a handkerchief out and, still with my eyes shut, cover my mouth with it. Then, stumbling along, I try to get through the loathsome cloud. I no longer even bother to pay any attention to the series of bursts from the volcano, being too anxious to get out of this hell before I lose grip entirely. I am getting pretty exhausted, staggering . . . The air filtered through the handkerchief just about keeps me going, but it is still too poisonous, and there is too little of it for the effort involved in making this agonising journey across rough and dangerous terrain. The gases are too concentrated, and the great maw that is belching them forth is too near.

A few steps ahead of me I catch a glimpse of the steep wall of the peak, or promontory, from the other side of which I started about a century ago, it seems to me now. The noxious mists are licking round the peak, which is almost vertical and twice the height of a man. It's so near! But I realise at once that I shall never have the strength to clamber up it.

In less than a second, the few possible solutions to this life-and-death problem race through my mind. Shall I turn my back to the crater and rush away down the outer slope, which is bombarded by the thickest barrages? No. About face and back along the ledge? Whatever I do, I must turn back. And then make my escape. By sliding down the northern slope? That is also under too heavy bombardment. And the worst of it would be that in making a descent of that sort there would be no time to keep a watch for blocks of lava coming down on one.

Only one possibility is left: to make my way back all along the circular ridge, more than a hundred yards of it, till I reach the eastern rim, where neither gas nor projectiles are so concentrated as to be necessarily fatal.

I swing round. I stumble and collapse on all fours, uncovering my mouth for an instant. The gulp of gas that I swallow hurts my lungs, and leaves me gasping. Red-hot scoriae are embedded in the palms of my hands. I shall never get out of this!

The first fifteen or twenty steps of this journey back through the acrid fumes of sulphur and chlorine are a slow nightmare; no step means any progress and no breath brings any oxygen into the lungs. The threat of bombs no longer counts. Only these gases exist now. Air! Air!

I came to myself again on the eastern rim, gasping down the clean air borne by the wind, washing out my lungs with deep fresh gulps of it, as though I could never get enough. How wide and comfortable this ledge is! What a paradise compared with the suffocating, torrid hell from which I have at last escaped! And yet this is where I was so anxious and so tense less than a quarter of an hour ago.

Several draughts of the prevailing breeze have relieved my agony. All at once, life is again worth living! I no longer feel that desire to escape from here as swiftly as possible. On the contrary, I feel a new upsurge of explorer's curiosity. Once more my gaze turns towards the mouth, out of which sporadic bursts of grape-shot are still spurting forth. Now and then there are bigger explosions and I have to keep a look-out for what may come down on my head, which momentarily interrupts the dance I keep up from one foot to the other, that *tresca* of which Dante speaks—the dance of the damned, harried by fire. True, I have come to the conclusion that the impact of these bombs is not necessarily fatal, but I am in no hurry to verify the observation.

The inner walls of the crater do not all incline at the same angle. To the north, west and south, they are practically vertical, not to say overhanging, but here on the east the slope drops away at an angle of no more than fifty degrees. So long as one moved along in a gingerly way, this might be an incline one could negotiate. It would mean going down into the very heart of the volcano. For an instant I am astounded by my own foolhardiness. Still, it's really too tempting . . .

Cautiously, I take a step forward . . . then another . . . and another . . . seems all right . . . it *is* all right. I begin the climb down, digging my heels as deep as I can into the red-hot scoriae. Gradually below me, the oval of the enormous maw comes nearer, growing bigger, and the terrifying uproar becomes more deafening. My eyes, open as wide as they will go, are drunken with its monstrous glory. Here are those ponderous draperies of molten gold and copper, so near—so near that I feel as if I, human being that I am, had entered right into their fabulous world. The air is stifling hot. I am right in the fiery furnace.

I linger before this fascinating spectacle. But then, by sheer effort, I tear myself away. It's time to get back to being 'scientific' and measure the temperatures, of the ground, and of the atmosphere. I plunge the long spike of the thermometer into the shifting scoriae, and the steel of it glitters among these brownish and grey screes with their dull shimmer. At a depth of six inches the temperature is two hundred and twenty degrees centigrade. It's amusing to think that when I used to dream it was always about polar exploration!

Suddenly, the monster vomits out another burst; so close that the noise deafens me. I bury my face in my arms. Fortunately almost every one of the projectiles comes down outside the crater. And now all at once I realise that it is I who am here—*alive* in this crater, surrounded by scorching walls, face to face with the very mouth of the fire. Why have I got myself into this trap, alone and without the slightest chance of help? Nobody in the world has any suspicion of the strange adventure on which I have embarked, and nobody, for that matter, could now do the slightest thing about it. Better not think about it . . .

Without a break the grim, steady growling continues to rise from the depths of that throat, only out-roared at intervals by the bellowing and belching of lava. It's too much; I can feel myself giving up. I turn my back on it, and try, on all fours, to scramble up the slope, which has now become incredibly steep and crumbles and gives way under my weight, which is dragging me down, down . . . 'Steady, now,' I say to myself. 'Keep calm for a moment. Let's work it out. Let's work it out properly. Or else, my boy, this is the end of *you*.'

Little by little, by immense exertions, I regain control of my movements, as well as the mental steadiness I need. I persuade myself to climb *calmly* up this short slope, which keeps crumbling away under my feet. When I reach the top, I stand upright for just a moment. Then, crossing the two glowing fissures that still intersect my course, I reach the part of the rim from where there is a way down into the world of ordinary peaceful things.

Fire and Ice

Some say the world will end in fire,
Some say in ice.
From what I've tasted of desire
I hold with those who favor fire.
But if I had to perish twice,
I think I know enough of hate
To say that for destruction ice
Is also great
And would suffice.

—Robert Frost

Most volcanoes lie in areas remote from centers of dense population. Those that do occur in inhabited areas are generally well studied and well known, their moods and tempers being, if not predictable, at least recognizable. One terrible exception is Mont Pelée, in the West Indies, which after more than 50 years of dormancy, provided signs of renewed activity in the Spring of 1902. On the morning of the 8th of May, a cloud of incandescent gas and glowing volcanic fragments poured down the sides of the cone at speeds as high as 60 meters per second, destroying buildings in its path, and bringing instant death to some 40,000 people in and around St. Pierre, the capitol of the island of Martinique. The account of this disaster by Fairfax Downey provides a vivid and terrible illustration of its effects on the lives of the members of the community.

LAST DAYS OF ST. PIERRE

Fairfax Downey

How graciously had fortune smiled on Fernand Clerc. Little past the age of forty, in this year of 1902, he was the leading planter on the fair island of Martinique. Sugar from his broad cane fields, molasses, and mellow rum had made him a man of wealth, a millionaire. All his enterprises prospered.

Were the West Indies, for all their beauty and their bounty, sometimes powerless to prevent a sense of exile, an ache of homesickness in the heart of a citizen of the Republic? Then there again fate had been kind to Fernand Clerc. Elected a member of the Chamber of Deputies, it was periodically his duty and his pleasure to embark and sail home to attend its sessions—home to France, to Paris.

Able, respected, good-looking, blessed with a charming wife and children, M. Clerc found life good indeed. With energy undepleted by the tropics, he rode through the island visiting his properties. Tall and thick grew the cane stalks of his plantation at Vivé on the slopes of Mont Pelée. Mont Pelée— Naked Mountain—well named when lava erupting from its cone had stripped it bare of its verdure. But that was long ago. Not since 1851 had its subterranean fires flared up and then but insignificantly. Peaceful now, its crater held the lovely Lake of Palms, whose wooded shores were a favorite picnic spot for parties from St. Pierre and Fort-de-France. Who need fear towering Mont Pelée, once mighty, now mild, an extinct volcano?

Yet this spring M. Clerc and all Martinique received a rude shock. The mountain was not dead, it seemed. White vapors veiled her summit, and by May 2nd she had overlaid her green mantle with a gown of gray cinders. Pelée muttered and fumed like an angry woman told her day was long past. Black smoke poured forth, illumined by night by jets of flame and flashes of lightning. The grayish snow of cinders covered the countryside, and the milky waters of the Rivière Blanche altered into a muddy and menacing torrent.

Nor was Pelée uttering only empty threats. On May 5th, M. Clerc at Vivé beheld a cloud rolling from the mountain down the valley. Sparing his own acres, the cloud and the stream of smoking lava which it masked, enveloped the Guerin sugar factory, burying its owner, his wife, overseer, and twenty-five workmen and domestics.

Dismayed by this tragedy, M. Clerc and many others moved from the slopes into St. Pierre. The city was crowded, its population of 25,000 swollen to 40,000, and the throngs that filled the market and the cafés or strolled through the gorgeously luxuriant Jardin des Plantes lent an air of added animation, of almost hectic gaiety. When M. Clerc professed alarm at the behavior of Pelée to his friends, he was answered with shrugs of shoulders. Danger? On the slopes perhaps, but scarcely here in St. Pierre down by the sea.

Thunderous, scintillant, Mont Pelée staged a magnificent display of natural fireworks on the night of May 7th. Whites and negroes stared up at it, fascinated. Some were frightened but more took a child-like joy in the vivid spectacle. It was as if the old volcano were celebrating the advent of to-morrow's fête day.

M. Fernand Clerc did not sleep well that night. He breakfasted early in the household where he and his family were guests and again expressed his apprehensions to the large group of friends and relatives gathered at the table. Politely and deferentially—for one does not jeer a personage and man of proven courage—they heard him out, hiding their scepticism.

The voice of the planter halted in mid-sentence; he half rose, his eyes fixed on the barometer. Its needle was actually fluttering!

M. Clerc pushed back his chair abruptly and commanded his carriage at once. A meaning look to his wife and four children, and they hastened to make ready. Their hosts and the rest followed them to the door. *Non, merci,* none would join their exodus. *Au revoir. A demain.*

From the balcony of their home, the American Consul, Thomas Prentis, and his wife waved to the Clerc family driving by. "Stop," the planter ordered and the carriage pulled up. Best come along, the planter urged. His American friends thanked him. There was no danger, they laughed, and waved again to the carriage disappearing in gray dust as racing hoofs and wheels sped it out of the city of St. Pierre.

Governor Mouttet, ruling Martinique for the Republic of France, glared up at rebellious Mont Pelée. This *peste* of a volcano was deranging the island. There had been no such crisis since its captures by the English, who always relinquished it again to France, or the days when the slaves revolted. A great pity that circumstances beyond his control should damage the prosperous record of his administration, the Governor reflected.

That miserable mountain was disrupting commerce. Its rumblings drowned out the band concerts in the Savane. Its pyrotechnics distracted glances which might far better have dwelt admiringly on the proverbial beauty of the women of Martinique. . . . Now attention was diverted to a cruder work of Nature, a sputtering volcano. *Parbleu!* It was enough to scandalize any true Frenchman.

Governor Mouttet sighed and pored over the reports laid before him. He had appointed a commission to study the eruption and get at the bottom of *l'affaire Pelée*, but meanwhile alarm was spreading. People were fleeing the countryside and thronging into St. Pierre, deserting that city for Fort-de-France, planning even to leave the island. Steamship passage was in heavy demand. The *Roraima*, due May 8th, was booked solid out of St. Pierre, one said. This would never do. Steps must be taken to prevent a panic which would scatter fugitives throughout Martinique or drain a colony of France of its inhabitants.

A detachment of troops was despatched by the Governor to St. Pierre to preserve order and halt the exodus. His Excellency, no man to send others where he himself would not venture, followed with Mme. Mouttet and took up residence in that city. Certainly his presence must serve to calm these unreasoning, exaggerated fears. He circulated among the populace, speaking soothing words. *Mes enfants,* the Governor avowed, Mont Pelée rumbling away there is only snoring soundly in deep slumber. Be tranquil.

Yet, on the ominous night of May 7th, as spurts of flame painted the heavens, the Governor privately confessed to inward qualms. What if the mountain should really rouse? Might it not then cast the mortals at its feet into a sleep deeper than its own had been, a sleep from which they would never awaken?

It was dark in the underground dungeon of the St. Pierre prison, but thin rays of light filtered through the grated opening in the upper part of the cell door. Enough so that August Ciparis could tell when it was night and when it was day.

Not that it mattered much unless a man desired to count the days until he should be free. What good was that? One could not hurry them by. Therefore Auguste stolidly endured them with the long patience of Africa. The judge had declared him a criminal and caused him to be locked up here. Thus it was settled and nothing was to be done. Yet it was hard, this being shut out of life up there in the gay city—hard when one was only twenty-five and strong and lusty.

Auguste slept and dozed all he could. Pelée was rumbling away in the distance—each day the jailer bringing him food and water seemed more excited about it—but the noise, reaching the subterranean cell only as faint thunder, failed to keep him awake. . . .

Glimmerings of the dawn of May 8th filtered through the grating into the cell, and Auguste stirred into wakefulness. This being a fête day, imprisonment was less tolerable. What merriment his friends would be making up there in the squares of St. Pierre! He could imagine the sidelong glances and the swaying hips of the girls he might have been meeting today. Auguste stared sullenly at the cell door. At least the jailer might have been on time with his breakfast.

The patch of light in the grating winked out into blackness. *Ai! Ai!* All of a sudden it was night again.

On the morning of May 8th, 1902, the clocks of St. Pierre ticked on toward ten minutes of 8 when they would stop forever. Against a background of bright sunshine, a huge column of vapor rose from the cone of Mont Pelée.

A salvo of reports as from heavy artillery. Then, choked by lava boiled to a white heat by fires in the depths of the earth, Pelée with a terrific explosion blew its head off.

Like a colossal Roman candle it shot out streaks of flame and fiery globes. A pall of black smoke rose thousands of feet in the air, darkening the heavens. Silhouetted by a red, infernal glare, Pelée flung aloft viscid masses which rained incandescent ashes on land and sea.

Then, jagged and brilliant as the lightning flashes, a fissure opened in the flank of the mountain toward St. Pierre. Out of it issued an immense cloud which rushed with unbelievable rapidity down on the doomed city and the villages of Carbet and Le Precheur.

In three minutes that searing, suffocating cloud enveloped them, and 40,000 people died!

Fernand Clerc, the planter, watched from Mont Parnasse, one mile east of St. Pierre, where he had so recently breakfasted. Shrouded in such darkness as only the inmost depths of a cavern afford, he reached out for the wife and children he could not see and gathered them in blessed safety into his arms. But the relatives, the many friends he had left a short while ago, the American consul and his wife, who had waved him a gay good-by—them he would never see alive again. . . .

In that vast brazier which was St. Pierre, Governor Mouttet may have lived the instant long enough to realize that Pelée had in truth awakened and that eternal sleep was his lot and his wife's and that of all those whose flight he had discouraged. . . .

Down in that deep dungeon cell of his Auguste Ciparis blinked in the swift-fallen night. Through the grating blew a current of burning air, scorching his flesh. He leaped, writhing in agony and screaming for help. No one answered.

Not until the afternoon of May 8th did the devastation of St. Pierre cool sufficiently to allow rescuers from Fort-de-France to enter. They could find none to rescue except one woman who died soon after she was taken from a cellar.

"St. Pierre, that city this morning alive, full of human souls, is no more!" Vicar-General Parel wrote his Bishop. "It lies consumed before us, in its windingsheet of smoke and cinders, silent and desolate, a city of the dead. We strain our eyes for fleeing inhabitants, for men returning to bury their lost ones. We see no one! There is no living being left in this desert of desolation, framed in a terrifying solitude. In the background, when the cloud of smoke and cinders breaks away, the mountain and its slopes, once so green, stand forth like an Alpine landscape. They look as if they were covered with a heavy cloak of snow, and through the thickened atmosphere rays of pale sunshine, wan, and unknown to our latitudes, illumine this scene with a light that seems to belong to the other side of the grave."

Indeed St. Pierre might have been an ancient town, destroyed in some half-forgotten cataclysm and recently partly excavated—another Pompeii and Herculaneum. Cinders, which had buried its streets six feet deep in a few minutes, were as the dust of centuries. Here was the same swift extinction Vesuvius had wrought.

Here was no slow flow of lava. That cloud disgorged by Pelée was a superheated hurricane issuing from the depths of the earth at a speed of ninety miles an hour. Such was the strength of the blast, it killed by concussion and by toppling walls on its victims. The fall of the fourteen-foot metal statue of Notre Dame de la Garde—Our Lady of Safety—symbolized the dreadful fact that tens of thousands never had a fighting chance for their lives.

Then four days after the catastrophe, two men walking through the wreckage turned gray as they heard faint cries for help issuing from the depths of the earth.

"Who's that?" they shouted when they could speak. "Where are you?"

Up floated the feeble voice: "I'm down here in the dungeon of the jail. Help! Save me! Get me out!"

They dug down through the debris, broke open the dungeon door, and released Auguste Ciparis, the criminal.

Some days later, George Kennan and August F. Jaccaci, American journalists arriving to cover the disaster, located Ciparis in a village in the country. They secured medical attention for his severe burns, poorly cared for as yet, and obtained and authenticated his story. When the scorching air penetrated his cell that day, he smelled his own body burning but breathed as little as possible during the moment the intense heat lasted. Ignorant of what had occurred, not realizing that he was buried alive, he slowly starved for four days in his tomb of a cell. His scant supply of water was soon gone. Only echoes answered his shouts for help. When at last he was heard and freed, Ciparis, given a drink of water, managed with some assistance to walk six kilometers to Morne Rouge.

One who lived where 40,000 died! History records no escape more marvelous.

And surely the mountain falling cometh to
        nought, and the rock is removed out of his place.
The waters wear the stones: thou washest away
        the things which grow *out* of the dust of
        the earth: and thou destroyest the hope of man.

Job 14, verses 18 and 19

The passage by Hans Cloos, "Beacons on the Passage Out," is of a different kind. Cloos was born in 1886 in Magdeburg, Germany. After finishing his geological studies at Freiburg, he went to southwest Africa. It was on the way out that he witnessed Vesuvius for the first time, and gave the account which we include in this section. This is not an account of a geologist at work. Cloos made no scientific readings and recorded no new observations as he gazed on Vesuvius for the first time. He already knew all the technical explanations of vulcanism and was a well-informed exponent of present earth processes as an adequate explanation for the former history of the earth. And yet, as he gazed on that volcano, he knew that in reality, he had really learned nothing, "because that concept of the physical world has not yet become a true possession of [his] own." This is a beautiful description of the difference between the acquisition of information and the possession of knowledge and understanding. Moments of real self-discovery, when the truth is revealed and assimilated on such a personal basis, are rare in our daily experience. But their very rarity, make them precious, and their significance is great, for they are major landmarks on our journey of personal intellectual inspiration. It is one such landmark that Cloos describes.

BEACONS ON THE PASSAGE OUT

Hans Cloos

When I arrived at Naples, it was a warm turmoil of noise and lights under a starless sky. From the hotel window I heard nothing but the soft lapping of the sea on the mole.

The next morning, however, I experienced the great moment which made me a real geologist. As I threw the shutters open, the whole splendor of the famous picture before me was unveiled like a vast triptych. From right to left; the sparkling bay hemmed in by mountainous shores; then, close by, the gloomy little fort on the beach; and finally, the colorful city rising landward in a thousand steps.

Upward the vision was cut off by a low cloud. Somewhat disappointed, I was about to turn back into the room when I saw a bright sheen above the clouds.

There the clear-cut triangular silhouette of the summit of Vesuvius seemed to be floating in the air, gleaming white with the new winter's snow, and from its sunken crater a little cloud of smoke idly detached itself. . . .

So it was really true!

Year after year I had read and learned little else but this: that our old earth had changed in countless ways during its endless history, and that the whole variegated mass of strata and rocks of the primeval world and of the present mountain ranges was but the result or relic of such changes; that the earth is still active today, living and working on its old material, adding new matter and energy to its old stores; that it is but an optical illusion to assume that the earth has reached a stability that provides an unshakeable foundation for 'uman planning, and that recent changes are only superficial.

These teachings I had heard and believed. I had defended them against he incredulous and recited them before stern judges in searching examinations. But now I had to realize in an unguarded moment that in reality I had learned nothing at all because that concept of the physical world had not yet become a true possession of my own—not till this unique and unforgettable moment when I became a geologist forever by seeing with my own eyes: *the earth is alive!*

Up there in those noble isolated heights above seas and crowds the monstrous happens: the earth, this permanent and time-honored stage of our growing, being, and dying, of our digging and building, thunders and bursts and blows acrid fumes from dark caverns into the pure air we breathe. It spits red fire and flings up hot rocks to let them crash to the ground and smash to bits. There, too, glowing waves rise and flow, burning all life on their way, and freeze into black, crusty rock which adds to the height of the mountain and builds land, thereby adding another day to the geological past.

So it is that the earth itself growing, grows and renews itself.

And as we look, the bright cloud fades into the blue sky and a second one gushes up from the peak. It swells and rises, parting easily, like a soap bubble, from the stony crater rim, which remains bound by terrestrial gravity.

A third cloud follows. Hundreds more; some larger, and more powerful, wrecking the peak; and others, more gentle ones, quietly covering the mountain with dust and ashes, building it step by step to new heights.

Countless eruptions have preceded them, building and destroying, piling up and shattering again. In such fashion, this mountain, softly and smoothly contoured, this cone within a cone, has become for people all over the world the symbol of another, hotter, and more active realm under the cooler one we know, an ever-growing product of a cosmic process of construction patiently carried on through the ages.

The student of the earth thus sees for the very first time the significant relationship between the shape, structure, and history of the terrestrial formation. Deeply moved, he senses the inseparable bonds of matter and space with time and history. And he learns that mountain- and land-forms are the handwriting of the earth, and that he who wanders attentively over the earth's surface can follow the traces of their history.

Naples has a famous museum, but far richer and more impressive to the student of geology is the natural museum of the city's volcanic ground. Landward, to the east, Mount Vesuvius rises solitary and dangerous. To the west huddle low-lying cones and craters, a miniature lunar landscape. Some of these cones appear in the suburbs of the city itself. None of them is large or dangerous. But their casual proximity to roads and buildings makes them exciting. Here the earth is alive everywhere! Like the temperamental people who live on it, the ground itself is noisy and moves, changes, and adds to itself constantly. The present terrain is no longer what it was in Roman times.

Wandering back toward Pozzuoli, the geologist looks for the Temple of Serapis between the road and the sea. Three pillars are still standing. The Temple is not famous in the history of art, but is sensational in the history of the earth. Each of these pillars is ringed to a height of about twenty-one feet by holes similar to those bored by marine mollusks in wood or stone. Between the time it was built and the present, the Temple must have been under water. Did the surface of the sea rise so high and retreat again in so short a period? Or did the little strip of land where the Temple stands rise after it had settled? A careful examination of the site has led to the latter conclusion. Not the ocean, but a rather small section of land has moved.

The height of sea level depends on the total amount of water, which could be increased by the melting of polar ice; on the capacity of the ocean basins; and on the rapidly changing elevation of the continents. Once in two thousand years the earth's crust has taken a deep breath. The eternal sea has spilled over and then slid back, as over the breathing chest of a man in a bath. Did the volcanic furnace first lower and then raise its plastic lid? Did it exhale the ashes of the volcanoes of the Phlegrean fields?

Or was it the pulsation of the whole earth itself? Did the lava yield to the movement of the earth's armor-plated crust itself?

Could it have been that in this little landscape the planet's two major forms of geological expression merged in a single, momentous rhythm? Which was the stronger, the more active force? Which the cause, and which the effect?

The sun has disappeared into Homer's "Okeanos." A cool breeze blows and night is falling fast. The craters below are black and menacing. Will the Temple at Pozzuoli again subside into the sea, and the Solfatara again awaken to life?

The next morning I heard that the ship for Africa would be two days late, which made a trip to Pompeii possible.

Pompeii is a southern city with straight sheets and light, rectangular walls. One can enter any of the buildings and find the sun pouring into the roofless rooms. I had a beautiful day. Even Vesuvius, the great villain, acted innocent. Leisurely I strolled into a *vineta* (an inn), whose floor is almost two thousand years old. All the vestiges of the living are still intact; there are even some traces of human form and gesture. I realized that these people felt and thought much as we do, yet I was also aware of the time-span separating their language and way of life from ours.

During the long period from Pliny the Elder to the days of Eleonora Duse, Vesuvius has remained always the same, has always been active and has always produced the same kind of lava and ashes. These two millennia have been but a day in the volcano's life, which had already begun millions of years before Pliny and Pompeii. And our own life is but the wink of an eye in the two-thousand-year existence of this fossil city.

Thoughtful and somewhat subdued, I rode back from the city of the dead to the city of the living on the beautiful bay.

The next afternoon the *Gertrud Woermann* arrived, and at eleven that night I embarked on my first long journey.

I stood at the stern and saw the city lights disappear. To the left and right beacons and channel-markers on cliffs and bars flashed their lights silently. Each performed its task punctually regardless of wind and weather.

Nature's lighthouses, however, rested. One I had visited three days before flared up only once. Epomeo and Vesuvius were somewhere, shadowy in the blackness of the night. When will they flare up again? The rhythm of their flashes is counted in centuries, perhaps millennia. As yet we do not know this rhythm, because our lives are too short to measure it. But some day it will be known. The beacons of the earth will be so scrupulously observed that they will yield the secrets of their construction and fuel supplies, so that periods of their eruptions and extinction will be known. And whoever solves the mystery of the earth's fiery breath will have caught and fettered the living earth itself. To do this was the hope and aim of the high science to which I had dedicated my life.

Silently the last lights ashore disappeared. The ship of my life bored its way into the black wall of the future.

The Poor World

The poor world is almost six thousands years old,

Rain added to a river that is rank
Perforce will force it overflow the bank.

—William Shakespeare

Our final description of a volcanic eruption concerns one which took place in the area of Öraefajökull, Iceland, in 1727. The account that we include is by Jon. Thorlaksson, the minister of Sandfell. The opening sentence of the extract sets the scene for a balanced, sober, and touching account of the impact of a catastrophe on the life of a local community.

ERUPTION OF THE ÖRAEFAJÖKULL, 1727

Jon. Thorlaksson

In the year 1727, on the 7th August, which was the tenth Sunday after Trinity, after the commencement of divine service in the church of Sandfell, as I stood before the altar, I was sensible of a gentle concussion under my feet, which I did not mind at first; but, during the delivery of the sermon, the rocking continued to increase, so as to alarm the whole congregation; yet they remarked that the like had often happened before. One of them, a very aged man, repaired to a spring, a little below the house, where he prostrated himself on the ground, and was laughed at by the rest for his pains; but, on his return, I asked him what it was he wished to ascertain, to which he replied, "Be on your guard, Sir; the earth is on fire!" Turning, at the same moment, towards the church door, it appeared to me, and all who were present, as if the house contracted and drew itself together. I now left the church, necessarily ruminating on what the old man had said; and as I came opposite to Mount Flega, and looked up towards the summit, it appeared alternately to expand and be heaved up, and fall again to its former state. Nor was I mistaken in this, as the event shewed; for on the morning of the 8th, we not only felt frequent and violent earthquakes, but also heard dreadful reports, in no respect inferior to thunder. Everything that was standing in the houses was thrown down by these shocks; and there was reason to apprehend, that mountains as well as houses would be overturned in the catastrophe. What most augmented the terror of the people was, that nobody could divine in what place the disaster would originate, or where it would end.

After nine o'clock, three particularly loud reports were heard, which were almost instantaneously followed by several eruptions of water that gushed out, the last of which was the greatest, and completely carried away the horses and other animals that it overtook in its course. When these exudations were over, the ice mountain itself ran down into the plain, just like melted metal poured out of a crucible; and on settling, filled it to such a height, that I could not discover more of the well-known mountain Lounagrupr than about the size of a bird. The water now rushed down the east side without intermission, and totally destroyed what little of the pasture-grounds remained. It was a most pitiable sight to behold the females crying, and my neighbours destitute both of counsel and courage: however, as I observed that the current directed its course towards my house, I removed my family up to the top of a high rock, on the side of the mountain, called Dalskardstorfa, where I caused a tent to be pitched, and all the church utensils, together with our food, clothes and other things that were most necessary, to be conveyed thither; drawing the conclusion that should the eruption break forth at some other place, this height would escape the longest, if it were the will of God, to whom we committed ourselves, and remained there.

Things now assumed quite a different appearance. The Jökull itself exploded, and precipitated masses of ice, many of which were hurled out to the sea; but the thickest remained on the plain, at a short distance from the foot of the mountain. The noise and reports continuing, the atmosphere was so completely filled with fire and ashes, that day could scarcely be distinguished from night, by reason of the darkness which followed, and which was barely rendered visible by the light of the fire that had broken through five or six cracks in the mountain. In this manner the parish of Oraefa was tormented for three days together; yet it is not easy to describe the disaster as it was in reality; for the surface of the ground was entirely covered with pumice-sand, and it was impossible to go out in the open air with safety, on account of the red-hot stones that fell from the atmosphere. Any who did venture out, had to cover their heads with buckets, and such other wooden utensils as could afford them some protection.

On the 11th it cleared up a little in the neighbourhood; but the ice-mountain still continued to send forth smoke and flames. The same day I rode, in company with three others, to see how matters stood with the parsonage, as it was most exposed, but we could only proceed with the utmost danger, as there was no other way except between the ice-mountain and the Jökull which had been precipitated into the plain, where the water was so hot that the horses almost got unmanageable: and, just as we entertained the hope of getting through by this passage, I happened to look behind me, when I descried a fresh deluge of hot water directly above me, which, had it reached us, must inevitably have swept us before it. Contriving, of a sudden, to get on the ice, I called to my companions to make the utmost expedition in following me; and by this means, we reached Sandfell in safety. The whole of the farm, together with the cottages of two tenants, had been destroyed; only the dwelling houses remained, and a few spots of the tuns. The people stood crying in the church. The cows which, contrary to all expectation, both here and elsewhere, had escaped the disaster, were lowing beside a few haystacks that had been damaged during the eruption. At the time the exudation of the Jökull broke forth, the half of the people belonging to the parsonage were in four nearly-constructed sheep-cotes, where two women and a boy took refuge on the roof of the highest; but they had hardly reached it when, being unable to resist the force of the thick mud that was borne against it, it was carried away by the deluge of hot water and, as far as the eye could reach, the three unfortunate persons were seen clinging to the roof. One of the women was afterwards found among the substances that had proceeded from the Jökull, but burnt and, as it were, parboiled; her body was so soft that it could scarcely be touched. Everything was in the most deplorable condition. The sheep were lost; some of which were washed up dead from the sea in the third parish from Oraefa. The hay that was saved was found insufficient for the cows so that a fifth part of them had to be killed; and most of the horses which had not been swept into the ocean were afterwards found completely mangled. The eastern part of the parish of Sida was also destroyed by the pumice and sand; and the inhabitants were on that account obliged to kill many of their cattle.

The mountain continued to burn night and day from the 8th of August, as already mentioned, till the beginning of Summer in the month of April the following year, at which time the stones were still so hot that they could not be touched; and it did not cease to emit smoke till near the end of the Summer. Some of them had been completely calcined; some were black and full of holes; and others were so loose in their contexture that one could blow through them. On the first day of Summer 1728, I went in company with a person of quality to examine the cracks in the mountain, most of which were so large that we could creep into them. I found here a quantity of saltpetre and could have collected it, but did not choose to stay long in the excessive heat. At one place a heavy calcined stone lay across a large aperture; and as it rested on a small basis, we easily dislodged it into the chasm but could not observe the least sign of its having reached the bottom. These are the more remarkable particulars that have occurred to me with respect to this mountain; and thus God hath led me through fire and water, and brought me through much trouble and adversity to my eightieth year. To Him be the honour, the praise, and the glory for ever.

CHAPTER 3

Controversy in Geology

Scientific knowledge grows partly by personal curiosity and partly by personal doubt. Doubt arises when existing explanations prove inadequate to account for personal experience, and especially when new data and new experiences impose an additional strain upon existing explanations. Doubt is resolved either by new experience, and especially the contrived experience that we call experiment, or by a comparison of individual experiences and competing explanations, involving at times intense debate, which in turn allows a modification of detailed explanation or general theory. One aspect of verification in science is that experiments are always described and published in such a way that they may be repeated by any trained observer who chooses to duplicate them. This opening of private experience to public scrutiny is the ultimate basis for the large body of agreement that characterizes the scientific world, as well as its classic disputes.

Because much of this debate takes place in public, the history of science is inevitably marked by a good deal of spirited controversy. Some of this has been deeply personal, for many scientists find it difficult to avoid a paternalistic attitude toward their own discoveries and explanations. Other bitter controversies have involved the non-scientific implications of scientific discoveries, and there have been relatively frequent debates between scientists and those of other groups. Still other controversies have wracked the whole scientific world, and their resolution has produced significant turning points in the history of scientific thought. The controversies regarding the nature of the universe following publication of Copernicus's work, and those concerning the organic world following publication of Darwin's *Origin of Species,* are cases in point. Because the effect of such disputes as these is often to change the whole scientific framework in which explanation is offered, such major controversies are rarely settled quickly; indeed some of them may extend over decades. The notion of continental drift, for example, was first suggested on the basis of scientific evidence in a map published by Antonio Snyder in 1855, but it was at first ignored and then later ridiculed, especially during the middle years of the 20th century. It was only the availability of new information in the form of paleomagnetic anomalies, and the development of the embracing concept of plate tectonics in the late nineteen sixties, that provided general acceptance for the theory of continental drift.

This example suggests that, even with the availability of new data, where two views compete for acceptance, it is relatively rare for either view to be ultimately fully proved or wholly disproved. What tends to happen is that one or the other explanation, or sometimes a wholly new one, slowly emerges as more general in its applicability and more economical or elegant in its explanation.

In the following extracts we trace one or two noteworthy geologic controversies. Some of these, such as that concerning footprints in sandstone, are of modest proportions, and were resolved by careful study and intelligent deduction. Others, such as the controversy between Cope and Marsh, were the results of private feuds and personal ambitions, while still others, such as the Oxford debate on evolution in 1860, had far deeper implications.

The Burlington Track

I have mentioned the Burlington (Iowa) track. It was an inspiring place. It was ballasted with glacial gravels where, by hard search, you discovered gems of agate and fossil coral which could, with infinite backaches, be polished on the grindstone. Their fine points came out wonderfully when wet, and you had to lick them with your tongue before each exhibit. I suppose that engineering has long since destroyed this inspiration to young geologists by mass-production of crushed rock.

—Herbert C. Hoover

Willy Ley, naturalist-author, provides an entertaining account of the discovery of fossil footprints in Triassic sandstone in the early 19th century from a quarry in the Hess Mountain, Germany. The debate which followed is typical of many others. The original discovery did not provide some key material or information (in this case, the remains of the animal that made the tracks). Conflicting interpretations were published, additional specimens were subsequently found, but the anomalies remained. One study, offered and disproved a major alternative explanation; then followed another careful study of the original material, and a challenge to a minor, but essential, assumption of the earlier descriptions. It was by this that the older ambiguities were resolved. The footprints in sandstone was a controversy of modest dimensions, conducted with restraint, but it is a useful model of a much more extensive process, by which knowledge is developed and established.

FOOTPRINTS IN RED SANDSTONE

Willy Ley

The first specimens of such footprints were found a little more than a hundred years ago and the case made its entry into scientific literature in the then customary form of a printed "Open Letter." Its author was a Professor F. K. L. Sickler and the letter was addressed to the very famous anatomist Johann Friedrich Blumenbach, professor of medicine at the University of Göttingen. It was published in 1834 and the title deserves to be quoted in full, not because it is a very good title but because it contained virtually all the information then available. It read: "Open Letter to Professor Blumenbach about the very strange reliefs of tracks of prehistoric, large, and unknown animals, discovered only a few months ago in the sandstone quarries of the Hess Mountain near the City of Hildburghausen."

I have not seen the originals which caused Professor Sickler to write his Open Letter, but I have seen the specimens that were in the collection of the Natural History Museum in Berlin. They made the sense of surprise and wonder which is apparent even in Sickler's deliberately stiff wording quite understandable. The dark red stone is flat, sometimes with a faint "wavy" contour which is more apparent to the touch of the hand than to the eye. And in the middle of such a flat stone there is suddenly the perfectly clear and rather deep print of a hand. Usually that handprint is somewhat too large for a human hand to fit well and when you try it you also find that the proportions are not quite the same. The fingers are much thicker and heavier, and so is the thumb, while the palm is too wide near the fingers and too narrow at the other end.

Still, the similarity, at first glance, is almost breathtaking. The differences do now show until one proceeds from general "looking" to detailed examination. In many of the better specimens, there appears a tiny handprint, like that of a child, immediately in front of the large handprint. And a good number of such slabs are overlaid with an irregular network of criss-crossing ridges—"ridges" because this is usually clearer on the "casts" on the upper slabs—which do not puzzle a trained observer for a minute. They can be seen almost anywhere now after a succession of hot days when puddles have dried out and the left-over mud has cracked. With enough material it was easy to find specimens where a print had been ripped apart by such a crack and other specimens where the print had been made over an already existing crack.

The general picture was clear almost from the outset. The area where workmen now quarried red sandstone from the flanks of the Hess Mountain had once been desert, wind-blown sandy desert. Either it rained occasionally, or else a river flooded the area periodically, as the Nile has done all through human history, and then there were places where a footprint could be preserved, at first temporarily, in hardening mud. And if wind-blown sand covered these prints deep enough so that the next moistening, whether a river flood or a rainfall, did not soften the old hardened mud again, the prints lasted to our time, the sand becoming sandstone under the pressure of other deposits piled on top of it.

Considering the shape of these tracks it is not surprising that discussion began at once, *furioso e fortissimo*. To Friedrich S. Voigt the whole case was perfectly simple. These were the tracks of a giant ape—let's call him *Palaeopithecus*. Others shook their heads, especially Alexander von Humboldt. Apes are rare and tropical and Hildburghausen was not in the tropics. More likely the tracks belonged to a large marsupial, something like the kangaroo of today. One Professor Link, who wrote in French, preferred to believe that it might have been a large toad—quite large since a large track is about 9 inches long. Then Voigt revised his opinion in part. He had a large track, presumably one where the imprint of the thumb had either broken off or happened not to show, and he declared that this must have been a bear, "possibly the famous cave bear itself." But he also had much smaller tracks of somewhat different shape; these must have been made by a monkey, "probably" a mandrill.

Nine different papers on the "Hess mountain quarry tracks" appeared in 1835, the year after the publication of Sickler's Open Letter. One of these nine, written by a Dr. Kaup, made the one lasting contribution of that year: he named the tracks, or rather the animal that had caused them. Of course nothing was really known about them except their shape. But that was enough for a name—after all, by order of the school commission one had spent six years learning Greek; now apply what had been learned. The tracks looked like hands; the Greek word for hand is *cheiros*. Unfortunately all safety stopped at that point. Did the animal that had made the tracks belong to the reptiles? If so, since *sauros* is Greek for "lizard," the name would be *Chirosaurus*. But if it was a mammal the name would have to be *Chirotherium*. Dr. Kaup believed with von Humboldt that it had probably been a marsupial and hence a mammal, but he still was careful. The title of his paper read: "Animals tracks of Hildburghausen, Chirotherium or Chirosaurus."

Later usage dropped the "chirosaurus" so that the tracks came to be called "chirotherium tracks" in all books in any language. It is unfortunate that it was the wrong term which was retained. We now know that the animal was a reptile, but the name still is chirotherium.

For a few years it seemed as if the mystery surrounding these tracks would remain centered on Hildburghausen. But in 1839 there appeared in the *London and Edinburgh Philosophical Magazine* a report by P. G. Egerton, with the title: "On Two Casts in Sandstone of the Impressions of the Hindfoot of a Gigantic Chirotherium, from the New Red Sandstone of Cheshire." Egerton's article dealt in the main with the tracks to which the title refers and which had been discovered near Storeton in 1838. But he could point out that chirotherium tracks had actually been discovered first in England. Some had been found in 1824 in the sandstone of Tarporley (near Chester, also in Cheshire) but had been neglected until the Storeton tracks came to light. Of course the Storeton find, two "handprints," each about 15 inches long, was impressive because of its size. The Tarporley tracks were smaller and apparently not very clear.

France was brought in when Daubée, in 1857, published a contribution in the *Comptes rendus* of the French Academy of Science, dealing with the *découverte de traces de pattes de quadrupèdes* in Triassic sandstones of Saint-Valbert, near Luxeuil (Haute Saône). Between these two dates, the second English and the first French, more had been found in Germany, in Hildburghausen as well as in other places. That the mysterious hand-animal had also lived in what is now Spain was reported for the first time in 1898 by Calderón in the *Actas de la Sociedad española de Historia natural*. Several years later chirotherium tracks were found in America; the reason they were discovered so late is clear from the place names occurring in the following short quotation, taken from a recent authoritative publication:

> Well represented in North America in Lower and Upper Moenkopi of Little Colorado River region by total of eight species. Chirotherium occurs as far east as Snowflake, Arizona, and as far west as Rockville, Utah, a lateral distribution of 250 miles. . . .

The main problem all through the history of the chirotherium tracks was, of course, "What kind of animal made those tracks? How did it look?" [There seemed to be several different kinds involved.] If the number of tracks is an indication of the abundance of a species, and not just an indication of some special habits which produced more tracks in places where they might be preserved, the most common variety was *Chirotherium barthi*, so named by Dr. Kaup. Sickler himself established a smaller form which he called *Chirotherium minus*. Later his own name was attached to still another species, *Ch. sickleri*, which is the smallest yet known. Its prints are only 3 inches long, as compared to the 9 inches of *Chi. barthi* and the 15 inches of Egerton's *Ch. herculis*.

All this, however, would have been prettier by far if the red sandstone had been kind enough to yield some fossil bones too. Scientists hovering around the quarries of Hess Mountain were quite certain that it would, sooner or later. After all, a chirotherium must have died somewhere, sometime, in the area where it had lived. Even if one assumed that it could defend itself most effectively against attackers, there was still old age. And there was thirst—remember it was a desert area.

The scientists who were patiently waiting for chirotherium bones from the Hess Mountain quarries knew quite well what desert conditions would or would not do and their hopes were not too farfetched. But unfortunately the hopes are still unrealized.

In addition to hoping one could guess a little more. It became clear just at that time that large mammals had not existed in Triassic days. That ruled out all guesses about apes, bears, and monkeys. But there was still a choice left. Chirotherium could have been a reptile, or it could have been a large amphibian. In the periods preceding the Triassic there had been large amphibians, grotesque, heavy, and large-headed salamanderlike monsters of crocodile size. And in 1841 the then very famous English Professor Richard Owen thought that he could point his finger at a specific group. Because of the strange labyrinthine structure of their usually enormous teeth one group of these ancient giant amphibians had been named labyrinthodonts. Remains of them had been found in England. Chirotherium tracks had been found in England too. They did not occur together, nor was it certain that tracks and bones belonged to the same reasonably short time interval. But Owen thought that it might be so and he announced that the originator of the chirotherium tracks was in all probability a labyrinthodont.

Some other paleontologists shook their heads and decided that they would not believe anything for a while. But the word of Owen happened to be the word of authority and many patiently tried to fit the known bones and the known tracks together. The whole episode culminated in a drawing published by Charles Lyell, showing a 6-foot toadlike labyrinthodont walking cross-legged. Some consecutive tracks which had been found indicated that the "thumb" was on the outside. Lyell therefore had to postulate that the right foot was put down to the left of the left foot in walking.

It was a bit strong for most informed observers, but the best they could do was to keep quiet. They had to admit that 6-foot labyrinthodonts had existed. They had to admit that some of them were generally toad-shaped. All they could really say against Lyell's picture was "We don't like it."

In 1925 there appeared a small book, only 92 pages, devoted to chirotherium tracks. Its author was Professor Wolfgang Soergel, then professor of geology and paleontology at the University of Tübingen, near Stuttgart. Professor Soergel, as became evident from his book, had spent years in a concentrated study of all the tracks he could find in museums and collections in southwest Germany. While he did not neglect other types he worked especially on tracks of *Chirotherium barthi*, the most common form and a rather large one.

The "thumb," Soergel saw soon, was on the outside of the foot and was a toe. Anatomists number fingers and toes in the same manner in which musicians mark piano scores, from the inside out, beginning with the thumb and ending with the little finger. The only difference between an anatomical picture and a piano score is that the piano score counts 1, 2, 3, 4, and 5, while the anatomical picture uses I, II, III, IV, and V. The "thumb" of chirotherium, therefore, was not I, it was V. This is not so surprising as it may sound. If you look at the foot of any lizard—well, almost any lizard; it does not apply to some—you'll see that the outside toe V is spread apart from the remainder of the foot to a fair degree, while I on the inside is usually more or less parallel to the others. Once you realize this the whole picture becomes clear; the long confusion had been due solely to the fact that everybody who saw such a track succumbed to the impulse of putting his own hand into it for comparison, putting the right hand into the left footprint.

Looking around among fossils for a skeleton of a reptilian foot which looked about the way the foot of chirotherium must have looked, Soergel found that there was one which "fitted" closely, except for size. In Triassic layers of South Africa the fossil foot of a small pseudo-crocodile (*Euparkeria capensis*) had been found. It was definitely, as one could tell from the bones, a right foot. Digits I to IV are roughly parallel, but digit V on the outside is spread away from the remainder of the foot. Soergel drew a picture of the probable footprint of an euparkeria; it resembles that of a chirotherium but not quite closely enough to be confused with one. And the foot length of euparkeria is only about 2 inches.

This settled, Soergel went on, looking for fine detail. Did chirotherium have claws? One print showed one clearly, on digit IV, but when others were examined closely faint marks of claw tips could be found, including one on digit V, the "thumb." In fact, chirotherium had possessed very strong and long claws which were carried in such a way that they were not worn by touching the ground. This strongly suggests carnivorous habits.

The next step was to examine the prints and their casts for skin structure. Folds of the "sole" were quite apparent on some pieces; scaly skin showed clearly on these and on many others. This proved that chirotherium was a reptile. A British zoologist, Richard Lydekker, had always insisted on that and had started writing "chirosaurus" in 1890, but unfortunately he had been unable to break the established habit.

Next step: check the impressions of a number of consecutive tracks for the probable sequence of movement. Soergel concluded that it must have been left foreleg, right hind leg, right foreleg, left hind leg—the normal gait of a four-legged animal. Next step after that was to measure the depth of impression, especially the relative depth of "foot" and "hand" of the same individual. It has been mentioned earlier that the "hand" is much smaller than the "foot." By measuring the depth of impression Soergel also found that the foot carried all the weight, while the hand just touched the ground.

From this fact alone the shape of the unknown animal could be reconstructed. If a four-legged reptile manages to carry virtually the whole weight of its body on its hind feet the body must be built in such a way that it almost balances. This means that chirotherium must have had a long and massive tail, heavy and stiff enough to serve as a counterweight for the body. For the same reason it cannot have had a very long neck, especially since a carnivorous reptile has to have a reasonably large head. This general shape—massive hind legs with large feet, weaker forelegs with small hands, fairly short neck with a relatively large head, and long and massive tail—is very well known to paleontologists. It is the shape of all the later dinosaurs that took to walking upright. Obviously chirotherium was evolving in the general direction of bipedal walk; it did not quite balance on two legs only, but it came close. Of the late forms of chirotherium which left tracks in British Keuper we can be virtually certain that they did walk upright. Of the earlier forms there exist a few suspicious tracks where the small prints of the front legs are unaccountably lacking. Maybe chirotherium could do what the Australian collared lizard (*Chlamydosaurus*) still demonstrates. When undisturbed it walks in the same manner as any other lizard, but when angered or frightened it will put on bursts of speed, holding the body stiffly curved backward and running on its hind legs only.

How big was chirotherium? Or, since there were a number of species, how big was *Ch. barthi*? It is easy to measure the length of the stride, but not quite as easy to draw conclusions from it. The legnth of the stride is determined not only by the length of the legs but also by their angle of "swing." Weighing carefully all the possibilities and especially all the possibilities of error, Professor Soergel concluded that the body of *Ch. barthi* must have been about 3 feet long. The tail, in order to balance the weight in front of the hind legs, must have been equally long or somewhat longer, while neck and head together measured probably a little less than the body. This, then, would make the over-all length 8 feet for *Ch. barthi,* while the total length of the smallest known varieties, *Ch. bornemanni* and *Ch. sickleri*, would work out to 14 inches.

Soergel's book was received with much surprise, much enthusiasm in selected circles, and without dissent. So it had been possible, by dint of hard work and sound reasoning, to make something out of these tracks which were never joined by any fossil bone.

More than ten years later an interesting addition to the story was provided by another German paleontologist, Friedrich Freiherr von Huene. He had been in Brazil in the early 1930s with another paleontologist, one Dr. R. Stahlecker. Their interests had been well known, of course, and one day a Brazilian named Vicentino Presto told them about a site of fossils near a place called Chinquá. The two Germans not only found the site promising, they could also determine its age. The fossils belonged to the Upper Rio-do-Rasto formation of Brazil, Upper Triassic and equivalent in age to the German Keuper. The main find was a rather large saurian, which von Huene named, in honor both of the original discoverer and of the place, *Prestosuchus chiniquensis*.

Of course one is quite careful with fossils in the field. Most especially nobody in his right mind will try to separate the fossil bone from the stony matrix to any larger extent than is absolutely necessary for recognition. Fossils are taken out with much of the surrounding stone sticking to them. They are then usually wrapped in burlap that has been soaked in fresh plaster of Paris, more plaster of Paris is smeared on, and the whole is nailed into stout boxes for transportation. The detail work of separating bone from matrix and of mounting the bones (if they are in a shape to be mounted) is decidedly an indoor job and may take years.

Doctors von Huene and Stahlecker packed their prestosuchus and returned home. When the fossil was in a sufficiently advanced state of preparation to be examined it was found to be a pseudo-crocodile, of a size far surpassing all the forms from the same group that had been found in Europe. Its total length was 15 feet 6 inches and it stood 3 feet 6 inches tall. The shape of prestosuchus was almost precisely what Professor Soergel had drawn as the "calculated" shape of chirotherium. And the foot of prestosuchus agreed with the foot of chirotherium as Soergel had reconstructed it from the prints.

Prestosuchus is considerably larger than *Ch. barthi* and it is also considerably later. And, of course, it lived in the Southern Hemisphere, while all chirotherium tracks known were found quite far north of the equator. But it does prove that the pseudo-crocodiles could not only attain the size required for chirotherium but even grow much larger.

As for fossil remains of the European chirotherium, it is still true that there are none. But this is not so disappointing any more, because we can be quite sure that if they are ever found they will merely confirm what could be deduced from patient detail studies and by careful thinking.

William Irvine's account of the "Oxford Meeting" describes an early round in a much larger controversy. Charles Robert Darwin (1809-1882) published The Origin of Species *on November 24, 1859. The first edition of 1,250 copies was sold out on the day of its publication, and the book soon reached an audience much wider than that of the scientific world to whom it was addressed. The book explained the use of artificial selection in domestic breeding and argued for the existence of a similar selection in nature, acting in an analogous way and favoring the survival of those organisms better adapted to the environment in which they lived. The book also dealt with what was then known of the laws of inheritance and variation, and the more general questions of evolution, or "descent with modification" as Darwin called it, grappling especially with difficulties in accepting the evolutionary origin of species.*

The book produced a controversy which extended into almost every area of human understanding. The controversy reached its height with the meeting of the British Association for the Advancement of Science in Oxford in 1860, at which Samuel Wilberforce, Bishop of Oxford, undertook to debate with T. H. Huxley, one of the leading zoologists of the 19th century, and the foremost champion of evolutionary theory, who was sometimes referred to as "Darwin's bulldog." The account that follows gives something of the flavor of their meeting.

APES, ANGELS, AND VICTORIANS

William Irvine

In June, 1860, the British Association met at Oxford. Science was not very much at home there, and neither was Professor Huxley. Beneath those dreaming spires, he always felt as though he were walking about in the Middle Ages; and Professor Huxley did not approve of the Middle Ages. At Oxford, he feared, ideas were as ivy-covered as the buildings, and minds as empty and dreamy as the spires and quiet country air. Professor Huxley's laboratory was set squarely in the middle of the nineteenth century, in the narrow downtown London thoroughfare of Jermyn Street, which was as crowded and busy as Professor Huxley's own intellect.

Reciprocally, Oxford did not feel in the least at home wich such people as Professor Huxley. In fact, she felt rather desperately at bay between a Tractarian past and a scientific future. Newman's conversion to Roman Catholicism had opened an abyss of conservatism on one side; now Mr. Darwin's patient and laborious heresy had opened an abyss of liberalism on the other. The ground of sanity seemed narrow indeed. But sanity can always be defended. After all, there was something obviously ridiculous in a heresy about monkeys.

Mr. Darwin himself was too ill to attend the meeting of the Association— as on a former even more important occasion, when Joseph Hooker and Sir Charles Lyell had acted in his place. A portentous absence from crucial events which deeply concerned him was already making Mr. Darwin a legend. It was just six months since his *Origin of Species* had appeared. Of course Darwinism was in everybody's mind. It was also on the program.

In Section D of the meeting, Dr. Daubeny of Oxford read a paper "On the final causes of sexuality in plants, with special references to Mr. Darwin's work on *The Origins of Species.*" Huxley was invited to comment by the president but avoided discussion of the vexed issue before "a general audience, in which sentiment would unduly interfere with intellect." Thereupon, Sir Richard Owen, the greatest anatomist of his time, rose and announced that he "wished to approach the subject in the spirit of the philosopher." In other words, he intended, as in his anonymous review of a few months before, to strike from under the cloak of scientific impartiality. "There were facts," he felt, "by which the public could come to some conclusion with regard to the probabilities of the truth of Mr. Darwin's theory." He then declared that the brain of the gorilla "presented more differences, as compared with the brain of man, than it did when compared with the brains of the very lowest and most problematical of the quadrumana."

Huxley rose, gave Owen's words "a direct and unqualified contradiction," pledged himself to "justify that unusual procedure elsewhere," and sat down. The effect was as though he had challenged Owen to a duel, and infinitely more dramatic than an immediate refutation, however convincing, would have been, though that duly appeared in the dignified pages of *The Natural History Review.*

Between Darwin and anti-Darwin the lines of battle were now drawn. The program of the Association encouraged peace on Friday, but the air was filled with rumors. A general clerical attack was to be made on Saturday, when a somewhat irrelevant American was to speak on the "Intellectual Development of Europe considered with reference to the views of Mr. Darwin." The Bishop of Oxford was arming in his tent and Owen was at his elbow, whispering the secret weaknesses of the enemy. Quite unaware that his larger destiny awaited him in a lecture room next day, Huxley had decided not to witness the onslaught. He was very tired, and eager to rejoin his wife at Reading. He knew that the Bishop was an able controversialist and felt that prevailing sentiment was strongly against the Darwinians. On Friday evening he met the much reviled evolutionist Robert Chambers in the street, and on remarking that he did not see the good of staying "to be episcopally pounded," was beset with such remonstrances and talk of desertion that he exclaimed, "Oh! if you are going to take it that way, I'll come and have my share of what is going on."

Perhaps revolutions often have their quiet beginnings in the classroom, but they seldom have their turbulent crises there. Saturday, June 30, 1860, was the exception. The Museum Lecture Room proved too small for the crowd, and the meeting was moved to a larger place, into which 700 people were packed. The ladies, in bright summer dresses and with fluttering handkerchiefs, lined the windows. The clergy, "shouting lustily for the Bishop," occupied the center of the room, and behind them a small group of undergraduates waited to cheer for the little known champions of "the monkey theory." On the platform, among others, sat the American Dr. Draper, the Bishop, Huxley, Hooker, Lubbock, and, as president of the section, Darwin's old teacher Henslow.

Dr. Draper had the compound misfortune to be at once a bore and the center of this exciting debate, having chosen to pick up the burning question of the day by its biggest annd hottest handle. His American accent added a quaint remoteness to his metaphysical fulminations. "I can still hear," writes one witness, "the American accents of Dr. Draper's opening address when he asked 'Air we a fortuitous concourse of atoms?'" But had he luxuriated in the combined gifts of Webster and Emerson, he would still have seemed an irrelevance. The audience wanted British personalities, not Yankee ponderosities; and they had already smelled blood.

Dr. Draper droned away for an hour, and then the discussion began. It was evident that the audience had tolerated its last bore. Three men spoke and were shouted down in nine minutes. One had attempted to improve on Darwin with a mathematical demonstration. "Let this point A be man and that point B be the mawnkey." He was promptly overwhelmed with cries of "mawnkey." And now there were loud demands for the Bishop. He courteously deferred to Professor Beale and then, with the utmost good humor, rose to speak.

Bishop Wilberforce, widely known as "Soapy Sam," was one of those men whose moral and intellectual fibers have been permanently loosened by the early success and applause of a distinguished undergraduate career. He had thereafter taken to succeeding at easier and easier tasks, and was now, at fifty-four, a bluff, shallow, good-humored opportunist and a formidable speaker before an undiscriminating crowd. His chief qualification for pronouncing on a scientific subject derived, like nearly everything else that was solid in his career, from the undergraduate remoteness of a first in mathematics.

Huxley listened to the jovial, confident tones of the orator and observed the marked hostility of the audience toward the Darwinians. How could he make an effective reply? He could hardly expound *The Origin of Species,* theory and evidence, in ten minutes. But Huxley was not the man to brood on disadvantages. He was encouraged to find that, though crammed to the teeth by Owen, the Bishop did not really know what he was talking about. Nevertheless, exploiting to the full the popular tendency to regard every novelty as an absurdity, he belabored Darwinism with such resources of obvious wit and sarcasm, saying nothing with so much gusto and ingenuity, that he was clearly taking even sober scientists along with him. Finally, overcome by success, he turned with mock politeness to Huxley and "begged to know, was it through his grandfather or his grandmother that he claimed his descent from a monkey?"

This was fatal. He had opened an avenue to his own vacuity. Huxley slapped his knee and astonished the grave scientist next to him by softly exclaiming, "The Lord hath delivered him into mine hands." The Bishop sat down amid a roar of applause and a sea of fluttering white handkerchiefs. Now there were calls for Huxley, and at the chairman's invitation he rose, a tall, slight, high-shouldered figure in a long black coat and an enormous high collar, which seemed to press the large, close-set features even more tightly together. His face was very pale, his eyes and hair were very black, and his wide lips were calculatingly, defiantly protruded. His manner, gauged with an actor's instinct, was as quiet and grave as the Bishop's had been loud and jovial. He said that he was there only in the interests of science, that he had heard nothing to prejudice his client's case. Mr. Darwin's theory was much more than an hypothesis. It was the best explanation of species yet advanced. He touched on the Bishop's obvious ignorance of the sciences involved; explained, clearly and briefly, Darwin's leading ideas; and then, in tones even more grave and quiet, said that he would not be ashamed to have a monkey for his ancestor; but he would be "ashamed to be connected with a man who used great gifts to obscure the truth."

The sensation was immense. A hostile audience accorded him nearly as much applause as the Bishop had received. One lady, employing an idiom now lost, expressed her sense of intellectual crisis by fainting. The Bishop had suffered a sudden and involuntary martyrdom, perishing in the diverted avalanches of his own blunt ridicule. Huxley had committed forensic murder with a wonderful artistic simplicity, grinding orthodoxy between the facts and the supreme Victorian value of truth-telling.

At length Joseph Hooker rose and botanized briefly on the grave of the Bishop's scientific reputation. Wilberforce did not reply. The meeting adjourned. Huxley was complimented, even by the clergy, with a frankness and fairness that surprised him. Walking back to lodgings with Hooker, he remarked that this experience had changed his opinion "as to the practical value of the art of public speaking," and that from this time forth he would "carefully cultivate it, and try to leave off hating it." Huxley had just enough of the sensitive romantic in him to imagine that he hated public speaking. How he actually felt at the time, he himself indicates in another sentence, "I was careful . . . not to rise to reply till the meeting called for me–then I let myself go."

Huxley's destiny had thus been captured by another man's book, and discovering almost with astonishment how many talents for action he possessed, this young professor of paleontology became the acknowledged champion of science at one of the most dramatic moments in her history. He defended Darwinian evolution because it seemed to constitute, for terrestrial life, a scientific truth as significant and far-reaching as Newton's for the stellar universe—more particularly, because it seemed to promise that human life itself, by learning the laws of its being, might one day become scientifically rational and controlled.

The Wreck of a Former World

In wandering through the Mauvaises terres, or "Bad Lands", it requires but little stretch of the imagination to think oneself in the streets of some vast ruined and deserted city. On ascending the butte to the east of our camp, I found before me another valley, a treeless barren plain, probably ten miles in width. From the far side of this valley butte after butte arose and grouped themselves along the horizon, and looked together in the distance like the huge fortified city of an ancient race. The utter desolation of the scene, the dried-up water-courses, and absence of any moving object, and the profound silence which prevailed, produced a feeling that was positively oppressive. When I then thought of the buttes beneath my feet, with their entombed remains of multitudes of animals forever extinct, and reflected upon the time when the country teemed with life, I truly felt that I was standing on the wreck of a former world.

—**Joseph Leidy**

Edward Drinker Cope (1840-1897), wealthy, talented, ambitious, prolific author, and brilliant scientist, studied at Princeton. He was one of the leading students of his time in herpetology and ichthyology and subsequently devoted his attention to fossil vertebrates. In 1870, he worked under F. V. Hayden with the U.S. Geological Survey, and discovered many new groups of fish and reptiles. He was deeply concerned, not only with systematic paleontology but also with the value of fossil vertebrates in correlation, and in the evolutionary mechanisms involved in the fossil record. He rejected Darwin's theory of natural selection and embraced a neo-lamarckian mechanism.

Othniel Charles Marsh (1831–1899) was born in Lockport, New York, and was a student at Andover and Yale. He subsequently traveled to Europe and, on his return to the United States, was appointed Professor of Paleontology at Yale. In 1870, he organized the first of four Yale scientific expeditions to the West and published extensively on the fossils that were collected on the expeditions. His millionaire uncle, George Peabody, founded the Peabody Museum at Yale, which, under Marsh's direction, became one of the leading museums in the world. Though of independent means, Marsh became vertebrate paleontologist at the U.S. Geological Survey in 1892, and it was during this period that his rivalry with E. D. Cope reached a climax. The present article, written by Howard S. Miller, Professor of History at the University of Missouri, St. Louis, describes the causes and effects of this controversy.

FOSSILS AND FREE ENTERPRISERS

Howard S. Miller

One summer day in 1868 the Union Pacific made an unscheduled stop at Antelope Station, a tiny dot on the rolling Nebraska prairies, just east of the Wyoming line. A dapper, obviously Eastern gentleman leapt from a parlor car and began searching through a mound of dirt beside a recently dug well. Othniel C. Marsh, a Yale paleontologist, had come in response to newspaper reports of fossil human bones unearthed by a well digger. He failed to find fragments of early man, but what he did find excited him hardly less. For scattered over the ground were remains of many animals, among them a diminutive horse which Marsh subsequently identified as an important link in the evolutionary chain. He left the scene only after the impatient conductor had flagged the train ahead, leaving the preoccupied scientist to run after the last car. "I could only wonder," Marsh later recalled, "if such scientific truths as I had now obtained were concealed in a single well, what untold treasures must there be in the whole Rocky Mountain region. This thought promised rich rewards to the enthusiastic explorer in this new field, and thus my own life work seemed laid out before me."

During the next three decades enthusiastic explorers, notably Joseph Leidy, Marsh himself, and Edward Drinker Cope, unearthed an extinct fauna of unimagined richness and variety. Before the three entered the field there were fewer than one hundred species of fossil vertebrates known in America. By the end of the century they had dug more than two thousand new species from the fossil cemeteries of the Great Plains. In the process they assembled impressive support for Darwinism, laid the foundations of modern paleontology in the United States, and became embroiled in one of the bitterest scientific and political controversies of the late nineteenth century.

But unveiling the past in the geological record was easier said than done, as Charles Darwin was the first to admit. Thus far fossil strata simply had not displayed the "finely graduated organic chain" demanded by Darwin's theory. Instead of orderly evolution, in fact, the rocks showed complex organisms suddenly appearing and disappearing without a trace of either ancestor or descendant. The evidence seemed to favor Progressionism, the respectable doctrine of Genesis and geology championed by Louis Agassiz. Undaunted, Darwin and the Darwinists had plunged ahead, blaming the missing links on the imperfection of the geological record. Fossilization only occurred under rare circumstances, they argued, and in any case only a fraction of the globe had been explored. The geological record was, in short, "a history of the world imperfectly kept, and written in a changing dialect," a book of which whole chapters had been lost, or never imprinted in the sediments of primeval seas.

The fossil strata of the Great Plains held many of Darwin's missing chapters. Fortunately for the scientific community the territory was just now becoming accessible, as westward expansion gained momentum after the Civil War. Soldiers and miners, cattlemen and homesteaders all quickly discovered the practical utility of a careful scientific reconnaissance of the public domain. Their practical needs plus natural curiosity added up to political pressure, so that by the late 1860's the federal government had launched elaborate scientific surveys of the western territories.

From the 1870's on O. C. Marsh and E. D. Cope contested for supremacy in vertebrate paleontology. Had the two men lived in different eras, each would undoubtedly have had a less controversial career. While Joseph Leidy seemed to value peace at any price, Marsh and Cope were prepared for war at any cost. Both were independently wealthy, scientific robber barons, fitting contemporaries of Daniel Drew and Commodore Vanderbilt. Joseph Leidy was forced to surrender, a casualty in the Cope-Marsh war. "Professors Marsh and Cope, with long purses, offer money for what used to come to me for nothing, and in that respect I cannot compete with them. So now . . . I have gone back to my microscope and my Rhizopods and make myself busy and happy with them."

Marsh owed his long purse—indeed much of his scientific career—to his uncle, George Peabody, a self-made man who had amassed a fortune in Anglo-American trade and finance.

Peabody was a doting bachelor uncle, who paid special attention to the education and well-being of his nieces and nephews. Routine family philanthropy prompted him to see O. C. Marsh through preparatory school and Yale, through two years graduate work in the Sheffield Scientific School, plus three more years of study abroad. Marsh set his sights on a professorship of paleontology at Yale. His mentors, James Dwight Dana and the younger Silliman, had also schooled him in the logistics of scientific enterprise, so that by 1862 Marsh was prepared to combine his request for further personal subsidy with a general plea for science at Yale. Marsh reported that Peabody had strong commitments to Harvard, and that he had already promised Edward Everett $100,000 for astronomy, or perhaps art. Still, there was hope. "As I have at present no interest in any institution except Yale," he assured Silliman, "I shall use all my influence with Mr. P. in her favor."

Silliman and Dana had long since envisioned a grand natural history museum and research center to fill out the Scientific School. Their protégé was to be the living bond between the plan and the patron. By May 1863 Marsh had secured a $100,000 Peabody legacy for Yale science. Three more years of complicated revision converted the legacy into a lifetime donation, and increased the amount to $150,000. In the meantime Marsh had also helped the Cambridge scientific corps redirect Peabody's proposed Harvard gift from a school of design to a museum of archeology and ethnology, and in the interest of institutional parity, had convinced his uncle to make equal gifts to the two universities.

George Peabody drew the formal trust deed in October 1866, establishing a museum specializing in zoology, geology, and mineralogy. It was, for all practical purposes, his nephew's private research institute. Several months before the grateful Yale Corporation had invited Marsh to fill a full professorship of paleontology created for the occasion. He had agreed to serve without salary (Uncle George provided a substantial allowance, and at his death in 1869 would leave him $100,000), so that the university could not demand that he teach under-graduate courses. Moreover, as a legal trustee of the museum as well as its director, Marsh could dictate research policy, control the collections, and publish his research at institutional expense. It was an enviable position for a fledgling scientist of thirty-five.

Under Marsh's direction the Peabody Museum quickly gained prominence. In the summer of 1870 he launched the first of a series of annual wild west fossil hunts. The Yale expeditions were technically private affairs, funded by the museum, but their familiarity with the government surveys, and their constant use of army facilities and personnel, clothed them at least partially in the public interest. The expeditions provided field training, augmented Marsh's collections, won him national publicity, and, at least to his own satisfaction, established squatter sovereignty over the fossil fields of the Great Plains.

In 1871 Edward Drinker Cope trespassed on Marsh's self-proclaimed preserve. "I am now in a territory which interests me greatly," he reported from Kansas. "The prospects are that I will be able to do something in my favorite line of Vertebrate Paleontology. . . . Marsh has been doing a great deal I find, but has left more for me." Cope had only recently committed himself to a full-time career in science. His wealthy Quaker father had intended that he become a farmer, and in the interests of scientific agriculture had provided Cope with a comprehensive if unorthodox scientific education at home and abroad. But Cope hated farming, and found the teaching of zoology at Haverford College little better. He had attached himself to the Hayden (and later the Wheeler) Survey, and began with a certain malicious enthusiasm to compete with Marsh for priority of discovery and publication. If Marsh had entrée to *Silliman's Journal* in New Haven, Cope had the Government Printing Office. "Hayden has fathered my Paleontological Bulletins. . . . A disagreeable pill for the Yale College People."

For the next twenty years the Cope-Marsh conflict sputtered intermittently, much to the embarrassment of the scientific community at large. It proved in the long run to be a war of attrition in which Marsh countered Cope's brilliant but erratic forays with a less spectacular but relentless marshaling of financial, personal, and institutional support. The turning point came in 1878, when Marsh succeeded Joseph Henry as President of the National Academy of Sciences. Marsh's commanding position in the academy carried prestige and power. But even more significant in 1878 was the fact that Congress had recently asked the academy, as official science advisor to the government, to recommend a unified scheme for the scientific surveys which would "secure the best results for the least possible cost." Fate had placed Marsh in a position to cut off his rival's source of support.

The practical outcome was quickly told. Marsh's National Academy report (adopted with the sole dissent of E. D. Cope, who understood that reform would in effect freeze him out), urged a scheme which would have provided for geodesy, the systematic disposition of public lands, and theoretical and practical geology. Congress, in keeping with the general tenor of the Gilded Age, ignored all but the latter when it authorized the creation of a new United States Geological Survey. After all, said Congressman Hewitt, "the science of geology and the science of wealth are indissolubly linked." Nations became great only as they developed "a genius for grasping the forces and materials of nature within their reach and converting them into a steady flowing stream of wealth and comfort." The Hayden and Powell surveys, once work in progress had been completed, were to be either absorbed or abolished. With Marsh's help Clarence King emerged as director of the new survey, only to be replaced by John Wesley Powell, when King left government service the following year. Largely in return for past favors, in 1882 Powell appointed Marsh official paleontologist of the Geological Survey, E. D. Cope, cut off from federal patronage, could only bewail his dwindling fortune, appeal for congressional and private support, and rage at his enemies, who he charged had created "a gigantic politico-scientific monopoly next in importance to Tammany Hall."

More was at stake than a personal feud and bureaucratic empire building. The survey reorganization of 1878–79 was of lasting significance as the beginning of a ten year debate over the proper relationship between science and the federal government. Except in special circumstances, notably Joseph Henry's Smithsonian battle of the 1850's, heretofore government science had been neither costly nor controversial enough to attract searching congressional criticism. But by the late 1870's it had grown too big to ignore, and politicians began to understand what years of salutory neglect had wrought.

A Psalm of Life

Lives of great men all remind us
We can make our lives sublime,
And, departing, leave behind us
Footprints on the sands of time...

—Henry W. Longfellow

Sir Archibald Geikie's book The Founders of Geology *contains a rare account of one of the decisive discoveries that ultimately led to the resolution of the great eighteenth and early nineteenth century debate between the Neptunists (those who believed that all rocks had been precipitated from aqueous solution) and the Plutonists (who regarded some of them as volcanic in origin). In 1788, Desmarest was appointed by the king of France as Inspector-General and Director of the Manufactures of France. His influence continued even after the Revolution. His studies of French industry were of enormous importance, but he was interested, too, in geology. His biographer records ". . . he made his journeys on foot, with a little cheese as all his sustenance. No path seemed impracticable to him, no rock inaccessible. He never sought the country mansions, he did not even halt at the inns. To pass the night on the hard ground of some herdsman's hut, was to him only an amusement."*

During these journeys, he found himself in the Auvergne in 1763, and was fascinated by the hexagonal pillars of basalt which characterize those lava fields.

THE FOUNDERS OF GEOLOGY

Archibald Geikie

Shortly after the middle of the eighteenth century, the Governments of Europe, wearied with ruinous and profitless wars, began to turn their attention towards the improvement of the industries of their peoples. The French Government especially distinguished itself for the enlightened views which it took in this new line of national activity. It sought to spread throughout the kingdom a knowledge of the best processes of manufacture, and to introduce whatever was found to be superior in the methods of foreign countries. Desmarest was employed on this mission from 1757 onwards. At one time he would be sent to investigate the cloth-making processes of the country: at another to study the various methods adopted in different districts in the manufacture of cheese. Besides being deputed to examine into the condition of the industries of different provinces of France, he undertook two journeys to Holland to study the paper-making system of that country. He prepared elaborate reports of the results of his investigations, which were published in the *Mémoires* of the Academie des Sciences, or in the *Encyclopédie Méthodique*. At the last in 1788 he was named by the King Inspector-General and Director of the Manufactures of France.

He continued to hold this office until the time of the Revolution, when his political friends—Trudaine, Malesherbes, La Rochefoucault, and others—perished on the scaffold or by the knife of the assassin. He himself was thrown into prison, and only by a miracle escaped the slaughter of the 2nd

September. After the troubles were over, he was once more called to assist the Government of the day with his experience and judgment in all matters connected with the industrial development of the country. It may be said of Desmarest that "for three quarters of a century it was under his eyes, and very often under his influence, that French industry attained so great a development."

Such was his main business in life, and the manner in which he performed it would of itself entitle him to the grateful recollection of his fellow-countrymen. But these occupations did not wholly engross his time or his thoughts. Having early imbibed a taste for scientific investigation, he continued to interest himself in questions that afforded him occupation and solace, even when his fortunes were at the lowest ebb.

"Resuming the rustic habits of his boyhood," says his biographer, "he made his journeys on foot, with a little cheese as all his sustenance. No path seemed impracticable to him, no rock inaccessible. He never sought the country mansions, he did not even halt at the inns. To pass the night on the hard ground in some herdsman's hut, was to him only an amusement. He would talk with quarrymen and miners, with blacksmiths and masons, more readily than with men of science. It was thus that he gained that detailed personal acquaintance with the surface of France with which he enriched his writings."

During these journeyings, he was led into Auvergne in the year 1763, where, eleven years after Guettard's description had been presented to the Academy, he found himself in the same tract of Central France, wandering over the same lava-fields, from Volvic to the heights of Mont Dore. Among the many puzzles reported by the mineralogists of his day, none seems to have excited his interest more than that presented by the black columnar stone which was found in various parts of Europe, and for which Agricola, writing in the middle of the sixteenth century, had revived Pliny's old name of "basalt." The wonderful symmetry, combined with the infinite variety of the pillars, the vast size to which they reached, the colossal cliffs along which they were ranged in admirable regularity, had vividly aroused the curiosity of those who concerned themselves with the nature and origin of minerals and rocks. Desmarest had read all that he could find about this mysterious stone. He cast longing eyes towards the foreign countries where it was developed. In particular, he pictured to himself the marvels of the Giant's Causeway of the north of Ireland, as one of the most remarkable natural monuments of the world, where Nature had traced her operations with a bold hand, but had left the explanation of them still concealed from mortal ken. How fain would he have directed his steps to that distant shore. Little did he dream that the solution of the problems presented by basalt was not to be sought in Ireland, but in the heart of his own country, and that it was reserved for him to find.